Nanotechnology and Neuroscience: Nano-electronic, Photonic and Mechanical Neuronal Interfacing

Massimo De Vittorio • Luigi Martiradonna
John Assad
Editors

Nanotechnology and Neuroscience: Nano-electronic, Photonic and Mechanical Neuronal Interfacing

Springer

Editors
Massimo De Vittorio
Dip. Ing. Innovazione - Università del Salento
Center for Biomolecular Nanotechnologies
Istituto Italiano di Tecnologia
Arnesano (LE), Italy

Luigi Martiradonna
Center for Biomolecular Nanotechnologies
Istituto Italiano di Tecnologia
Arnesano (LE), Italy

John Assad
Center for Neuroscience and Cognitive Systems
Istituto Italiano di Tecnologia - Trento, Italy

Department of Neurobiology
Harvard Medical School
Boston, MA, USA

ISBN 978-1-4939-4765-2 ISBN 978-1-4899-8038-0 (eBook)
DOI 10.1007/978-1-4899-8038-0
Springer New York Heidelberg Dordrecht London

Printed on acid-free paper

Springer is part of Springer Science+Business Media (www.springer.com)

Preface

Understanding the function and organization of billions of neurons and trillions of connecting synapses in the human brain is a tremendous challenge requiring coordinated efforts from scientists from many different research areas. This belief has motivated both the European Union and the United States to invest billions of dollars in long-term, multidisciplinary projects—the "Human Brain" flagship project and the "Brain Initiative," respectively—involving hundreds of scientists and technologists to build a comprehensive picture of the human brain. To achieve this goal, new technologies and approaches for interrogating neuronal circuits will be needed, as well as innovative approaches to handle an exponentially increasing amount of physiological data.

Strategies for recording and stimulating neurons in the central and peripheral nervous systems have been investigated since the eighteenth century, with Luigi Galvani's first electrophysiology experiments. After more than two centuries, the current methodologies and tools allow simultaneous interaction with only a relatively small number of cells. As a result, although it has been possible to elucidate the mechanisms and roles of specific groups of neurons and areas of the brain, we have no available techniques to systematically monitor thousands of functional links that each neuron forms with other neurons. Thus, we lack a clear understanding of large-scale integration in the brain.

Neuroscientists need new tools that enable recording or stimulation at high spatial and temporal resolution and for extended periods of time. Ideally, such devices should be able to interface—with high fidelity and low invasiveness—a large number of neurons to record, perturb, and control their electrical activity. Nanoscience and nanotechnology can play a key role in improving the performance of existing tools and in developing new ideas and experimental approaches to create detailed maps of the human brain. Indeed, reducing the size of electrophysiology probes down to the nanoscale minimizes their invasiveness and increases their spatial resolution; furthermore, more sophisticated electrical, mechanical, and optical functionalities can be merged in miniaturized devices thanks to advanced nanofabrication protocols.

This book provides an overview of the different ways in which the "nano-world" can be beneficial for neuroscientists. The volume encompasses the latest developments in the field of micro- and nanotechnology applied to neuroscience, discussing technological approaches applied to both in vitro and in vivo experiments. A variety of different nanotechnologies are presented that include nanostructured electrodes and their electrical, mechanical, and biochemical properties, active and passive 2D and 3D multielectrode arrays (MEAs), nanoscale transistors for subcellular recordings, and an overview on methods, tools, and applications in optogenetics.

Advances in electrode technology including improvements in the mechanical adhesion of cells on electrodes and the realization of novel, minimally invasive extracellular and intracellular recording approaches are reported. Electrodes are still the most common tools in neuroscience for bidirectional communication with the brain because of their simple fabrication, long-term stability, and large bandwidth, enabling the registration of action potentials as well as low-frequency field potentials (LFPs). Nanotechnology has dramatically improved their spatial resolution by exploiting surface micromachining and technologies derived from CMOS integrated circuits (IC). In addition, three-dimensional shaping, optimized electrode–electrolyte–neuron interfaces, surface functionalization by means of inorganic (e.g., carbon nanotubes) or organic materials, and cell-engulfment strategies for in-cell recording and stimulation have significantly increased their signal-to-noise ratio.

Arrays of electrodes have also become a key methodology in neuroscience for studying the function of neuronal networks and for the development of brain–machine interfaces (BMIs) and neural prostheses. Advances in semiconductor technologies and surface- and bulk-micromachining techniques have doubled the number of electrodes on the same substrate and the number of neurons that can be recorded simultaneously every 7 years. Nanotechnologies and nanostructuring capabilities have opened novel perspectives for multielectrode arrays. In fact, electrode arrays have progressed towards massively parallel single-unit recordings, at the same time taking advantage of low-power microelectronics to implement onboard circuits for conditioning, multiplexing, and wireless transmission of the recorded signals. Real-time operation is now reported for novel *active* multielectrode arrays based on CMOS technologies composed of several thousand electrodes. The impressive fabrication approaches of 2D and 3D MEAs on different substrates and their in vitro and in vivo applications developed during the last decades are reviewed in this book.

At the nanoscale, it is also possible to probe neurons with subcellular resolution, to follow the propagation of electrical signals in axons and dendrites. Such applications have been demonstrated with nanowire-based field effect transistors (FETs). The volume reports on 3D-distributed self-organized nano-FETs produced by bottom-up approaches and on their use in subcellular and high-frequency recordings. Substrate-free nanoscale electrical probes that could penetrate into living tissues or artificial tissue constructs, allowing electrical recording and mapping of cellular activities in a 3D microenvironment, are reported.

Defined as one of the breakthroughs of the decade, optogenetics is still at its infancy, but its capability for controlling neural activity rapidly and reversibly is already revolutionizing behavioral and neurophysiological analysis in animal models, both for understanding the function of the healthy brain and in brain disease. Nanotechnologies are having a major impact in this fast-growing field; indeed, novel micro- and nanotools that target light-sensitive proteins in multiple single neurons are making possible investigations of causal relationships in neural networks. In the last chapter of the book, new optogenetic approaches are discussed, including new ways in which nano- and microtechnologies can advance optogenetic approaches.

The book focuses specifically on fabrication strategies, to offer a comprehensive guide for developing and applying micro- and nanostructured tools for neuroscientific applications. It is intended as a reference both for neuroscientists and nanotechnologists on the latest developments in neurotechnological tools.

Massimo De Vittorio
John Assad
Luigi Martiradonna

Contents

Chapter 1
Carbon Nanotubes for Neuron–Electrode Interface with Improved Mechanical Performance

David Rand and Yael Hanein

Abstract The capacity of neuronal cells to elicit and propagate action potentials in response to electrical stimulation is harnessed in neuro-prosthetic devices to restore impaired neuronal activity. Recording and stimulating electrodes are accordingly one of the major building blocks of these systems, and extensive investigations were directed to build better performing electrodes. The electrochemical properties of the electrodes have clearly gained a lot of attention in securing an electrode technology suitable for high signal-to-noise recordings as well as low-power and high-efficacy stimulation. In addition to electrochemical considerations, the design of the electrodes has to take into account multitude of other concerns ranging from surface chemistry, electrode stability, biocompatibility, mechanical properties, to manufacturability. It is now widely accepted that the neuron–electrode interface is considerably impacted by physical cues and that the mechanical properties of the electrode have to be carefully addressed to achieve optimal performances. Mechanical properties affect the manner neurons proliferate, adhere, and possibly operate. In this chapter, we will focus on the mechanical properties of the neuron–electrode interface. We begin by reviewing neuronal mechanics and its relevance to electrode design and performance. In particular, we will address surface properties such as roughness and shape as important properties in the realm of neuronal electrodes. The ultimate aim and focus of this chapter will be to introduce carbon nanotube electrodes as a powerful system for improved mechanical performances and to discuss their unique properties.

D. Rand • Y. Hanein (✉)
School of Electrical Engineering, Tel Aviv University, Tel Aviv, Israel
e-mail: YaelHa@tauex.tau.ac.il

M. De Vittorio et al. (eds.), *Nanotechnology and Neuroscience: Nano-electronic, Photonic and Mechanical Neuronal Interfacing*, DOI 10.1007/978-1-4899-8038-0_1,

1.1 Introduction

Neurons have unique electrical properties which underlie their capacity to process and propagate information. In a healthy functional neural system, the entire neural pathway is intact, allowing proper input and output from and to the peripheral nervous system. In a diseased or traumatized system however, the information flow pathway is broken and input may not reach the brain (e.g., vision or hearing loss) or may not activate the limbs (as in the case of amyotrophic lateral sclerosis (ALS)). Improper neuronal activity in the central nervous system may cause undesired interference between different brain regions causing severe disabling medical conditions (e.g., Parkinson's disease). Neuronal stimulation and recording with electrodes have long been proposed to address such conditions (for comprehensive review see Cogan S.F. [3]) by providing a new input as in the case of cochlear [5], retinal [2, 6], and deep brain stimulation (DBS) [1] implants, or by activating a machine with signals recorded and analyzed from the brain [7]. Neuronal electrodes are also very important for in vitro applications where artificial cultured networks can be studied over extended duration of weeks, allowing basic investigations [8], bio-sensing [9], as well as pharmacological studies [10].

While neurons are recognized foremost for their electrical properties and their responsiveness to electrical stimulations, they are in fact very sensitive to the mechanics of their environment [11]. Moreover, mechanical signaling appears to be an important factor in the development and maturation of neuronal systems [12].

What are the implications of these phenomena in determining the neuron–electrode interface? Several aspects are directly and indirectly impacted. Foremost is cellular adhesion, which appears to be affected by surface morphology. Adhesion and proliferation of neurons appear to be markedly different on smooth versus rough surfaces [13]. Neuronal adhesion is also associated with morphological and other biological responses, effectively affecting the electrical coupling between the electrode and the cells [4]. The cellular response can be strong enough to turn the coupling to be ohmic-like, by affecting the manner in which its membrane engulfs the electrode surface. A similar effect can be associated with the capacity of cells to impale themselves on nano-sized pikes, establishing intracellular interface with electrodes [14]. These effects can also affect the electrical coupling between neurons and nano-based electrodes, as was suggested by some recent studies [15]. Overall, neuronal mechanics is critically important in determining the neuron–electrode interface, and surfaces with different mechanical properties appear to contribute to differences in the neuron–electrodes interface.

This general theme motivated many recent studies to explore surface morphology as a viable mean to affect the neuron–electrode interface. Several electrode categories were developed in this regard including porous silicon, dendritic polymer coatings, and, recently, carbon nanotube (CNT)-based electrodes. The latter in particular offer electrical conductivity, chemical stability, and mechanical durability, making them very attractive as a neuronal interface. In this chapter, we aim to summarize the accumulated know-how pertaining to the use of CNT as an improved mechanical interface. To do so, we will first introduce basic concepts in neuromechanics.

We will then introduce the various properties of CNT surfaces, emphasizing their mechanical properties. We will conclude by discussing the CNT potential as a novel and improved interface in neuron–electrode technology.

1.2 Neuronal Mechanics

Cell and tissue mechanics is critically important in cell function and organ development as multitudes of studies have demonstrated over the past several decades [16, 17]. While neuronal mechanics has received relatively limited attention, recent investigations have clearly demonstrated the important role of mechanical forces in neuronal development and activity [11, 18, 19]. These mechanical forces also play an important role in neuronal interaction with surfaces and accordingly are relevant in the general context of neuron–electrode interactions [20].

From their very early developmental stages, neurons apply and sense mechanical queues. When axons grow, they rapidly extend in length and diameter while also developing internal tension [12]. This tension in turn may be used to apply a force on the soma, affecting its position [21] and various other biological mechanisms. Pulling forces applied by neurons are apparent in various conditions, most conspicuously when attempting to culture them with predefined positions [22].

The significance of mechanical forces in neuronal development was demonstrated by several studies using both indirect and direct methods. Primarily, the work of Bray and coworkers postulated the existence of such forces and demonstrated that the morphology of cultured neuronal networks is consistent with a network of balanced pulling axons [23, 24]. Cultured neuronal networks, in particular from invertebrate models, appear to exhibit a clear morphology of tensed axons with markedly straight processes. At every branching point, it is possible to associate the measured angles with the axon diameter, where the diameter of the axon is assumed to scale linearly with tension. The linear relation is thought to emerge from the circumference of the axon with its corresponding layer of interacting actin and myosin molecules, which give rise to the observed force. Bernal and coworkers used a calibrated bendable tip to directly measure neurite tension in PC12 cells [25]. Using this approach, it was possible to demonstrate that axons not only have passive viscous–elastic properties but that there is an active tension buildup, probably owing to the action of the motor protein mentioned above. Similar results were obtained from a related study in which axonal mechanics was explored using a microfluidic device [26].

While these investigations revealed the existence of a mechanical tension in axons, recent studies also explored the possible role of such a tension force in shaping the structure and function of neuronal systems [27]. As we briefly reviewed above, neuronal mechanics affects many cellular processes. Adhesion and proliferation are two primary ones with particular significance in the realm of neuronal electrodes.

Owing to the important role of mechanical forces in neuronal development, neurons have an incredible capacity to respond to mechanical cues associated with artificial surfaces. This capacity is best manifested in the manner neurons proliferate on surfaces with different morphologies. Interestingly, surfaces with similar chemistry but with different morphology elicit entirely different response. These differences can be exploited for the development of culturing substrates for neurons, for the design of novel regrowth templates, as well as for the investigation of cellular processes [19]. The sensitivity of neurons to different surface roughness was demonstrated using a multitude of techniques, including roughened silicon dioxide as well as coated carbon nanotubes and various other engineered substrates.

The general field of neuron–surface interactions is very broad and was reviewed in many publications [28]. Here we will focus on specific issues with direct relevance to neuron–electrode interactions. Two primary issues are of particular interest in this regard. The first is the direct interaction between cell processes and surfaces, allowing neurons to preferentially and tightly adhere to rough surfaces. Such entanglement is very helpful in ensuring intimate contact between the cells and the electrodes, which may be effective at building improved neuron–electrode electric coupling and also circumvent glia cell screening of the electrodes by providing neuronal processes enough contact area with the electrodes. The second issue concerns with proteins and other extracellular matrix adhesion molecules, securing strong anchoring of the tissue to the rough surface. In the next section, we further review these issues.

1.3 Neuronal Adhesion to Rough Surfaces

In the previous section, we alluded to two mechanical–neuronal mechanisms which ultimately contribute to strong neuronal adhesion to rough surfaces. The first is the capacity of neuronal processes to extend, or be guided, along elongated structures, and the second is the capacity of the same processes to apply a tension force, allowing them, and the cell soma, to have improved mechanical anchoring.

Elongation and tension are apparently two contradicting mechanisms. Yet, careful investigations have revealed that the two can coexist in concert: An axon can continually extend in length owing to buildup of internal pressure through its microtubule core. At the same time, the same elongation can cause stress buildup. In addition to a passive process of tension buildup due to axonal elongation, active actin–myosin interaction can generate internal tension directed towards axonal shortening. Using CNT islands as anchor points and locust neurons as a model system, it was recently shown how these two processes work together to allow axonal development and consolidation (Fig. 1.1). When neurites grow, they constantly sprout new segments which keep elongating as long as they can secure their adhesion to the surface. When this adhesion is overtaken by tension buildup, processes lose their adhesion to the surface and quickly consolidate into the parent process.

An additional effect observed over many systems is the capacity of neurites to extend along ridges and to twist and curl, ultimately anchoring themselves and their somas to the surface. It is important to keep in mind that the neurites not only extend

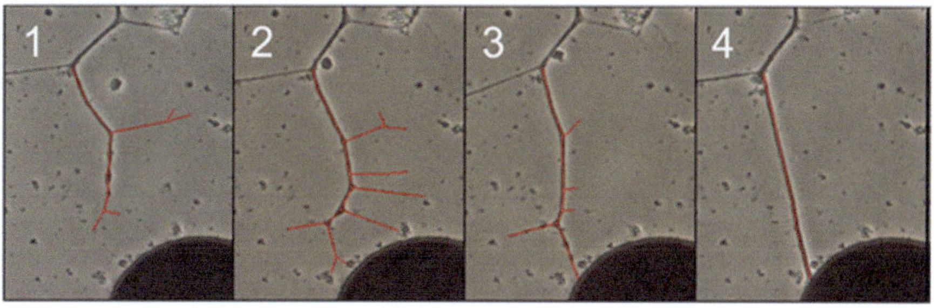

Fig. 1.1 Time-lapse images and analysis of locust neurite development and neurite branch pruning in culture. The data demonstrate the role of tension in developing neuronal systems in response to neurite attachment to a CNT island (appears as a section of a *dark disk* at the *bottom*). Reproduced with permission from [21] © (2009) Elsevier

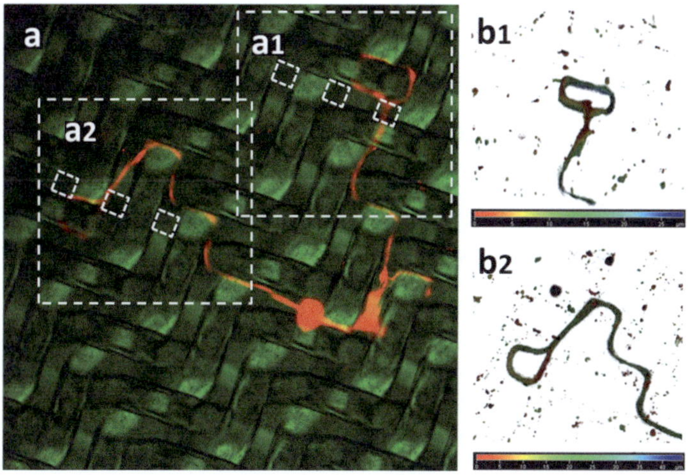

Fig. 1.2 (**a**) Fluorescent confocal image of a locust neuron cultured on a mesh. Three-dimensional rendering clearly demonstrates the tendency of these neurons not only to follow straight paths but also to twist and curl providing the cell with mechanically balanced anchoring to the surface. Field of view = 460 μm. (**b1** and **b2**) show a zoom in into the regions marked (**a1**) and (**a2**). *Small dashed square* indicate holes in the grid. Credit: Anava S, Ayali A, and Hanein Y (unpublished)

as they develop but they also build up internal tension, further establishing the anchoring. This effect is nicely demonstrated in Fig. 1.2, showing a fluorescent confocal image of locust frontal ganglion neurons cultured on an artificial grid. Note that the cell processes not only extend along the grid network but also twist and engulf the grid lines. Moreover, the soma, the two major processes are aligned in parallel. These mechanical effects, which are easily visualized with the relatively large neurons of the locust neurons, are also observed in other types of neurons, with some differences associated with the dimensions and the corresponding forces and flexibility.

The effect demonstrated in Fig. 1.2 can be reproduced on many different surfaces, including pillars, holes, and various other mechanical structures, all reflecting the same response to surface mechanics, accompanied by innate capacity to curl while ultimately developing and applying tension [29].

Before the emergence of nanotubes and nanowires as attractive substrates for neuronal culturing, great attention was directed towards the investigation of neuronal adhesion and proliferation on engineered surfaces with different adhesion patterns [13]. The realization that pristine and modified nanotubes and nanowires can be used in neuronal engineering has opened up a new front directed towards the identification of optimal materials with preferred properties for recording, stimulation, and proliferation of neurons and neuronal tissue [27, 30].

Among the various surfaces and systems that can be used to harness this behavior, CNT surfaces are particularly attractive owing to their chemical stability and electrical conductivity. The fact that CNTs can be readily integrated into complex devices is an additional major advantage [31]. In the next sections, we will review how neuronal response to rough surfaces can be utilized for various applications while using CNT surfaces. We will focus our attention at three related topics: cellular adhesion to CNT surface, tissue response to the CNT surfaces, and the possible contribution of the three-dimensional CNT surface to prevent clogging and retain the beneficial low-impedance properties of the electrodes for long-term applications.

1.4 Neurons on CNT Surfaces

A multitude of recent studies explored how cultured neurons respond to CNT surfaces [15, 32]. Special effort was directed to compare neuronal adhesion and neurite development on different CNT preparation (i.e., tube length, diameter, surface chemistry). Surface chemistry was often considered as a primary parameter, but some data from these investigations hint that mechanical factors may also be at play. Specifically, it was reported in Zhang et al. [32] that guided neurite growth was formed preferably on long vertical MWCNTs compared to short ones. It is possible that long nanotubes are flexible and undergo deformation to accommodate the proliferating neuritis, ultimately affecting the overall response.

The mechanical nature of neuron–CNT interaction and its impact on network development are best visualized and analyzed when exploring the organization of cultured neurons on surfaces with isolated CNT structures (see Fig. 1.3). Such an investigation was reported first by Gabay et al., who showed how disassociated rat cortical neurons that were cultured on quartz substrate with isolated CNT island self-organized into engineered networks over several days [33]. The self-organization of the neurons is coupled with preferred adhesion to the CNT surfaces. Neuronal processes appear to strongly entangle into the CNT surfaces, stabilizing the networks in patterns, which tightly correspond with the CNT island pattern [28]. Extensive soma migration and neurite network consolidation result in replacement of the uniform or unorganized clustered organization with exclusive attachment of

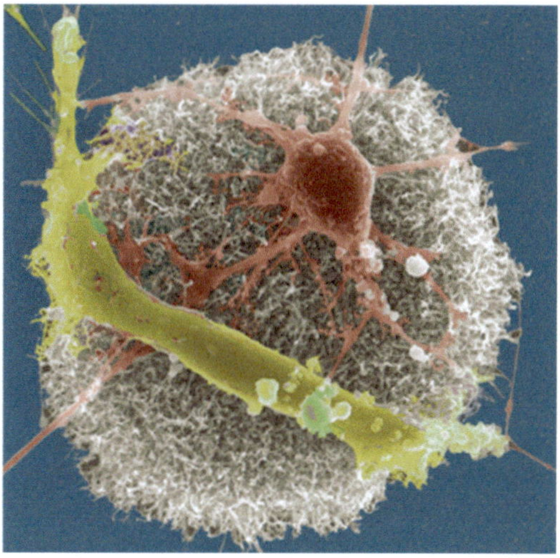

Fig. 1.3 A false-colored SEM image of fixed rat neuronal cells cultured on carbon nanotube islands. The carbon nanotube islands were grown using the chemical vapor deposition method directly on quartz support (for further details, see [21]). Width of image = 34.5 μm

clusters on the CNT surfaces, with distinctly taut neurite bundle bridges retaining axonal connectivity with neighboring clusters. Additional investigation with locust neuronal cultures revealed that the same effect can be also demonstrated in invertebrate networks. In such cultures, a single cell can anchor to specific locations on a CNT island electrode, allowing the formation of networks with one-to-one correspondence between the neuron and the recorded activity from the electrodes [34]. Both in vertebrates and invertebrates, axons appear to be taut corresponding with internal buildup tension within the axons or the axonal bundles. It is clear then that the morphology of the neuronal culture is dramatically impacted by the CNT structures. This mechanism is so robust that it can be utilized systematically to pattern stable cultured networks [28, 35].

The origin of what appears to be a preferred adhesion to CNT surfaces is a direct result of the mechanical forces: As neurons are cultured on a surface, they continually extend processes while establishing anchor contact with the surface. The rougher the surface, the more effective is the ability of the cells to form this anchoring. In conjunction, cells are continually forming connections with other cells, resulting with a constant application of force [27]. The fate of the cell is now dependent on the competition between cell–cell and cell–surface interactions. Ultimately, rough surfaces provide the best anchoring, winning over smooth surfaces. The apparent organization of neurons with preferential adhesion to CNTs is clearly associated with mechanical sensitivity to these rough surfaces which is akin to neuronal response to other rough surfaces.

The ability of neuronal processes to entangle into a CNT matrix facilitates a very strong bond between the neurons and the surface which guarantees both mechanical stability and improved electrical coupling. Interestingly axonal diameter plays an important role as an entanglement necessitates the right correspondence between the CNT diameter and the diameter of the neuronal processes [28]. Indeed, as

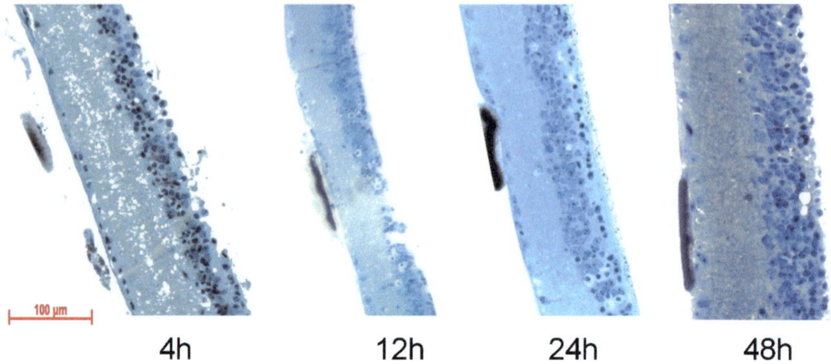

Fig. 1.4 Transverse mouse retinal sections through CNT islands at increasing incubation times. With time, CNTs become more intimately coupled to the retina. Adapted with permission from [37] © (2012) The Author(s)

mentioned above, the difference between axonal diameters of rat versus locust neurons can explain the difference in the adhesion profile between CNT and the two neuron types.

As stated above, CNT–neuron coupling results not only with improved anchoring of the cells to the surface but also improved electrical coupling (for a comprehensive review, see [30]), as it guarantees a closer contact between the neurons and the electrode surface. Evidence for this improved electrical coupling was provided by experiments with whole mount retinas conducted using CNT-based multielectrode arrays. Interestingly, the CNT MEA not only provides excellent signal-to-noise ratio recording, but it was also observed that the signal-to-noise ratio improved over time, an effect which is not revealed in conventional TiN electrodes. Similar investigation was performed focusing on the retinal stimulation thresholds [36].

Even more tale-telling are recent results obtained by using CNT islands loosely grown on SiO_2 substrates. When retinas were kept on these surfaces for 48 h (Fig. 1.4), it was possible to remove the retina with the CNT islands totally embedded in the retina. Apparently, the retina is capable to engulf the CNTs in such a way that it can faithfully remove it from the support surface. Despite having CNT islands embedded in the retina, the retina seems to be perfectly healthy, demonstrating normal electrical activity in MEA recordings [37].

These results demonstrate that a neuronal tissue, such as a retina, appears to adhere exceptionally well to CNT surfaces. Here the effect can be mediated by cellular as well as molecular processes which facilitate the adhesion of the tissue to the CNT surface, making the CNTs act effectively as a Velcro film.

Three-dimensional electrodes are clearly very attractive for neuronal stimulation and recording applications. The enlarged surface area reduces the electrode impedance, allowing reduced noise as well as lower stimulation thresholds for neuronal activation [22]. Several different schemes have been developed to achieve roughness while also maintaining the mechanical integrity of the electrodes. Extremely rough

TiN is an excellent non-Faradaic electrode material often used both for recording and stimulation [38]. Optimized sputtering approach has been developed to yield a TiN electrode coating with preferred performances. Platinum black is another example for a commonly used porous material achieved using rapid electroplating coupled with prolonged sonication to establish both roughness and mechanical stability [22]. Electroplating has been used to deposit other materials such as IrO with improved performances. Despite the great performances of rough surfaces in saline characterization as well as in short-term investigations, the low impedance of the electrodes, associated with the high porosity, appears to quickly deteriorate under long-term in vivo investigations. It is likely that protein clogging in the pores and overall electrode engulfing by glia cells (e.g., gliosis) render the total impedance of the electrode to be significantly below the values obtained through surface enhancement.

Carbon nanotube electrodes offer a dramatic improvement in the electrical performances of the electrodes, when studied in vitro or ex vivo. Owing to the special manner by which neuronal cells interface with CNT (i.e., process entanglement), these surfaces may be beneficial for long-term applications. This issue however will have to be studied further to validate the true potential of CNT electrodes in this regard.

1.5 Summary

Cells appear to have the capacity to distinguish between surfaces of different mechanical properties and respond to these surface cues by modifying their proliferation, adhesion, and overall activity [39]. Specifically, neurons are sensitive to different surfaces and are known to preferentially adhere to rough surfaces. The origin of this effect appears to be associated with the manner by which neurites develop preferentially along ridges and grooves. Neuronal fasciculation, the collection of axons into a bundle, is readily apparent in many in vivo and in vitro scenarios. The most obvious in vivo examples, the optic and auditory nerves in which axonal navigation along a preexisting path, is an important developmental tool allowing guided growth of these systems. Many in vitro studies have shown how axons can be guided along certain directions using mechanical objects such as pillar arrays and ridges. Both in the in vivo and the in vitro cases, neurites will not continue with a predestined path but will ultimately deviate from it. By forming such twists and turns, a neuron can arbor itself very securely to a patterned substrate. It appears that rough surfaces generally provide neurons with optimal conditions for extensive elongation and anchoring.

Surface roughening can be achieved using various techniques with different materials. CNTs offer a unique combination of an extremely rough surface coupled with chemical and mechanical durability as well as bio-inertness and excellent electrochemical properties. All these considerations place CNT as a superior interface for neuronal applications.

To conclude, in this chapter, we reviewed various topics underlying the mechanical neuron–electrode interface and demonstrated that mechanical factors are ubiquitous and important. They may affect how the cells and the tissue form at the interface, possibly affecting short- and long-term properties such as adhesion, electrical coupling, and overall biological response. In vitro examinations have already revealed the many benefits of CNT surfaces [40, 41]. Some recent studies have begun exploring the long-term in vivo manifestations of these effects [42]. Further in vivo investigations are needed to validate the capacity of CNT electrode to form an improved electrical as well as mechanical interface.

References

1. Rosin, B., Slovik, M., Mitelman, R., Rivlin-Etzion, M., Haber, S. N., Israel, Z., Vaadia, E., and Bergman, H.: Closed-loop deep brain stimulation is superior in ameliorating Parkinsonism, Neuron **72**, 370–384 (2011)
2. Zrenner, E., Bartz-Schmidt, K. U., Benav, H., Besch, D., Bruckmann, A., Gabel, V.-P., Gekeler, F., Greppmaier, U., Harscher, A., Kibbel, S., et al.: Subretinal electronic chips allow blind patients to read letters and combine them to words. Proceedings Biological Sciences/The Royal Society **278**, 1489–1497 (2011)
3. Cogan, S. F.: Neural stimulation and recording electrodes. Annual Review of Biomedical Engineering **10**, 275–309 (2008)
4. Hai, A., Shappir, J., and Spira, M. E.: In-cell recordings by extracellular microelectrodes. Nature Methods **7**, 200–202 (2010)
5. Eshraghi, A. a, Gupta, C., Ozdamar, O., Balkany, T. J., Truy, E., and Nazarian, R.: Biomedical engineering principles of modern cochlear implants and recent surgical innovations. The Anatomical Record (Hoboken) **295**, 1957–1966 (2012)
6. Fernandes, R.A., Diniz, B., Ribeiro, R., and Humayun, M.: Artificial vision through neuronal stimulation. Neuroscience Letters **519**, 122–128 (2012)
7. Schwartz, A. B., Cui, X. T., Weber, D. J., and Moran, D. W.: Brain-controlled interfaces: movement restoration with neural prosthetics. Neuron **52**, 205–220 (2006)
8. Shahaf, G., and Marom, S.: Learning in networks of cortical neurons. The Journal of Neuroscience **21**, 8782–8788 (2001)
9. O'Shaughnessy, T. J., Gray, S. A., and Pancrazio, J. J.: Cultured neuronal networks as environmental biosensors. Journal of Applied Toxicology **24**, 379–385 (2004)
10. Johnstone, A. F. M., Gross, G. W., Weiss, D. G., Schroeder, O. H.-U., Gramowski, A., and Shafer, T. J.: Microelectrode arrays: a physiologically based neurotoxicity testing platform for the 21st century. Neurotoxicology **31**, 331–350 (2010)
11. Rajagopalan, J., Tofangchi, A., and Saif, M. T. A.: Drosophila neurons actively regulate axonal tension in vivo. Biophysical Journal **99**, 3208–3215 (2010)
12. Ayali, A.: The function of mechanical tension in neuronal and network development. Integrative Biology: Quantitative Biosciences from Nano to Macro **2**, 178–182 (2010)
13. Dowell-Mesfin, N. M., Abdul-Karim, M.-A., Turner, A. M. P., Schanz, S., Craighead, H. G., Roysam, B., Turner, J. N., and Shain, W.: Topographically modified surfaces affect orientation and growth of hippocampal neurons. Journal of Neural Engineering **1**, 78–90 (2004)
14. Shalek, A. K., Robinson, J. T., Karp, E. S., Lee, J. S., Ahn, D.-R., Yoon, M.-H., Sutton, A., Jorgolli, M., Gertner, R. S., Gujral, T. S., et al.: Vertical silicon nanowires as a universal platform for delivering biomolecules into living cells. Proceedings of the National Academy of Sciences of the United States of America **107**, 1870–1875 (2010)
15. Mazzatenta, A., Giugliano, M., Campidelli, S., Gambazzi, L., Businaro, L., Markram, H., Prato, M., and Ballerini, L.: Interfacing neurons with carbon nanotubes: electrical signal transfer

and synaptic stimulation in cultured brain circuits. The Journal of Neuroscience: the Official Journal of the Society for Neuroscience **27**, 6931–6936 (2007)

16. Mammoto, T., and Ingber, D. E.: Mechanical control of tissue and organ development. Development (Cambridge, England) **137**, 1407–1420 (2010)
17. Rajagopalan, J., and Saif, M. T. A.: MEMS sensors and microsystems for cell mechanobiology. Journal of Micromechanics and Microengineering: Structures, Devices, and Systems **21**, 54002–54012 (2011)
18. Leong, W. S., Wu, S. C., Pal, M., Tay, C. Y., Yu, H., Li, H., and Tan, L. P.: Cyclic tensile loading regulates human mesenchymal stem cell differentiation into neuron-like phenotype. Journal of Tissue Engineering and Regenerative Medicine **6** Suppl 3, s68–79 (2012)
19. Baranes, K., Kollmar, D., Chejanovsky, N., Sharoni, A., and Shefi, O.: Interactions of neurons with topographic nano cues affect branching morphology mimicking neuron-neuron interactions. Journal of Molecular Histology **43**, 437–447 (2012)
20. Brunetti, V., Maiorano, G., Rizzello, L., Sorce, B., Sabella, S., Cingolani, R., and Pompa, P. P.: Neurons sense nanoscale roughness with nanometer sensitivity. Proceedings of the National Academy of Sciences of the United States of America **107**, 6264–6269 (2010)
21. Anava, S., Greenbaum, A., Ben Jacob, E., Hanein, Y., and Ayali, A.: The regulative role of neurite mechanical tension in network development. Biophysical Journal **96**, 1661–1670 (2009)
22. Maher, M., Pine, J., Wright, J., and Tai, Y.-C.: The neurochip: a new multielectrode device for stimulating and recording from cultured neurons. Journal of Neuroscience Methods **87**, 45–56 (1999)
23. Bray, D.: Mechanical tension produced by nerve cells in tissue culture. Journal of Cell Science **37**, 391–410 (1979)
24. Bray, D.: Axonal growth in response to experimentally applied mechanical tension. Developmental Biology **102**, 379–89 (1984)
25. Bernal, R., Pullarkat, P., and Melo, F.: Mechanical properties of axons. Physical Review Letters **99**, 018301 (2007)
26. Bernal, R., Melo, F., and Pullarkat, P.: Drag force as a tool to test the active mechanical response of PC12 neurites. Biophysical Journal **98**, 515–523 (2010)
27. Hanein, Y., Tadmor, O., Anava, S., and Ayali, A.: Neuronal soma migration is determined by neurite tension. Neuroscience **172**, 572–579 (2011)
28. Sorkin, R., Greenbaum, A., David-Pur, M., Anava, S., Ayali, A., Ben-Jacob, E., and Hanein, Y.: Process entanglement as a neuronal anchorage mechanism to rough surfaces. Nanotechnology **20**, 015101 (2009)
29. Xie, C., Hanson, L., Xie, W., Lin, Z., Cui, B., and Cui, Y.: Noninvasive neuron pinning with nanopillar arrays. Nano Letters **10**, 4020–4 (2010)
30. Bareket-Keren, L., and Hanein, Y.: Carbon nanotube-based multi electrode arrays for neuronal interfacing: progress and prospects. Frontiers in Neural Circuits **6**, 122 (2012). doi: 10.3389/fncir.2012.00122
31. Gabay, T., Ben-David, M., Kalifa, I., Sorkin, R., Abrams, Z. R., Ben-Jacob, E., and Hanein, Y.: Electro-chemical and biological properties of carbon nanotube based multi-electrode arrays. Nanotechnology **18**, 035201 (2007)
32. Zhang, X., Prasad, S., Niyogi, S., Morgan, A., Ozkan, M., and Ozkan, C.: Guided neurite growth on patterned carbon nanotubes. Sensors and Actuators B: Chemical **106**, 843–850 (2005)
33. Gabay, T., Jakobs, E., Ben-Jacob, E., and Hanein, Y.: Engineered self-organization of neural networks using carbon nanotube clusters. Physica A: Statistical Mechanics and Its Applications **350**, 611–621 (2005)
34. Greenbaum, A., Anava, S., and Ayali, A.: One-to-one neuron–electrode interfacing. Journal of Neuroscience Methods **182**, 219–224 (2009)
35. Shein, M., Greenbaum, A., Gabay, T., Sorkin, R., David-Pur, M., Ben-Jacob, E., and Hanein, Y.: Engineered neuronal circuits shaped and interfaced with carbon nanotube microelectrode arrays. Biomedical Microdevices **11**, 495–501 (2009)
36. Shoval, A., Adams, C., David-Pur, M., Shein, M., Hanein, Y., and Sernagor, E.: Carbon nanotube electrodes for effective interfacing with retinal tissue. Frontiers in Neuroengineering **2**, 4 (2009)

37. Eleftheriou, C. G., Zimmermann, J., Kjeldsen, H., David-Pur, M., Hanein, Y., and Sernagor, E.: Towards the development of carbon nanotube based retinal implant technology: electrophysiological and ultrastructural evidence of coupling at the hybrid interface. Proceedings of the 8th International MEA Meeting on Substrate Integrated Microelectrode Arrys. Reutlingen, Germany (2012)
38. Janders, M., Egert, U., Stelzle, M., and Nisch, W.: Novel thin film titanium nitride microelectrodes with excellent charge transfer capability for cell stimulation and sensing applications. Proceedings of 18th Annual International Conference of the IEEE Engineering in Medicine and Biology Society (Amsterdam: IEEE), 245–247 (1996)
39. Shein Idelson, M., Ben-Jacob, E., and Hanein, Y.: Innate synchronous oscillations in freely-organized small neuronal circuits. PLoS One **5**, e14443 (2010)
40. Sirivisoot, S., Yao, C., Xiao, X., Sheldon, B. W., and Webster, T. J.: Greater osteoblast functions on multiwalled carbon nanotubes grown from anodized nanotubular titanium for orthopedic applications. Nanotechnology **18**, 365102 (2007)
41. Gabriel, G., Gómez-Martínez, R., and Villa, R.: Single-walled carbon nanotubes deposited on surface electrodes to improve interface impedance. Physiological Measurement **29**, S203–212 (2008)
42. Minnikanti, S., and Peixoto, N.: Carbon nanotubes applications on electron devices. In: J. M. Marulanda (ed.), Implantable electrodes with carbon nanotube coatings. InTech: Croatia (2011)

Chapter 2
Nanoscale Field-Effect Transistors for Minimally Invasive, High Spatial Resolution, and Three-Dimensional Action Potential Recording

Xiaojie Duan

2.1 Introduction

This chapter is devoted to the development and application of nanoscale field-effect transistors (FETs) for neural and cardiac activity recording. Compared to optical methods, the electrical recording of action potentials has high signal-to-noise ratio (SNR) and temporal resolution. But the need for electrodes limits its spatial resolution and also poses perturbation on the biological system under investigation. One way to overcome these problems is to develop electrical recording devices with nanometer size and high-density scaling-up ability. Microfabricated metal electrodes can be readily patterned into arrays, but it is hard to decrease the size of these electrodes to nanometer scale because of the necessity to ensure a reasonable impedance value at the electrode/electrolyte interface for sufficient SNR. Field-effect transistors (FETs), on the other side, can sense the potential of the solution independently on the device/electrolyte interface impedance and hence allow for the miniaturization of the probes to nanometer scale, which is important for minimally invasive, high spatial resolution electrical recording and mapping of neuronal activities, as will be discussed in this chapter.

In a standard FET, the conductance of the semiconductor channel between the source (*S*) and (*D*) drain electrodes is modulated by a third gate (*G*) electrode capacitively coupled through a thin dielectric layer to the semiconductor. In the case of a p-type semiconductor, applying a positive gate voltage depletes majority of charge carriers (positive holes) and reduces the conductance, whereas applying a negative gate voltage leads to an accumulation of carriers and a corresponding increase in conductance (Fig. 2.1) [1, 2]. If the FET is immersed into an electrolyte solution, the solution can act as analog of metallic gate electrode in the conventional FET

X. Duan (✉)
Peking University, Beijing, China
e-mail: xjduan@pku.edu.cn

M. De Vittorio et al. (eds.), *Nanotechnology and Neuroscience: Nano-electronic, Photonic and Mechanical Neuronal Interfacing*, DOI 10.1007/978-1-4899-8038-0_2,

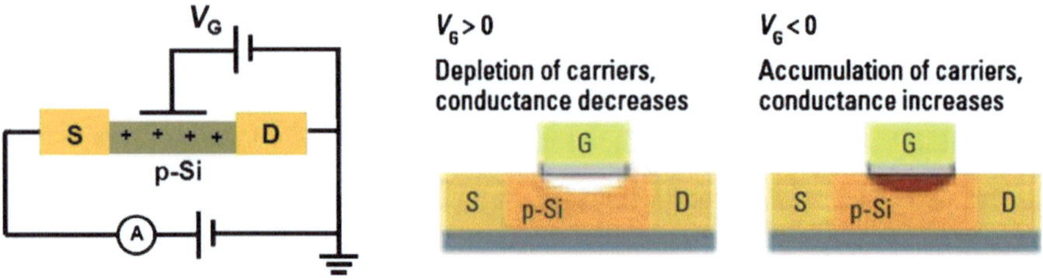

Fig. 2.1 FET voltage sensors. *S*, source; *D*, drain; and *G*, gate electrodes; V_G, gate voltage. Reproduced with permission from [2], © (2006) American Chemical Society

configuration and thus is termed a water gate. The dependence of the conductance on gate voltage makes FETs natural candidates for electrically based voltage sensing.

There are various ways to make the nanoscale FETs (nanoFETs) to fulfill the purpose of voltage sensing and action potential recording with minimal invasiveness and high spatial resolution. In this chapter, we choose to mainly talk about the nanoFETs which use freestanding, one-dimensional (1D) semiconducting nanomaterials, such as silicon nanowires (SiNWs), as the active channels of the FETs. Compared to their planar counterparts, the 1D nanoscale morphology of nanowires leads to depletion or accumulation of carriers through the entire cross section of the device from the potential change on the surface of a nanowire versus only a shallow region near the surface of a planar device. This leads to higher sensitivity for voltage sensing by using nanowire FETs [3]. Furthermore, the bottom-up paradigm used to fabricate nanowire FETs allows nanoscale FETs to be fabricated on nearly any type of surface, including those that are typically not compatible with standard CMOS processing, such as flexible plastic substrates. And it also enables the fabrication of flexible, three-dimensional (3D) freestanding devices, which can be utilized as the macroporous scaffold for synthetic tissue constructs and used for neuronal and cardiac activity monitoring throughout the 3D cellular networks inside the constructs, capabilities that are hard to achieve with conventional probes or even recently developed flexible electronics. Finally, the structure, morphology, physical properties, and corresponding functions of semiconducting nanowires can be readily controlled by encoded synthesis, which makes them ideally suited for hierarchical design of various extracellular and intracellular probes, as will be talked about later in this chapter.

In the next section, we will briefly go over some basics related to nanowire FET, including the general method for nanowires preparation and nanowire FETs fabrication. Then we will discuss the development and application of novel SiNW FET devices as high spatial resolution and minimally invasive extra- and intracellular action potential recording probes. Lastly, we will talk about the recent progress of developing 3D macroporous, freestanding nanoFETs for 3D electrical interfacing with living biological systems, such as synthetic neural and cardiac tissue constructs.

2.2 Nanowires and Nanowire FETs

Semiconductor nanowires exhibit a conductivity change in response to variations in the electric field or potential at their surface, which can be configured as FET channels and used for potential sensing. SiNWs are one of the best characterized examples of semiconducting nanowires which can be prepared as single crystal structures with diameters as small as 2–3 nm [4]. They can be prepared as p- or n-type materials and configured as FETs that exhibit electrical performance characteristics comparable to or better than those achieved in the microelectronics industry for planar silicon devices. Semiconductor nanowires are generally synthesized by employing metal nanoclusters as catalysts via a vapor–liquid–solid (VLS) process [5, 6]. In this process, the metal nanoclusters are heated above the eutectic temperature for the metal–semiconductor system of choice in the presence of a vapor-phase source of the semiconductor, resulting in a liquid droplet of the metal/semiconductor alloy. The continued feeding of the semiconductor reactant into the liquid droplet supersaturates the eutectic, leading to nucleation of the solid semiconductor. The solid–liquid interface acts as a sink causing the continued semiconductor incorporation into the lattice and, thereby, the growth of the nanowire with the alloy droplet riding on the top [6] (Fig. 2.2).

The gaseous semiconductor reactants can be generated through decomposition of precursors in a chemical vapor deposition (CVD) process. In CVD-VLS growth, the metal nanocluster serves as a catalyst at which site the gaseous precursor decomposes, providing the gaseous semiconductor reactants. In the case of SiNWs growth

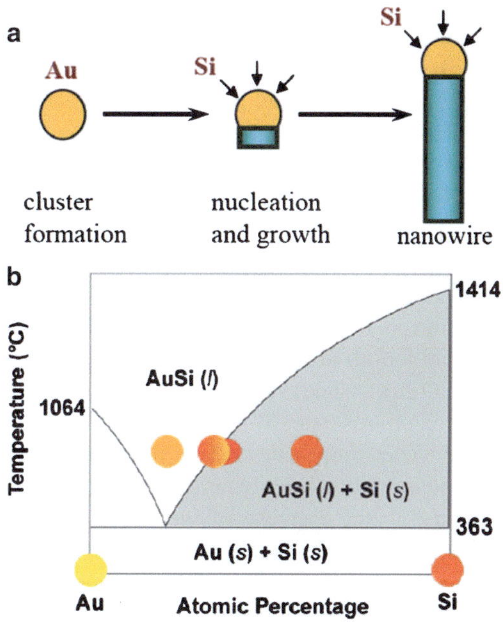

Fig. 2.2 Schematic of VLS growth of SiNWs. (**a**) A liquid alloy droplet AuSi is first formed above the eutectic temperature (363 °C) of Au and Si. The continued feeding of Si in the vapor phase into the liquid alloy causes oversaturation of the liquid alloy, resulting in nucleation and directional nanowire growth. (**b**) Binary phase diagram for Au and Si illustrating the thermodynamics of VLS growth. Reproduced with permission from [6], © (2006) IOP Publishing

(Fig. 2.2), silane (SiH_4) and Au nanoparticles are normally used as the precursor and catalysts, respectively [6]. The diameter of the nanowire is determined by that of the starting nanocluster in CVD-VLS growth, and nanoclusters with diameters down to a few nanometers are now commercially available. The crystallographic orientation of the nanowires during VLS growth is chosen to minimize the total free energy, as the process is thermodynamically driven. In reality, it is determined by the diameter and growth condition such as temperature and pressure.

The VLS process offers one key advantage compared with other approaches such as vapor–solid growth or solution-based liquid–solid growth, in that the heterostructures can be achieved at the individual device level in a controlled fashion. Axial heterostructures, in which sections of different materials or same materials with different dopant level of the same diameter are grown along the nanowire axis, can be readily realized by switching the reactant during the growth [7]. By doing VLS first and then vapor–solid growth which homogeneously deposits a conformal shell on the preformed nanowire surface versus reactant addition at the nanoparticle catalyst, radial heterostructures with core/shell and core/multi-shell can form along the radial direction [8]. 3D hierarchical branched nanowire structures with same or different materials for backbones and branches can be realized by multistep VLS growth [9]. In addition, the morphology of nanowires can also be controlled by adding disturbance during VLS growth to obtain, for example, the kinked nanowires [10]. These added controls over morphology, composition, and structure can play a critical role in creating novel nanoscale FET devices for extra- and intracellular action potential recording.

To make nanowire FET devices, the nanowires can be randomly dispersed from solution by drop casting or assembled by using contact printing onto substrates such as SiO_2/Si wafers. *S* and *D* metal electrodes are then defined on nanowires by using photolithography or electron-beam lithography. To get nanowire FETs with good electrical performance, the metal used as *S* and *D* needs to match the work function of the semiconducting materials, same as in the conventional planar devices. Besides, for the application of action potential recording, other considerations, such as the metal corrosion in electrolyte and biocompatibility issue, need to be taken as well. For SiNWs, palladium (Pd) with thin layer of titanium (Ti) as adhesion has been shown as a good candidate for *S* and *D* electrodes. Lastly, passivation on the metal connection lines using insulating layers such as SiN_x or SU-8 is needed for the nanowire FET potential sensing. In this chapter, the diameter of SiNWs used in action potential recording probes is normally in the range of 10–100 nm, and the channel length of the nanowire FETs, which is the nanowire length between the *S* and *D* electrodes, is in the range of 50–2,000 nm. Figure 2.3a shows the scanning electron microscope (SEM) image of a typical SiNW FET. An example of conductance versus water-gate voltage traces for three representative p-type SiNW FETs is shown in Fig. 2.3b, where all three FETs exhibit a conductance decrease in response to the increase of the solution potential [3]. For voltage sensing with FETs, the recorded voltage values are normally calibrated from the conductance change and sensitivity of the FETs, and the sensitivity can be modulated by changing the water-gate offset.

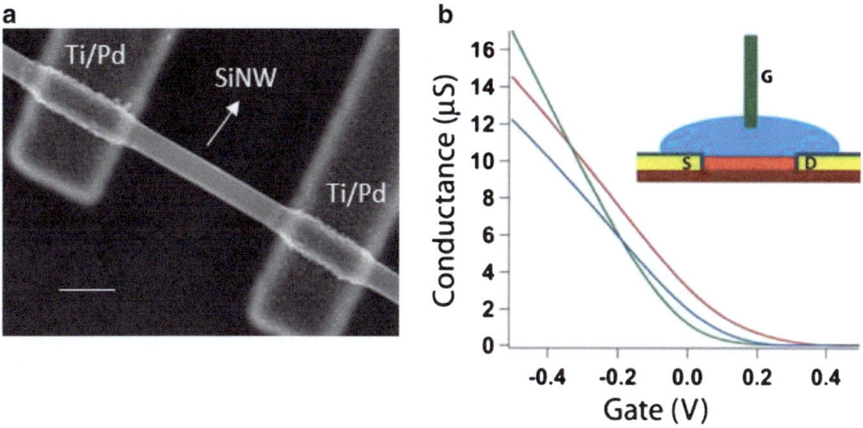

Fig. 2.3 SiNW FET voltage sensors: (**a**) SEM image of a SiNW FET. Here the *S* and *D* electrodes are made of palladium with titanium as adhesion layer. Scale bar is 100 nm. (**b**) Conductance versus water-gate voltage traces for three representative p-type SiNW FET devices. *Inset*: scheme representing experimental setup, (*red*) nanowire, (*yellow/navy*) passivated contact electrodes, (*blue*) electrolyte solution, and (*green*) Ag/AgCl reference/gate electrode [3]

2.3 Extracellular Recording with Nanoscale FETs

The potential change outside of excited cells associated with the transmembrane change can be recorded to monitor the electric activity of the cells. Compared to the intracellular electrical recording, which normally places the probes inside the cells, extracellular recording is much less invasive and provides ready access to simultaneous multisite recording. To date, the most widely used technique for extracellular recording of neuronal activities is the multielectrode arrays (MEAs) [11]. The size of the metal electrodes used in MEAs is normally in the range of 5–100 μm. This size ensures a reasonable impedance value at the electrode/electrolyte interface and gives sufficient SNR for action potential detection. With the use of FETs, the size of the extracellular recording devices can be decreased to the nanometer scale. This increase in spatial resolution will allow for the recording from small size critical subcellular structures, such as axons and dendrites of neurons, which is hard to achieve with MEAs. In MEA recording, the large size of metal electrodes also makes cell-to-electrode registration challenging because the measured signals for a given electrode typically are due to contributions from several nearby cells. As a result, identification of specific cellular signals generally requires complicated post-processing such as the spike sorting. With much smaller size, the nanoscale FET recording devices enable easy and accurate cell-to-electrode registry by decreasing the probability of coupling to multiple cells. Besides, higher-density arrays can be made from nanoscale FETs, which can increase the spatial resolution in neural activity mapping.

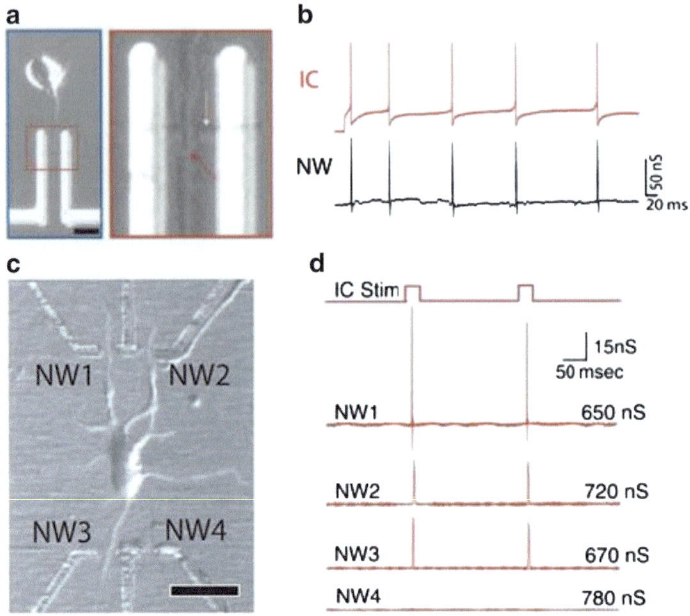

Fig. 2.4 Extracellular action potential recording from cultured neurons. (**a**) *Left*: optical image of a cortical neuron aligned across a SiNW FET; scale bar is 10 μm. *Right*: high-resolution image of the region where axon (*red arrow*) crosses a SiNW (*yellow arrow*). (**b**) *Red trace*: intracellular potential of a cortex neuron (after 6 days in culture) recorded by glass micropipette during stimulation; *black trace*: time-correlated signal from axon measured using a p-type SiNW FET. (**c**) Optical image of a cortical neuron interfaced to three of the four functional SiNW FETs in an array; scale bar is 20 μm. (**d**) Trace of intracellular current stimulation by a glass micropipette in the soma and the resulting electrical responses from the four SiNW FETs in (**c**). NW1 ~ NW4 means silicon nanowire 1–4 used in the four FETs [12]

2.3.1 *Extracellular Recording from Cultured Cells*

The cellular and subcellular interfaces between FETs and living excitable cells can be realized by either directly culturing cells on top of the FET devices or culturing cells on a separate elastic substrate, such as polydimethylsiloxane (PDMS) sheet, and then manually bringing the cell sheet and devices chip together for extracellular recording. For direct culturing, surface patterning of polylysine can be used to guide the growth of neuron cells, including the cell body and neurites, to ensure a high yield of neuron/FET junctions and hence an efficient interfacing. With this method, SiNW FETs have successfully recorded extracellular action potential signals from neuron axons and dendrites, where the signal was in good temporal correlation with the intracellular action potentials recorded by a glass micropipette (Fig. 2.4a, b). This direct correlation indicates that the depolarization of cell membrane during action potential firing results in negative charging of the extracellular space around

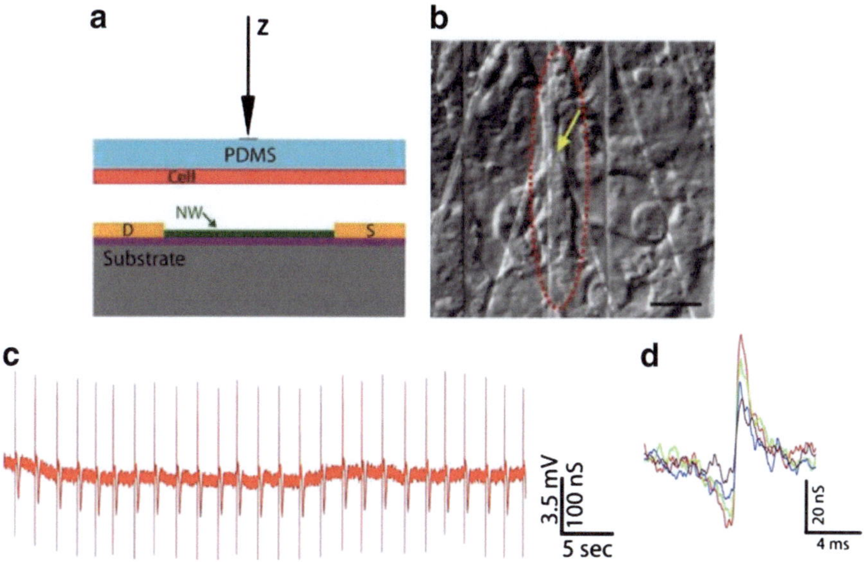

Fig. 2.5 Building interface between SiNW FETs and spontaneously firing cardiomyocytes. (**a**) Schematic illustration of manipulating the PDMS/cell substrate to make the interface with an NWFET device. NW means nanowire, same as below. (**b**) Distinct patch of beating cells (*red dashed oval*) over a SiNW device (*yellow arrow*). Scale bar is 20 μm. (**c**) A representative trace recorded by a SiNW FET from a spontaneously firing cardiomyocyte at displacement value close to cell failure. (**d**) High-resolution comparison of single spikes recorded with increasing displacement values (from *purple* to *red*) [13]

the nanowire, which is consistent with the fact that the membrane expresses a relatively high density of Na^+ ion channel. Multiplexed recording from different dendrites and axon of a single neuron was also realized by using high-density SiNW FETs array (Fig. 2.4c, d), which highlighted the advantage of higher spatial precision and resolution for the nanoscale-size SiNW FET probes. For the recording with SiNW FETs, the active junction area for nanowire/axon interfaces was 0.01–0.02 μm^2, which is at least two orders of magnitude smaller than microfabricated metal electrodes and previous planar FETs [12].

The second method of interfacing nanoFETs with living cells allows for separate design and optimization of the nanoFETs array and cell culture such that the two key components are brought together under precise manipulation only during the final measurement phase. Compared to direct culture, this approach enables people to identify, register to and record from specific cellular and subcellular regions with respect to nanoFET devices, and carry out multiplexed recording from well-defined multicellular configurations with overall subcellular resolution [3]. In addition, with this approach, the amplitude and SNR of the recorded signals can be tuned by manually adjusting the coupling strength between the cells and nanoFETs. As shown in Fig. 2.5, embryonic chicken cardiomyocytes were cultured on 100–500 μm thick, optically transparent and flexible PDMS pieces to form cell monolayers. By bringing the cells closer to the SiNW FET devices, the amplitude of recorded extracellular action potential spikes can reach a value of 10.5 mV from ~1 mV and a maximum SNR of 25 before making irreversible changes and cessation of the spontaneous cell beating [13]. It should be noted that within this range of PDMS/cell displacement value, the spike amplitude changes are reversible, and the SiNW FET/cell interface is stable for different displacements of the cells toward the devices. The enhanced signal amplitudes can be attributed to a decrease in the gap between the cell membrane and SiNW FET devices, although further studies will be needed to quantify such junction changes.

This new interfacing strategy can also be used for multiplexing measurements, as different devices can all form tight junctions with the cells and hence simultaneously give signals with high SNR. The time shift between devices derived from the cross-correlation analysis of data traces recorded in this manner provides information about the action potential propagation direction and speed. The flexibility of interfacing with specific cellular areas for high-density multiplexed recording using SiNW FETs represents a powerful platform to study the effects of cell monolayer inhomogeneity on action potential propagation, enabling both intra- and intercellular propagation to be characterized in details for well-defined cellular structures.

The small size of nanoFETs allows for more localized and higher spatial precision recording than achievable with other larger size detectors, as reflected by the different signal shapes. As shown in Fig. 2.6, in the extracellular recording from spontaneously firing chicken cardiomyocytes, the signal recorded by the a SiNW FET with ~30 nm diameter and ~2.3 μm channel length has a peak-to-peak width of 760 ± 40 μs. This value is smaller than 1,310 ± 40 μs obtained from a graphene FET (Gra-FET) which uses graphene of 20.8 μm × 9.8 μm large as the active channel for FET (the Gra-FETs were fabricated side by side on the same chip with the SiNW FETs) [14]. This result indicates that the signals recorded with the large Gra-FET (Gra1-FET in Fig. 2.6) do not represent a localized detection but rather an average of the extracellular potential from different parts of the beating cell or even from different cells, which caused the broadening of the peak. However, the signal recorded by another Gra-FET (Gra2-FET in Fig. 2.6) with active channel of 2.4 μm × 3.4 μm, which is 16 μm away from the SiNW FET, gives peak-to-peak width of similar value, 730 ± 40 μs, although the SiNW FET has ~100× smaller active detection area [14]. This can be explained by the micrometer-scale channel length of the SiNW FET detector. Although the detection can be localized in nanometer scale in the radial direction, there will still be average from the micrometer long axial detection, which leads to the broadened peaks detected by the SiNW FET compared to the intrinsic values expected for sodium ion channels.

To make real "point-like" nanoscale detectors, the active channel length needed to be shrunk down to the size comparable to the nanowire diameter. One way of making this short-channel nanowire FET is to put the source and drain electrode close to nanoscale distance. However, these metal electrodes will physically limit cell access and also electrostatically screen the active nanowire channel, making it less sensitive to the potential change of the solution around it [15]. With VLS grown nanowires, the "point-like" ultrasmall detector can be realized in a unique way of

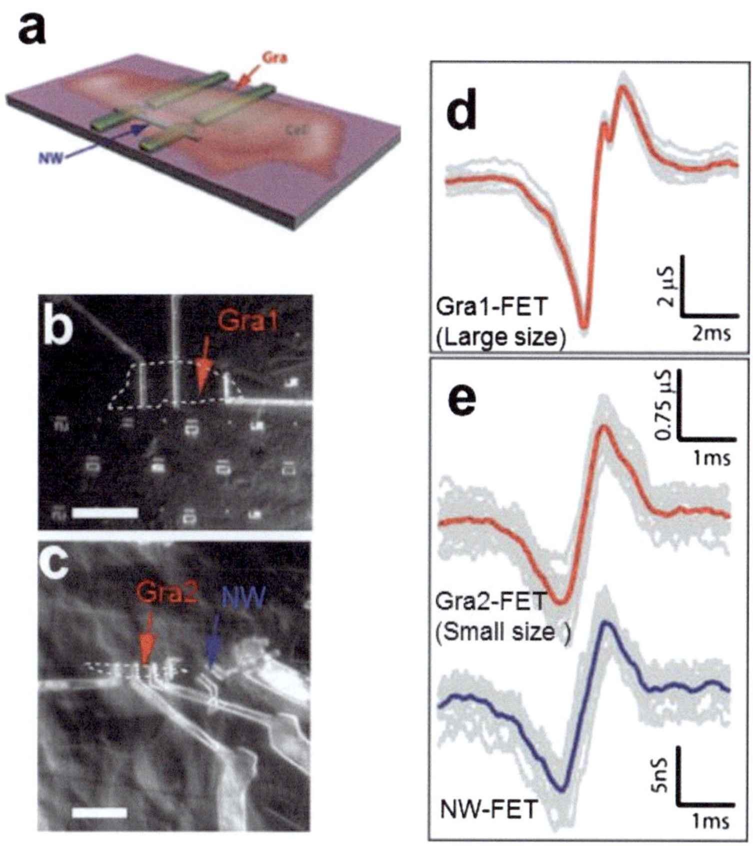

Fig. 2.6 Extracellular action potential recording from FET probes with different size of active channels. (**a**) Schematic illustrating the chip design incorporating graphene and SiNW FET devices. (**b**) Optical microscope image of PDMS/cells interfaced with large flake graphene FET. Scale bar is 30 μm. (**c**) Optical microscope image of PDMS/cells interfaced with smaller flake graphene FET and SiNW FET. Scale bar is 13.6 μm. Graphene flake outline is marked by *white dashed line*, measured graphene device is marked by *red arrow*, and measured SiNW device is marked by *blue arrow*. (**d**) Recorded averaged peak (*red*) and raw data (*gray traces*) for the Gra1-FET and cell in (**b**). (**e**) Thirteen raw signal peaks (*gray traces*) from the Gra2-FET (*upper* data) and SiNW FET (*lower* data) devices marked by *red* and *blue arrows*, respectively, in (**c**). The average of the peaks was plotted in *red* and *blue*, respectively. Reproduced with permission from [14] © (2010) American Chemical Society

dopant modulation along axial direction. By changing the dopant ratio in the feeding sources during nanowire growth, short length of lightly doped nanowire segment with two heavily doped nanowire segments connected to its two ends can be synthesized. This short lightly doped nanowire segment will be the FET active channel and two heavily doped arms function as nanoscale *S*/*D* electrode arms. Metal interconnects are then placed on these two arms, thus ensuring an intimate contact between the cells and detectors.

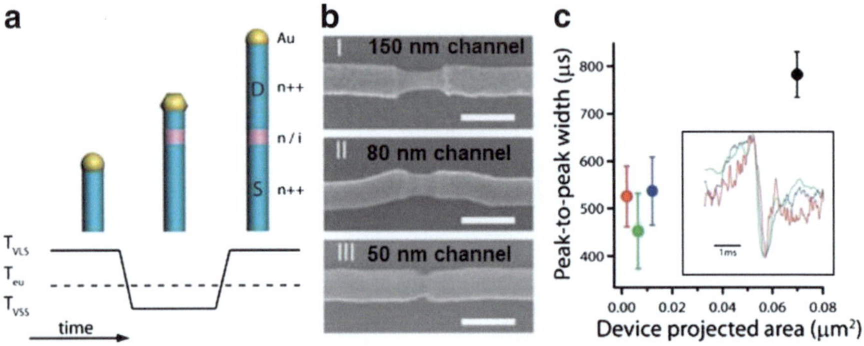

Fig. 2.7 Extracellular action potential recording with short-channel SiNW FETs. (**a**) Illustration of Au-nanocluster-catalyzed nanowire growth with well-controlled axial dopant profile introduced during VSS growth. (**b**) Short-channel $n^{++}/i/n^{++}$ SiNWs with channel lengths of 150, 80, and 50 nm, respectively. Scale bars are 150 nm. The Au nanoclusters were ~80 nm in diameter, and nanowires were selectively etched to reveal the active channel. (**c**) Summary of the peak-to-peak widths for the 150, 80, and 50 nm channel length devices. In addition, a previously published 2.3 μm channel length SiNW FET device (*black*) is shown for comparison [14]. *Inset*, an example of single peaks from each of the short-channel devices. Reproduced with permission from [16] © (2010) American Chemical Society

Except for VLS mode, nanowire growth can also use a vapor–solid–solid (VSS) mechanism; in this case, the growth temperature is below the eutectic temperature, and instead of liquid metal–semiconductor alloy droplets, a solid alloy forms and acts as the nanowire growth interface [17]. In VSS growth, the nanowire growth rate is at least 10–100 times lower than for VLS growth. By switching to VSS for the lightly doped nanowire segment (while maintaining VLS for the most part of the heavily doped nanowire arms growth to keep a reasonable growth time in total, Fig. 2.7a), short-channel nanowire devices can be realized in a way which enables control of the lightly doped nanowire channel segment on a 10 nm scale and abrupt lightly heavily doped nanowire junctions. SEM imaging of selectively etched lightly doped nanowire segments (Fig. 2.7b) showed that nanowires with channel lengths of 150, 80, and 50 nm could be synthesized using this approach [16].

Extracellular recording from spontaneously firing cardiomyocytes using SiNW FETs configured from these new short-channel SiNWs (Fig. 2.7c) yielded action potential signals with peak-to-peak widths of 520 ± 40, 450 ± 80, and 540 ± 50 μs for the 150, 80, and 50 nm devices, respectively [16]. These widths are smaller than the peak-to-peak widths of 750–850 μs recorded with the conventional nanowire with micrometer-scale active channels discussed above, which indicates the advantage of the recording with "point-like" detectors to avoid extrinsic temporal broadening. Interestingly, the time scale reported for Na^+ channel conduction is about 500 μs [18], which is consistent with the peak widths measured from the short-channel SiNW FETs here. This suggests that these short-channel SiNW FET devices may be able to study ion channel on the length and time scale of single ion channel events in the future.

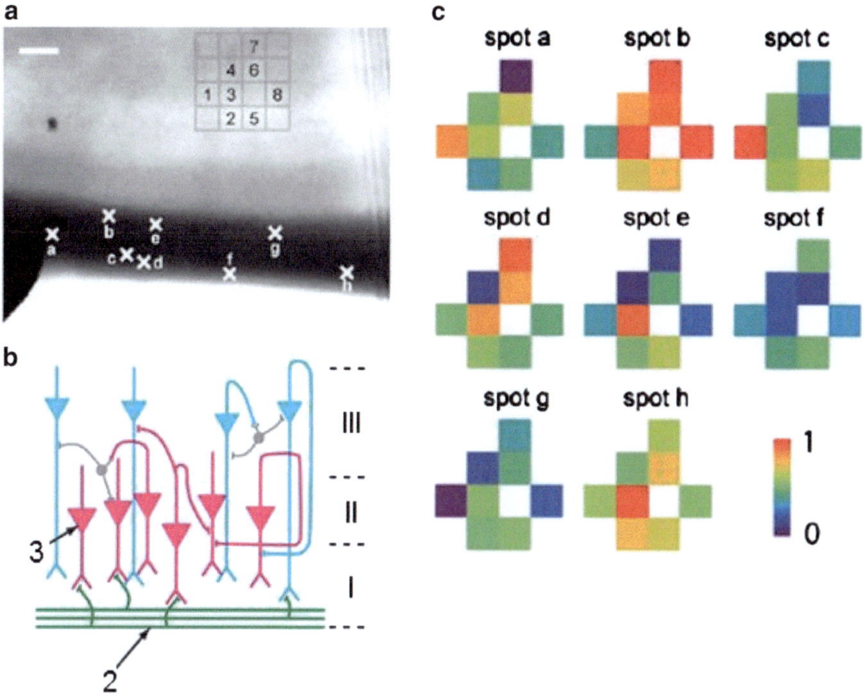

Fig. 2.8 Extracellular recording from acute brain slices with SiNW FETs. (**a**) Optical image of an acute slice over a 4×4 SiNW FET array. Crosses along the LOT fiber region of the slice mark the stimulation spots *a–h*. Scale bar is 100 μm. (**b**) Laminar organization and input circuitry of the piriform cortex (layers I–III). (**c**) Maps of the relative signal intensity or activity for devices 1–8 [19]

2.3.2 Extracellular Recording from Tissue Slices and Organs

In the brain, neural circuits are organized through synaptic connections into hierarchical networks operating on spatial and temporal scales that span multiple orders of magnitude. From this perspective, electrical recording with high spatiotemporal resolution from populations of neurons using nanoscale FET arrays presents unique opportunities in neural activity mapping of the brain tissue. On the other hand, electrical recording in vitro and in vivo from whole hearts is important in areas ranging from basic studies of cardiac function to patient healthcare. The flexible and transparent recording chips enabled by the bottom-up paradigm used in nanowire FETs fabrication provide a powerful way to make conformal contact to the 3D soft tissue of heart, as will be discussed in this section.

The high-resolution recording capability of SiNW FET arrays was exploited to map the neural connectivity in the olfactory cortex of a rat acute brain slice as shown in Fig. 2.8. An optical image of an oriented brain slice (Fig. 2.8a) shows the lateral olfactory tract (LOT) (*dark band*) and the pyramidal neuron layers. A schematic of the organization and circuit of the slice (Fig. 2.8b) highlights the LOT and

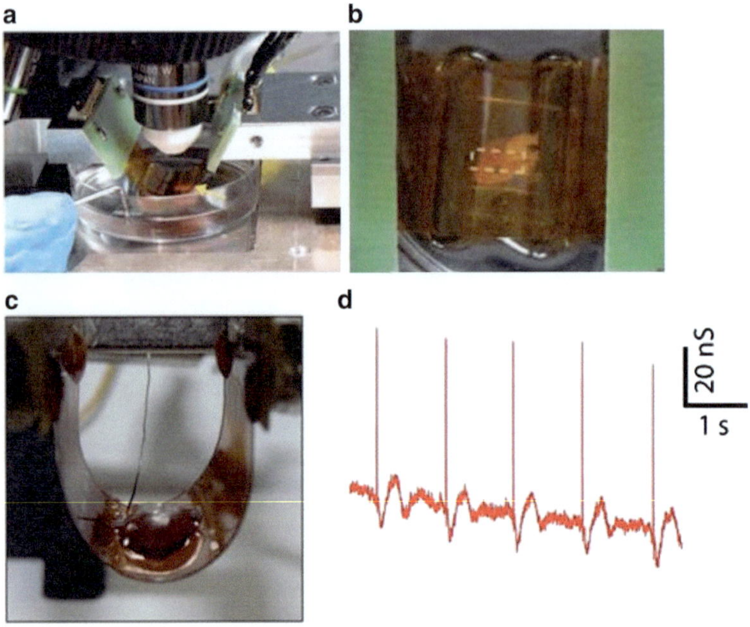

Fig. 2.9 Extracellular recording from whole hearts with SiNW FETs fabricated on flexible plastic substrate. (**a**) Photograph of a flexible SiNW FETs chip interfaced with a chicken heart in concave configuration. *Yellow arrow* marks the location of the heart. (**b**) Top-down photograph of same system, which enables overall registration between heart and lithographically defined markers on the substrate. (**c**) Photograph of a flexible SiNW FETs chip interfaced with a chicken heart in convex configuration. (**d**) A representative recorded conductance data from a SiNW FET interfacing with chicken heart in bent configuration. Reproduced with permission from [20] © (2009) American Chemical Society

synaptic connections (*layer I*) with the pyramidal cells in *layers II* and *III*, which are oriented over the SiNW FET array. Following stimulation at eight different spots or different sets of axon fibers in the LOT (a–h, Fig. 2.8a) using a metal electrode or glass micropipette, eight SiNW FETs (1–8, Fig. 2.8a) showed distinct responses in terms of the amplitude for the recorded postsynaptic signal, as shown in the 2D maps in Fig. 2.8c. These 2D maps from the SiNW FET array resolved clearly the heterogeneous activity of the neural circuit [19]. Note that the distance between adjacent devices, 60 μm, in which distinct activity was recorded is already better than the 100 μm scale resolution achieved in MEA recording of brain slices, although the mapping resolution can be substantially improved by using higher-density nanowire FET arrays.

Silicon nanowire FETs were fabricated on thin polymer substrates and interfaced to beating embryonic chicken hearts in bent chip configuration, as shown in Fig. 2.9. The use of flexible and transparent SiNW FETs chips enables simultaneous optical imaging and electronic recording. A bent device chip with concave surface facing a beating heart (Fig. 2.9a) shows that the SiNW FET/heart can be readily examined with microscope. This allows for visual orientation of the device array to the heart and higher-resolution imaging through the transparent substrate (Fig. 2.9b). Recording from the whole heart in a convex configuration (Fig. 2.9c), where the heart/device interface is on the convex side of the bent chip, gives extracellular signals with excellent SNR (Fig. 2.9d). The signal correlates with the spontaneous beating of the heart and shows an initial sharp peak followed by a slower phase, where these two phases can be attributed to transient ion channel currents and mechanical motion, respectively [20].

The above studies on acute brain slices and whole hearts demonstrate the powerful capability of nanoFETs to interface with hierarchically organized tissue and organs with cellular and subcellular spatial resolution and submillisecond temporal resolution. The advantage of conformal interfaces with soft and irregularly shaped tissues and organs enabled by the flexible nanowire FET chips represents unique opportunities in cellular activity mapping, which demonstrates the potential of nanowire FETs as tools to understand the functional connectivity and address critical biological problems in the study of neural and cardiac systems.

2.4 Intracellular Recording with Nanoscale FETs

Intracellular recording has several advantages over extracellular recording. Firstly, it measures true transmembrane potential of the cells, instead of the derivative form of the cell transmembrane potential change reflected by the extracellular recording, which makes the extracellular signal much smaller in amplitude and distinct in shape/time scale from real action potentials. As a result, extracellular recording reflects the time of occurrence of action potentials but is not able to record action potentials with the details needed to explore the properties of ion channels. Second, intracellular recording can measure subthreshold events and DC or slowly changing potentials across the cell membrane associated with synaptic interactions, which is important for neural network activity but difficult to measure with extracellular probes. Third, intracellular recording enables clear cell-to-electrode registry. Extracellular signals often represent an average over several cells located at the vicinity of the probe. As discussed above, nanoscale extracellular probes, such as nanowire FETs, can overcome this issue to some extent by decreasing the probability of coupling to multiple cells. But there is still uncertainty when a nanoscale extracellular probe is close to the boundary between cells or interfaced with subcellular structures, such as dendrites. Intracellular recording, on the other hand, can register the position and signal to specific cell/electrode interfaces unambiguously.

The fact that a direct physical contact between probes and the interior of cells is needed for intracellular recording makes it more invasive. For example, commonly used glass micropipettes typically have open tip diameter of ~0.2–5 μm, which can be a substantial fraction of the cell size, and during recording, there will be mixing of cell cytosol and exogenous filling solution (through the open tip) that cause irreversible changes to cells and make long-term measurements difficult [21].

In addition, the large probe size also poses a challenge to record from small subcellular structures, such as axons and dendrites. Another limitation of conventional intracellular recording comes from its difficulty in simultaneous recordings at a large number of sites with high spatial resolution, which is faced by many intracellular recording probes, such as glass micropipettes, sharp metal electrodes, and carbon-based microelectrodes, which all needs a micromanipulator for the recording.

By developing novel nanoscale probes with active FET detection elements, intracellular electrical recording with minimal invasiveness and toxicity, high spatial precision and resolution, and high-density, large-scale scaling-up capability can be realized. We will talk about the recently developed FET-based intracellular probes and their applications in intracellular action potential recording in this section.

2.4.1 Novel FET-Based Intracellular Probes

2.4.1.1 Kinked Nanowire FET Probes

The main challenge to realize the FET-based intracellular probes is to couple the active channel to the interior of cells in a minimally invasive manner. Unfortunately, the conventional FET geometry in which the active channel connected in a linear geometry to large *S* and *D* electrodes precludes insertion without disruption of cells. One efficient way of solving this challenge is to use kinked nanowires in which the voltage sensitive active channel—a lightly doped segment—is encoded synthetically at or close to the tip of the kink (Fig. 2.10a, b). The kink tip can be inserted into cells for intracellular interfacing. In this structure, two heavily doped nanowire arms of the kink serve as synthetically defined nanoscale *S*/*D* electrodes. By connecting the nanowire *S*/*D* arms with strained microscale metal interconnects, the device is bent up and forms a 3D, freestanding probe with the kink tip presented to the 3D space to access the cell interior (Fig. 2.10c, d) [22]. The heavily doped arms effectively prevent the metal electrodes from disrupting the cell. Since the synthetically defined active part of the kinked nanowire is localized at the tip region, the entire active channel of the FET can be coupled to the interior of the cell, thus ensuring highly sensitive transmembrane potential recording.

The variation of reactant pressure during the Au-nanocluster-catalyzed VLS growth of SiNWs can introduce reproducible 120° kinks at defined positions [22]. By purging the reactant for a short time and then reintroduce the reactants, the growth of the SiNWs experiences a pause and resume process. And the heterogeneous nucleation during the resume step causes a change of the growth orientation which results in a 120° kink in the nanowire. An intracellular recording probe can be this 120° kink or a 60° kink formed with two *cis*-linked kinked units (Fig. 2.10 a, b). The dopant modulation can be conducted right before or after the introduction of the kink or between the two kinks for the 60° probe to make the short active channel close to or at the topologically defined kink tip (the pink segment in Fig. 2.10a).

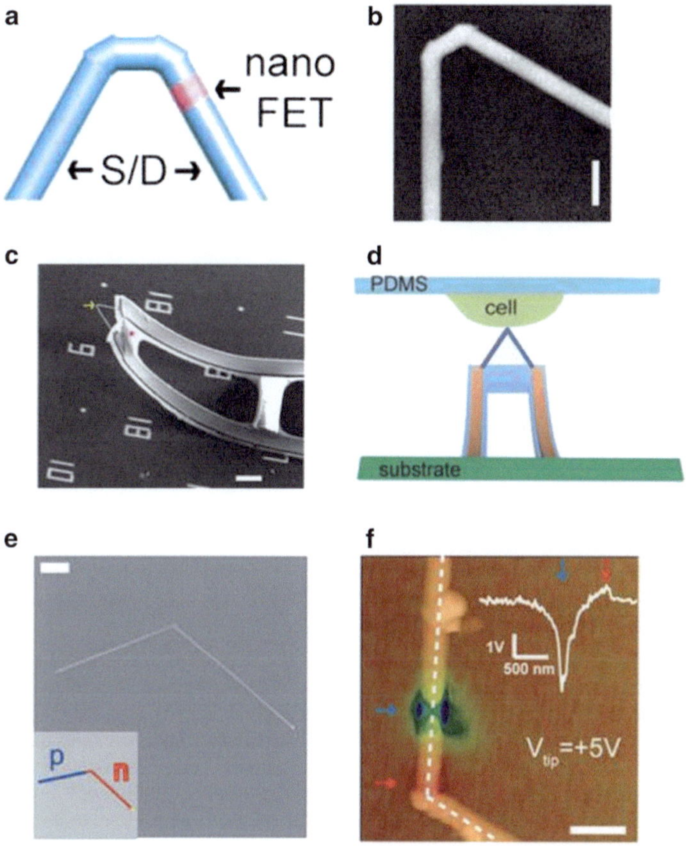

Fig. 2.10 Kinked nanowire FET probes. (**a**) Schematic of a kinked nanowire probe with encoded active region (*pink*). *Blue regions* are nanowire S/D. (**b**) SEM image of a doubly 60° kinked nanowire with a *cis* configuration. Scale bar, 200 nm. (**c**) A 3D, freestanding kinked nanowire FET bent-up probe. The *yellow arrow* and *pink star* mark the nanoscale FET and SU-8 (a polymer used to passivate metal electrodes), respectively. Scale bar, 5 μm. (**d**) Schematic of intracellular recording using kinked nanowire probes. (**e**) Representative SEM image and schematic (*inset*) of a kinked p–n junction SiNW with 120° tip angle. Scale bar, 1 μm. (**f**) Superposition of tmSGM images on AFM (atomic force microscopy) topographic images of a representative kinked p–n nanowire device under V_{tip} of +5 V. Scale bar is 0.5 μm. The *blue*/*red arrows* indicate the p-type and n-type depletion/accumulation regions, respectively. *Inset*: line profile of the tmSGM signal along the *white dashed lines*. (**a–c**) Reproduced with permission from [22] © (2010) The Author(s). (**d–f**) Reproduced with permission from [23] © (2012) American Chemical Society

Another approach we have implemented for making active kinked nanowire probes involves incorporation of a p–n junction near the probe tip by synthesis (Fig. 2.10e) [23]. They can be realized by changing the dopant source, instead of doping ratio during the growth. The active channel is then localized at the depletion region of the p–n junction, where the theoretical width of the depletion region could be as small as 10–30 nm [24], thus allowing potentially very high-resolution

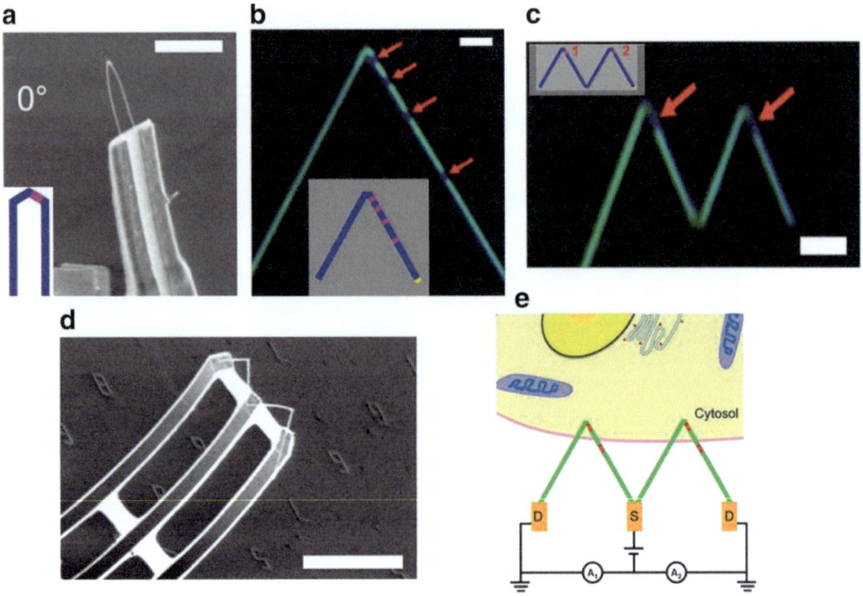

Fig. 2.11 Kinked nanowire superstructures. (**a**) SEM image of a 3D probe device fabricated using a 30 nm diameter U-shaped kinked nanowire. Scale bar, 3 μm. *Inset*, schematic of a U-shaped kinked nanowire with tip constructed from three 120° *cis*-linked kinks. The lightly doped n-type nanoFET element (*pink*) is encoded at the tip and connected by heavily doped n^{++} S/D arms (*blue*). (**b**) Dark-field optical microscopy image of a KOH-etched kinked nanowire with four nanoFETs. The dark segments correspond to the four lightly doped nanoFET elements (*red arrows*). Scale bar, 2 μm. *Inset*, schematic of the probe design. (**c**) Dark-field optical microscopy image of KOH-etched W-shaped kinked nanowire. The two *dark* color segments correspond to the lightly doped nanoFET elements (*red arrows*) near the two tips. Scale bar, 2 μm. *Inset*, schematic of the probe design. (**d**) SEM image of W-shaped kinked nanowire bent-up probe. Scale bar, 20 μm. (**e**) W-shaped kinked nanowire with multiple nanoFETs (*red*) illustrated as a bioprobe for simultaneous intracellular/extracellular recording. *Green* indicates heavily doped (n^{++}) S/D nanowire electrode arms, and *red* highlights the point-like active nanoFET elements. Reproduced with permission from [25] © (2013) American Chemical Society

recording. Tip-modulated scanning gate microscopy (tmSGM) measurements from a typical p–n junction kink nanowire (Fig. 2.10f) show that the synthetically defined p–n junction region near the kink exhibits a p-type gate response (because of the lower dopant concentration for the p-arm that the n-arm). The length of the p-depletion region, which defines the spatial resolution of the device, was estimated from the full width at half maximum (FWHM) of the SGM line profiles to be 210 nm [23]. This resolution is lower than the theoretical limit of 10–30 nm and represents an area where future improvements could be realized.

The strategy of encoding well-defined FET active segments into geometrically controlled nanowire superstructures for 3D, freestanding devices can be extended to prepare a variety of functional bioprobes. Zero-degree kinked nanowire probes, which have two parallel heavily doped arms in U-shape and the active nanoscale FET (nanoFET) channel located at the tip of the “U” (Fig. 2.11a), were prepared by encoding three *cis*-kinks. The 60° V-shaped kinked nanowire probes with multiple nanoFETs encoded in series along one arm from the tip (Fig. 2.11b) can be realized by adding additional dopant modulation. Furthermore, superstructures in which two-kink nanowire devices were juxtaposed in a single W-shape with nanoFETs integrated at the tip of each of the kinked regions (Fig. 2.11c, d) were synthesized [25]. By encoding multiple nanoFETs in these complex structures and precisely controlling the probe/cell interface, these probes offer high-density multisite intracellular recording and/or simultaneous recording of both intra- and extracellular signals (Fig. 2.11e), a capability truly unique to these bottom-up nanowire structures.

2.4.1.2 Branched Intracellular Nanotube and Active Nanotube FET Probes

The kinked nanowire presents a unique design for FET-based intracellular probes. To further reduce the size of the probes and make them more amenable to large-scale high-density parallel recording, some other designs using nanotube channels to bridge between the inside of cells and FET detector elements were developed. The first design, termed as the branched intracellular nanotube FET (BIT-FET), involves the use of a vertical or nearly vertical electrically insulating SiO_2 nanotube which is integrated on top of the FET channel (e.g., a SiNW channel). After the nanotube tip penetrates the cell membrane, the cytosol fills in the nanotube and gates the underlying FET, thus enabling the recording of the intracellular transmembrane potential change or action potentials (Fig. 2.12a) [26]. This BIT-FET design has several unique advantages: (1) It uses the tip of controlled diameter nanotube to interface to and probe intracellular regime, and together with an "impedance-free" FET detector, it allows for much smaller probe size, thus adds the capability for interfacing with small subcellular structure such as neuronal dendrites. (2) This design is compatible with the large-scale high-density planar nanoFET arrays, which provides a powerful way for parallel recording from large numbers of sites with spatial resolution that exceeds other probes.

The BIT-FET device was realized using bottom-up synthesis together with more conventional top-down processing, which enables an easy control over the critical probe dimensions. In short, germanium nanowire (GeNW) branches were grown on top of SiNWs, coated with a conformal, controlled thickness SiO_2 layer by atomic layer deposition (ALD), followed by selective removal of the topmost part of the SiO_2 shell and etching of the GeNW core to yield the a hollow SiO_2 nanotube on the SiNW FET as shown in the inset of Fig. 2.12b [26]. This design and fabrication scheme is not limited to SiNW FETs but can be used on any other nanoFETs including the planar ones. The etching step used to remove the upper portion of SiO_2 results in a controlled taper at the tip due to isotropic etching of the SiO_2 shell, which is particularly advantageous for further decreasing probe size.

The BIT-FETs respond selectively and with high sensitivity to the potential change of the solution inside the nanotubes rather than that outside and thus meet the requirements for intracellular recording outlined schematically in Fig. 2.12a. In terms of temporal resolution, modeling shows that the BIT-FETs with typical

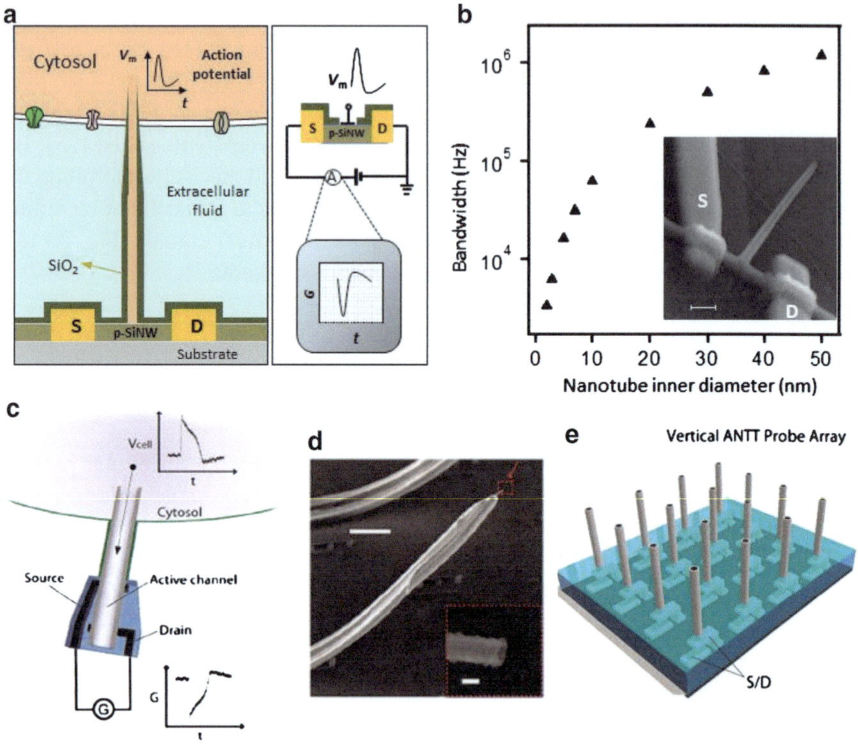

Fig. 2.12 Nanotube coupled FET probes. (**a**) Schematic illustrating the working principle of the BIT-FET. (**b**) Calculated bandwidth of the BIT-FET device versus the inner diameter of the nanotube for fixed nanotube length of 1.5 μm. *Inset*, SEM image of a BIT-FET device. Scale bar, 200 nm. (**c**) Schematic illustration of the working principle for the ANTT probe. (**d**) SEM image of an ANTT probe. Scale bar, 10 μm. *Inset*, zoom of the probe tip from the dashed red box. Scale bar, 100 nm. (**e**) Schematic of chip-based vertical ANTT probe arrays for enhanced integration. (**a, b**) Reproduced from [26] © (2012) The Author(s). (**c–e**) Reproduced with permission from [27] © (2012) American Chemical Society

nanotube dimension (inner diameter, 50 nm; SiO_2 wall thickness, 50 nm; length, 1.5 μm) have a bandwidth in the MHz scale, which is higher than necessary for recording the neuronal action potentials. The bandwidth decreases with decreasing nanotube size but can still maintain a ≥6 kHz value for nanotube inner diameters as small as 3 nm according to the modeling [26] (Fig. 2.12b).

The nanotube used to couple to the cell interior can also act as the active channel of the FET detector as shown schematically in Fig. 2.12c. In this alternative nanotube-based intracellular probe design, the *S*/*D* electrodes are fabricated on one end of the nanotube while leaving the other end free for cell membrane penetration. The cytosol filling the nanotube after membrane penetration can gate the FET from inside the nanotube. This design is termed as *a*ctive silicon *n*ano*t*ube *t*ransistor (ANTT) [27]. The ANTT probes were realized by synthesizing Ge/Si core/shell nanowires and then etching away the Ge core with H_2O_2 to leave a Si nanotube. Similar to the kinked nanowire probes, strained metal electrodes are used to lift up the nanotube FET and make it accessible to cells (Fig. 2.12d). The use of microscale metal electrodes to orient the ANTT probe limits its application in large-scale high-density multisite recording. However, this limitation could be overcome in the future by preparing vertical nanotube FET arrays in a manner similar to work on vertical nanowire FETs (Fig. 2.12e) [28].

2.4.2 *Intracellular Action Potential Recording with FET-Based Probes*

The cell membrane penetration and high-resistance sealing between the membrane and nanoprobes are critical for faithful, stable, and minimally invasive intracellular recording. Mechanical suction or insertion has been used successfully by many conventional recordings. In the studies of the above nanoFET probes, a different approach using phospholipid bilayer modification was applied to facilitate the cell membrane penetration which shows advantages compared to the mechanical method. As shown in Fig. 2.13, when bringing a dissociated HL-1 cell (a mouse cardiac muscle cell line) into contact with a kinked nanowire probe, where the intracellular potential of the cell was clamped at a certain value with a glass micropipette, the probe modified with the phospholipid bilayer showed clear change in calibrated potential to a value close to the intracellular potential of the cell, which is indicative of the cell membrane penetration. But the unmodified probes exhibited only baseline fluctuations ($<\pm 1$ mV) [22]. These results show that the phospholipid bilayer modification helped the cell membrane penetration by the kinked nanowire FET probe, even when there is no external force applied. While there is no detailed study on this membrane penetration process or the nature of the membrane/probe interface, a speculation of biomimetic lipid fusion [29] was proposed to play an important role. The fusion of the lipid bilayer on the probe with the cell membrane could result in a biomimetic and spontaneous membrane penetration by the kink tip and also a stable, high-resistance probe/membrane seal, which will be important for long-term and stable intracellular recording.

With phospholipid bilayer modification, kinked nanowire FET probe, BIT-FET, and ANTT recorded intracellular action potentials successfully from spontaneously firing cardiomyocyte cells cultured on the PDMS sheets in a setup as shown schematically in Fig. 2.5a. Representative data from a BIT-FET recording shows that approximately 45 s after gentle contact of the cell with the device, there was a dramatic change from extracellular spikes to intracellular action potential peaks with a concomitant baseline shift of approximately −35 mV (Fig. 2.14a). Because the SiNW used in this BIT-FET is p-type, the recorded intracellular conductance peaks were inverted, although the calibrated potential for these very stable peaks has the standard polarity, shape, amplitude, and duration for intracellular cardiac action potential. These intracellular action potential peaks show five characteristic

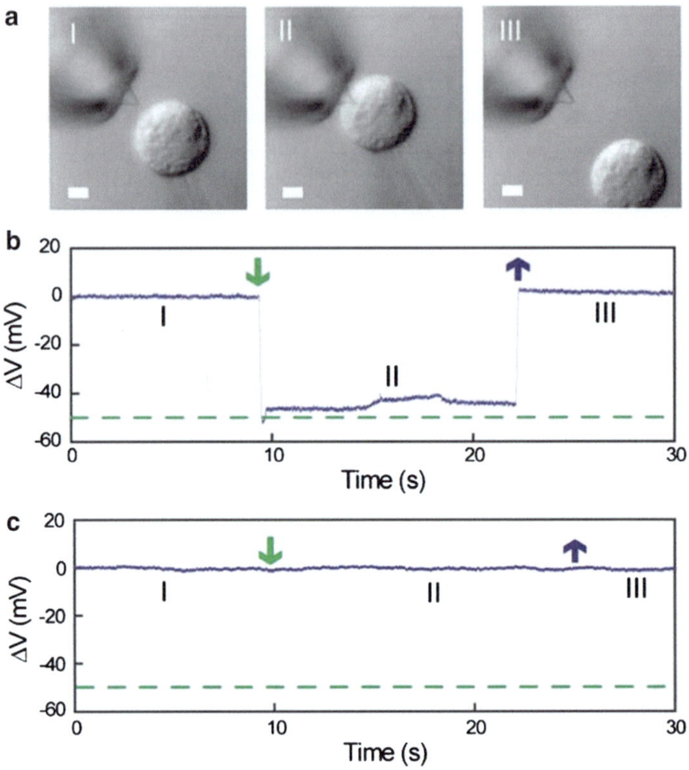

Fig. 2.13 Cellular entry by a kinked nanowire FET probe. (**a**, **b**) Differential interference contrast (DIC) microscopy images (**a**) and electrical recording (**b**) of an HL-1 cell and 60° kinked nanowire probe as the cell approaches (*I*), contacts and internalizes (*II*), and is retracted from (*III*) the nanoprobe. A glass micropipette was used to manipulate and voltage clamp the HL-1 cell. The *dashed green line* corresponds to the micropipette potential. Scale bars, 5 μm. (**c**) Electrical recording with a 60° kinked nanowire probe without phospholipids surface modification. *Green* and *blue arrows* mark the beginnings of cell penetration and withdrawal, respectively. Reproduced with permission from [22] © (2010) The Author(s)

phases of a cardiac intracellular potential, including (I) resting state, (II) rapid depolarization, (III) plateau, (IV) rapid repolarization, and (V) hyperpolarization (Fig. 2.14b) [26].

In the above action potential recording, and other ones using lipid modified nanoFET probes, the transition from extra- to intracellular signals, which is indicative of the cell membrane penetration, occurs without application of external force. This is consistent with the phospholipid-assisted spontaneous, biomimetic cell membrane penetration. And we believe that the small probe sizes can be beneficial for the lipid fusion and biomimetic cell membrane penetration to happen. As a result of this biomimetic penetration, a stable, high-resistance sealing between probes and cell membrane forms, as reflected by the full amplitude of the action potential signals, ~75–100 mV [26], which is recorded without the need for circuitry to

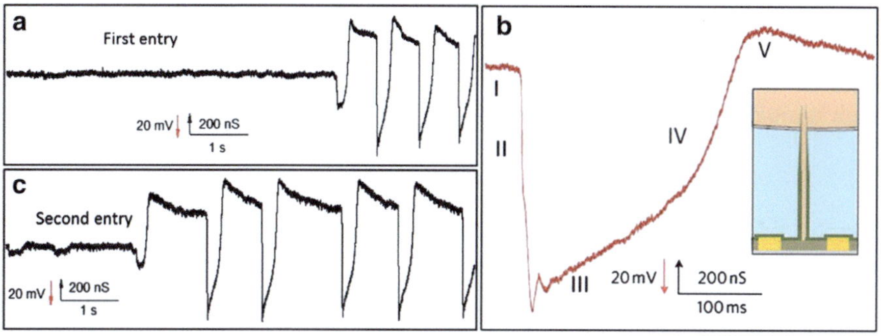

Fig. 2.14 Intracellular action potential recording by a BIT-FET. (**a**) Representative trace (conductance vs. time) reflecting the transition from extracellular to intracellular recording. (**b**) Magnified view of an intracellular peak. *I–V* correspond to the five characteristic phases of a cardiac intracellular potential. *Inset*, schematic for BIT-FET intracellular recording. (**c**) Trace corresponding to the second entry of the BIT-FET nanotube at approximately the same position on the cell as in (**a**). Reproduced from [26] © (2012) The Author(s)

compensate for putative probe/membrane leakage. In fact, control experiments carried out without phospholipid modification on the BIT-FETs required external forces to achieve the transition to intracellular action potential signals, which proved the critical role of lipid modification in this spontaneous cell membrane penetration.

Another unique characteristic of the nanoFET recording, which distinguishes them from conventional intracellular probes, is that they can be retracted from the cell and reenter approximately the same position on the same cell to record intracellular action potentials for multiple times without affecting the cell (Fig. 2.14c) [26]. This capability, a cooperative result from the small size of the nanoFET probes and the use of lipid modification, allows for long-term stable recording on scale of hours and demonstrates the reliability, robustness, minimally invasiveness of the nanoFET recording.

Simultaneous, multisite intracellular recording of action potentials from both single cells and cell networks has been demonstrated by interfacing the independently addressable nanoFET probes with cells [26, 27]. The use of phospholipid modification is especially advantageous for multiplexed recording in that it can help to achieve stable, tight sealing between multiple nanoprobes and a cell membrane (or membranes of multiple cells) at the same time. Different nanoFET probes may penetrate the cell membrane at different time, but eventually all of them yield stable, full-amplitude action potential recording thus demonstrating the possibility of large-scale parallel measurements for neural and cardiac activity mapping.

Due to their potential in high spatiotemporal resolution and large-scale brain activity mapping, nanoscale tools have received considerable attention [30]. The small size of the FET-based nanoprobes discussed here not only makes it possible for an unprecedented high-density device arrays which are critical for cellular and even subcellular resolution mapping but also yields minimal perturbation of cells and/or tissues being studied. We believe that these nanoprobes represent great candidates for use in brain activity mapping and related research. With further development and application, they can also extend substantially the scope of fundamental and applied electrophysiology studies and contribute to areas such as high-throughput drug screening.

2.4.3 Metal Electrode-Based Intracellular Probes

In addition to our active FET-based intracellular probes, several groups have been exploring the on-chip metal electrodes for intracellular action potential recording. The first example is the micrometer-sized protruding gold-spine microelectrode array [31, 32]. The mushroom-shaped Au electrodes here were functionalized with a peptide containing multiple Arg-Gly-Asp (RGD) repeats, named engulfment-promoting peptide (EPP9), to facilitate the engulfment of the gold-spine electrodes by the membrane of the cultured neurons thus forming a high-resistance seal. Following glass micropipette stimulation of nearby or the same cells, the engulfed gold-spine electrodes recorded subthreshold synaptic and action potential signals from individual neurons. From a representative measurement, the amplitude of the "raw" action potentials recorded by different spine electrodes was 0.1–25 mV, which is smaller than the value simultaneously recorded with a glass micropipette. Differences in the signal amplitude from different electrodes could arise from variations of the electrode impedance and/or the electrode/cell membrane coupling [31].

In another design, multiple vertical nanowires/nanopillars of metal, or silicon capped with metal, were used together as an electrode for intracellular recording. An example, named as vertical nanowire electrode arrays (VNEA) by the authors, uses a 3 × 3 array of vertical SiNWs (diameter of ~150 nm, length of ~3 μm) capped with a sputter-deposited metal tip for each intracellular probe [33]. Measurements from HEK293 and rat cortical neurons, which were cultured on the probes, suggested that some nanowire electrodes spontaneously penetrated HEK293 cell membranes during culture, although for others, a large voltage pulse (~±3 V; duration, 100 ms) was needed to yield intracellular signals. The signals recorded with this SiNWs array electrode were 10× smaller in amplitude than that simultaneously recorded with the glass patch pipette [33]. Another similar example uses multiple vertical platinum nanowires (PtNWs, diameter of ~150 nm, length of ~1.5 μm) for each intracellular probe, which is termed as nanopillar electrode. The direct recording using these nanopillar probes from HL-1 cells cultured on them initially yielded small ~100–200 μV extracellular spikes, but following voltage pulse (2.5 V; 200 μs), the signal transitions to one similar to an intracellular action potential albeit with a reduced amplitude of ~11.8 mV [34]. The authors suggest that the voltage pulse induces nanometer-sized pores thereby providing access to the intracellular potential. This electroporation mechanism was thought as an underlying mechanism for the electrical pulse-induced intracellular recording by the VNEA as well.

Compared to the nanoFET probes, the metal-based intracellular electrodes discussed here have an advantage in that the electrode can deliver stimulatory pulses to evoke action potentials and act as a stimulator, in addition to potential detector. However, the dependence of the recording on electrode/electrolyte interface impedance limits the probe size to be in the micrometer scale (gold-spine electrode) or requires multiple nanowires per electrode (VNEA and nanopillar probes) which also results in overall size of electrodes in micrometer scale. In addition, nanoFET probes usually give larger signal amplitude and SNR than the metal-based electrodes. The overall comparison of different intracellular recording probes is summarized in Table 2.1 [35].

2.5 Merging Nanoelectronics with Artificial Tissues Seamlessly in Three Dimensions

Implementing electronic sensors in 3D and the capabilities for monitoring cells throughout the 3D microenvironment of tissues are critical for functional neural activity mapping and understanding physicochemical changes relevant to living organisms. Previous works have been limited in terms of merging electronics with tissues throughout 3D space with minimal tissue disruption, because the 2D support structures and the electronic sensors are generally on much larger scale than the extracellular matrix (ECM) and cells. The bottom-up paradigm from separately synthesized nanomaterials used for nanoFETs fabrication provides a unique opportunity to make freestanding, substrate-free nanoscale electrical probes that are distributed in 3D space. By merging these 3D probes with living tissues or artificial tissue constructs, electrical recording and mapping of cellular activities in 3D from inside the tissues could be realized. In this section, we will review the effort toward realizing the seamless merging between electronic system and artificial tissues through the fabrication of 3D macroporous, freestanding nanoscale devices.

2.5.1 Three-Dimensional Macroporous Nanoelectronic Network

Conceptually, the seamless merging between nanoelectronic network and artificial tissue in 3D can be achieved in three basic steps as follows (Fig. 2.15) [36, 37]: (1) fabrication of the nanoelectronic network/nanoprobe arrays in 2D with underlying sacrificial layers and substrate support (step A, Fig. 2.15); (2) removal of the sacrificial layers to release the nanoelectronic network to yield 3D, freestanding nanoelectronic scaffolds (nanoES), where the nanoES is used alone or combined with traditional tissue scaffold materials (step B, Fig. 2.15); and (3) cell seeding and culture on the nanoES to yield 3D nanoelectronic-tissue hybrids (step C, Fig. 2.15) [37]. In this new paradigm, the use of nanoES is the critical advance, which addresses key constraints as follows: (1) the electronic network must be

Table 2.1 Electrical and methodological specifications of the various modalities of intracellular electrical recordings:

	Electrode impedance (Ω)	Primary electrode material	Size	Seal (Ω)	Maximum AP (mV)	Maximum EPSP (mV)	Cell type studied	Coupling coefficient	Insertion method	Coating	Duration of stable recording	Ref.
Patch-clamp electrodes (whole cell)	10^6	Glass micro-pipette	0.2–5 μm	$>10^9$	[a]	[a]	Multiple	1	Mechanical	None	Hours	21
Kinked silicon nanowire FET	Irrelevant (FET device)	Si	Approximately several hundreds of nanometers	N/A	Full amplitude, 75–100	None	Cardiomyocytes	~1	Mechanical or spontaneous	Phospholipids	Hours	22
Branched intracellular nanotube FET	Irrelevant (FET device)	Si/SiO_2	~30–150	N/A	Full amplitude, 75–100	None	Cardiomyocytes	~1	Mechanical or spontaneous	Phospholipids	Hours	26
Active silicon nanotube transistor	Irrelevant (FET device)	Si	~30–150	N/A	Full amplitude, 75–100	None	Cardiomyocytes	~1	Mechanical or spontaneous	Phospholipids	Hours	27
Gold mushroom-shaped MEA (g MμE)	10^{11}	Au	1–2 μm	10^6	25	5	Neurons (aplysia)	~0.5	Biological endocytosis	RGD motif peptide	>2 days	31
Vertical nanowire electrode arrays (VNEA)	3×10^8 (with application of bias volt-age = 1.5 V)	Si with Au tip	Multiple nanowires spread in a size of several micrometers	$>10^8$	4	None	None Neurons (cortical rat)	~0.3	Bias voltage	Silanization	10 min	33
Nanopillar electroporation	6×10^6 to 18×10^6 (depending on the number of pillars)	Pt	Multiple nanowires spread in a size of several μm	N/A	12	None	HL-1 cell line	~0.1	Electroporating current	Fibronectin	10 min per daily ses-sion, 3 days	34
gMμE electroporation	10^{11}	Au	1–2 μm	10^8	6	None	Neurons (aplysia)	~0.1	Electroporating current	Poly-L-lysine	5 min	32

[a]Glass micropipettes provide maximal (one-to-one) AP (action potential) and EPSP (excitatory postsynaptic potential) amplitudes

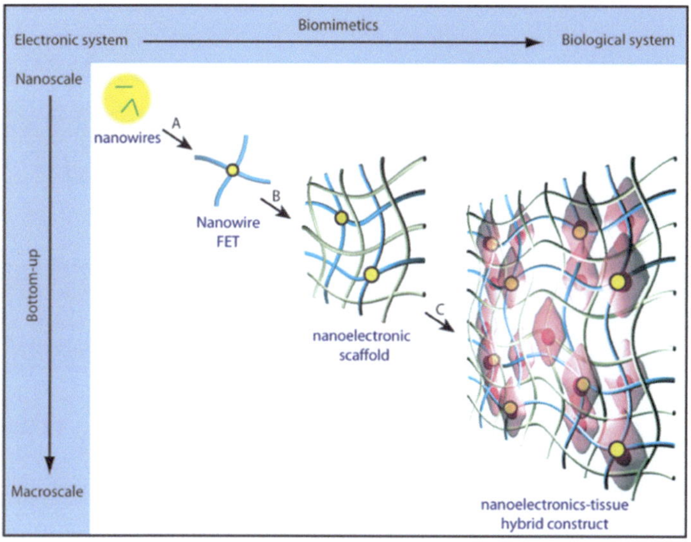

Fig. 2.15 Merging nanoelectronics with artificial tissues seamlessly in three dimensions [36]

macroporous (e.g., >99 % porosity), not planar, to enable 3D interpenetration with biomaterials and cells; (2) structural dimensions in nanometer to micrometer scale to match the natural ECM and mimic the well-studied scaffold materials used for tissue engineering and also to minimize the invasiveness; (3) 3D interconnectivity and addressability of the electronic devices; and (4) mechanical properties more compatible with the natural tissue (e.g., much softer than normal electronics).

Two basic types of nanoES have been developed and exploited for the creation of innervated tissues with neurons, cardiomyocytes, and smooth muscle cells. The first one, named reticular nanoES, uses stress in the bi- or tri-metallic interconnects to induce the self-bending and self-organization of the nanoelectronic network /nanoprobes array, in order to yield a 3D scaffold with interconnected and addressable nanowire FET sensors embedded in the different positions of the 3D space (Fig. 2.16a). Reconstructed 3D confocal fluorescence images of a typical reticular scaffold (Fig. 2.16b) shows clearly the 3D interconnected structure, with a magnified SEM image of one of the kinked SiNW FET sensor elements in the nanoES shown in Fig. 2.16c [36]. The second basic class of nanoES, the mesh nanoES, is prepared upon the use of a 2D macroporous nanoelectronic network sheet (Fig. 2.16d). A regular nanoelectronic devices array with structural backbone (e.g., made of Su-8) is patterned on the supporting substrate combining nanowire assembly and conventional 2D lithography (Fig. 2.16e). After release from the substrate with the removal of underlying sacrificial layer, the freestanding and flexible 2D macroporous nanoelectronic network sheet can be organized into 3D macroporous structures by either manual manipulation such as rolling (Fig. 2.16f) or stress-induced self-assembly [36, 37]. Usually, the manual manipulation gives more flexibility in making nanoES of large size than the stress-induced self-organization.

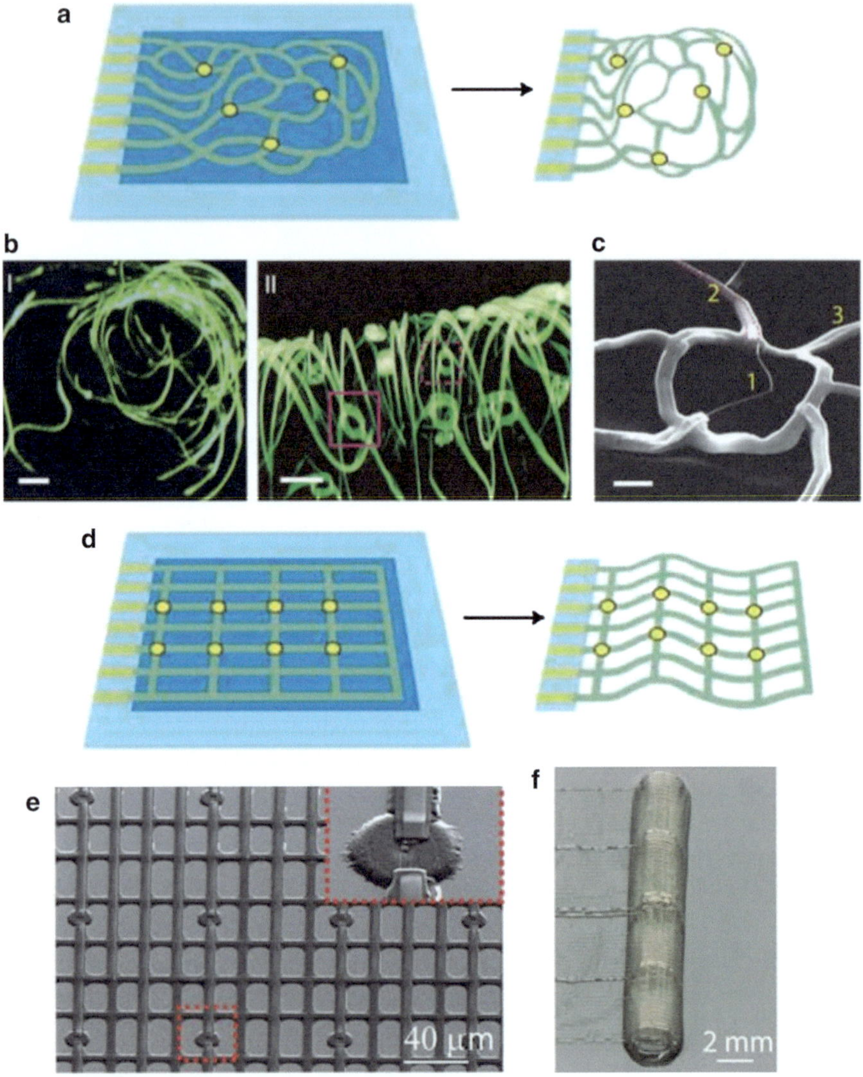

Fig. 2.16 Reticular and mesh nanoES: (**a**) device fabrication schematics for reticular nanoES. *Light blue*, silicon oxide substrates; *blue*, nickel sacrificial layers; *green*, nanoES; and *yellow dots*, individual nanowire FETs. (**b**) 3D reconstructed confocal fluorescence micrographs of a reticular nanoES viewed along the y (*I*) and x (*II*) axes. *Solid* and *dashed open magenta squares* indicate two SiNW FET devices located on different planes along the x axis. Scale bars, 20 μm. (**c**) SEM image of a single kinked SiNW FET within a reticular scaffold, showing (*1*) the kinked SiNW, (*2*) metallic interconnects (*dashed magenta lines*), and (*3*) the SU-8 backbone. Scale bar, 2 μm. (**d**) Device fabrication schematics for mesh nanoES. The color designation is same as (**a**). (**e**) SEM image of a 2D macroporous nanoelectronic network before release from the substrate. (*Inset*) Zoom-in of the region enclosed by the small *red dashed box* containing a single SiNW FET device [37]. (**f**) Photograph of a manually scrolled-up 3D macroporous mesh nanoES [37]. (**a**–**d**) [36]

In the nanoES, the 3D networks can have porosities larger than 99 %, ready for 3D interpenetration with cells. It can also contain hundreds of addressable nanowire devices and have feature sizes from the 1–10 μm scale (for electrical and structural interconnections) to the 10 nm scale (for device elements) [36, 37]. This size represents minimal invasiveness to the tissues, as reflected by the cytotoxicity tests which showed minimal difference in cell viability for culture in the scaffold with and without nanoES. Importantly, typical 3D macroporous nanoelectronic networks are reported to have a very low effective bending stiffness from 0.0038 to 0.0378 nN/m [37]. These values, which can be readily tuned by changing the materials and size of the nanoES components, are comparable with synthetic and natural ECMs, thus making them an ideal scaffold for innervating synthetic neural and cardiac tissue.

2.5.2 Three-Dimensional Nanoelectronics/Tissue Hybrids

Before cell seeding and culture, the reticular and mesh nanoES are always merged with conventional macroporous biomaterials. Specifically, gel casting, lyophilization, and electrospinning were used to deposit and construct macroporous collagen, alginate, and poly(lactic-co-glycolic acid) (PLGA), respectively, around nanoES. Embryonic rat hippocampal neurons were cultured on the reticular nanoES/Matrigel scaffold [36]. Reconstructed 3D confocal micrographs from a 2-week culture (Fig. 2.17a, b) show clearly neurons with a high density of spatially interconnected neurites that penetrate seamlessly the reticular nanoES (Fig. 2.17a), sometimes passing through the ring structures supporting individual SiNW FET sensors (Fig. 2.17b) [36]. Nanoelectronics/cardiac tissue was constructed from a hybrid mesh nanoES/PLGA scaffold following seeding and culture of cardiomyocytes. Confocal fluorescence microscopy of a typical cardiac 3D culture (Fig. 2.17c) revealed a high density of cardiomyocytes in close contact with the nanoES components. Epifluorescence micrographs of cardiac cells on the surface (Fig. 2.17d) further show striations characteristic of cardiac tissue [36]. Significantly, the original SiNW FET device characteristics were retained after the nanoES 3D organization, scaffold hybridization, and cell culture up to at least 12 weeks. This capability of the nanoES for long-term culture and monitoring is important for 3D recording and mapping of neural and cardiac activities. It also enables a number of in vitro studies, including drug screening assays with these synthetic neural and cardiac tissues, where 3D tissue models will provide a more robust link to in vivo disease treatment than the 2D cell cultures.

The recording of extracellular action potentials by the SiNW FET devices within the nanoES/cardiac tissue hybrid demonstrated the 3D monitoring capabilities of the nanoES. As shown in Fig. 2.17e, a SiNW FET ~200 μm below the construct surface gave regularly spaced spikes with a frequency of ~1 Hz, a calibrated potential change of ~2–3 mV, a SNR ≥3, and a ~2 ms width, which is consistent with the standard extracellular signals from cardiomyocytes. Following dosage of the construct with noradrenaline (also known as norepinephrine), which is a drug that

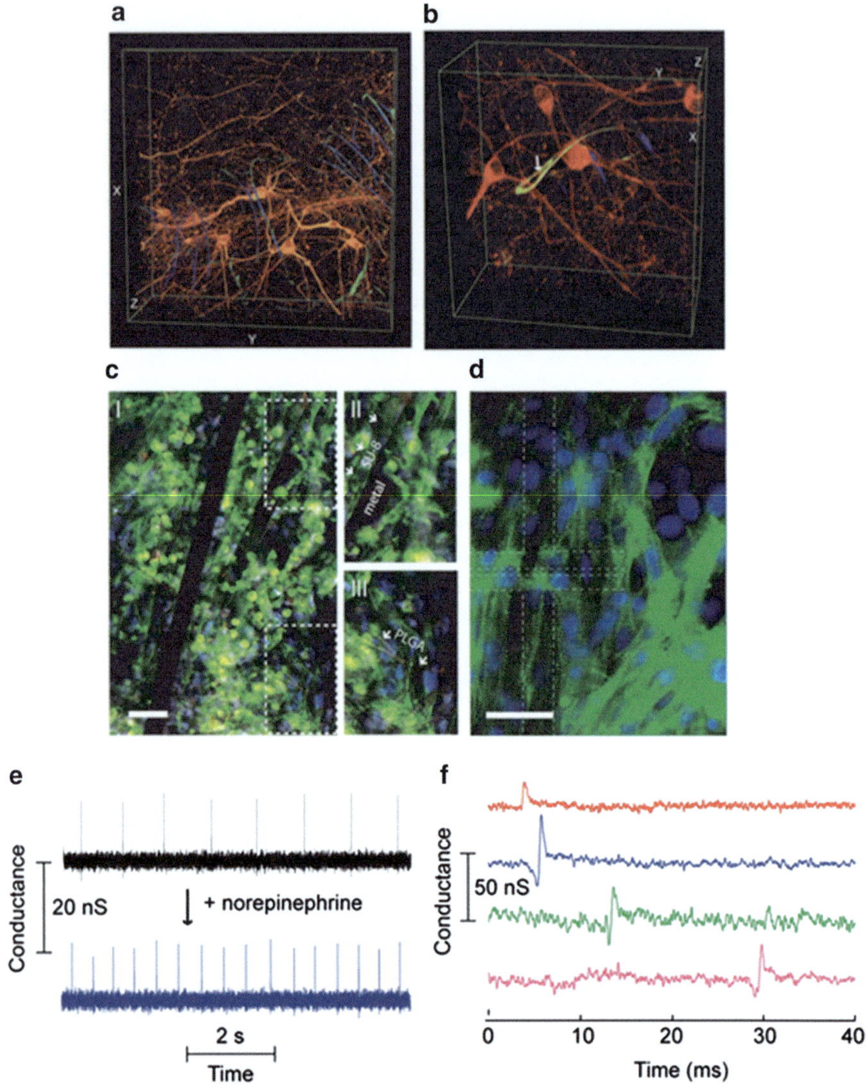

Fig. 2.17 3D nanoelectronics/tissue hybrids: (**a**, **b**) 3D reconstructed confocal images of rat hippocampal neurons after a 2-week culture on a reticular nanoES/Matrigel scaffold. The *white arrow* highlights a neurite passing through a ringlike structure supporting a nanowire FET. Dimensions in (**a**) x, 317 μm; y, 317 μm; and z, 100 μm and in (**b**) x, 127 μm; y, 127 μm; and z, 68 μm. (**c**) Confocal fluorescence micrographs of a synthetic cardiac patch. (*II* and *III*), Zoomed-in view of the *upper* and *lower dashed regions* in *I*, scale bar, 40 μm. (**d**) Epifluorescence micrograph of the surface of the cardiac patch. *Green*, α-actin; *blue*, cell nuclei. The *dashed lines* outline the position of the S/D electrodes. Scale bar, 40 μm. (**e**) Conductance versus time traces recorded from a single SiNW FET before (*black*) and after (*blue*) applying noradrenaline. (**f**) Multiplexed electrical recording of extracellular action potentials from four SiNW FETs at different depths in a nanoES/cardiac hybrid [36]

stimulates cardiac contraction, the recorded signal showed a twofold increase in action potential firing frequency [36]. It was reported that SiNW FETs at different position in the cardiac patch showed response to the noradrenaline application at different time scale. In addition, multiplexed recording from a coherently beating nanoES/cardiac construct demonstrated submillisecond temporal resolution from the four SiNW FETs with separations up to 6.8 mm within the 3D innervated cardiac tissue sample (Fig. 2.17f) [36].

The studies on the nanoelectronics/tissue hybrid and 3D neural and cardiac activity mapping are still at their infancy. However, the concept and strategy of making 3D macroporous nanoES discussed here provides a promising way toward this direction. The bottom-up paradigm started from separately synthesized nanomaterials to make the nanoprobes is the key for the 3D, freestanding nanoES. And the small size of these nanoprobes represents minimal invasiveness, which is critical for long-term and stable 3D recording. Further development of the nanoES could not only have a profound impact on areas such as 3D brain activity mapping and in vitro pharmacological studies but also suggest possibilities for implantable "cyborg" tissues enabling closed-loop monitoring and treatment of diseases.

2.6 Summary

This chapter reviews comprehensively the recent development of novel nanoscale probes, mainly FET-based nanoprobes, as extra- and intracellular action potential recording probes for biological systems ranging from single cultured cells to tissue slices and whole organs, through 3D artificial tissue constructs. The small size of these nanoscale probes minimizes the invasiveness and also increases the spatial resolution for the recording. The enabled large area, high-density extra- and intracellular recording of action potentials with cellular and subcellular resolution will be important for neural and cardiac activity mapping. Furthermore, the bottom-up paradigm used to fabricate these nanodevices allows for the development of 3D, freestanding nanoprobes which adds the capability for recording throughout the 3D space inside the artificial tissue constructs and opens up exciting possibilities for 3D neural activity mapping. All of these results represent great advances in nanoelectronics-biology interfacing and neural activity recording and could serve as foundations for many new fundamental studies and novel biomedical applications in neuroscience [38, 39].

References

1. Sze, S.M., Ng, K.K.: Physics of Semiconductor Devices, 3rd ed., Wiley Interscience, New York (2006)
2. Patolsky, F., Zheng, G., Lieber, C.M.: Nanowire-based biosensors. Anal. Chem. **78**, 4260–4269 (2006)

3. Timko, B.P., Cohen-Karni, T., Qing, Q., Tian, B., Lieber, C.M.: Design and implementation of functional nanoelectronic interfaces with biomolecules, cells, and tissue using nanowire device arrays. IEEE Trans. Nanotechnol. **9**, 269–280 (2010)
4. Lieber, C.M.: Nanoscale science and technology: building a big future from small things. MRS Bull. **28**, 486–491 (2003)
5. Wagner, R. S.: Whisker Technology, Wiley, New York (1970)
6. Lu, W., Lieber, C.M.: Semiconductor nanowires. J. Phys. D.: Appl. Phys. **39**, R387–R406 (2006)
7. Gudiksen, M.S., Lauhon, L.J., Wang, J., Smith, D.C., Lieber, C.M.: Growth of nanowire superlattice structures for nanoscale photonics and electronics. Nature **415**, 617–620 (2002)
8. Lauhon, L.J., Gudiksen, M.S., Wang, C.L., Lieber, C.M.: Epitaxial core–shell and core–multishell nanowire heterostructures. Nature **420**, 57–61 (2002)
9. Jiang, X., Tian, B., Xiang, J., Qian, F., Zheng, G., Wang, H., et al.: Rational growth of branched nanowire heterostructures with synthetically encoded properties and function. Proc. Natl. Acad. Sci. U S A **108**, 12212–12216 (2011)
10. Tian, B., Xie, P., Kempa, T.J., Bell, D.C., Lieber, C.M.: Single-crystalline kinked semiconductor nanowire superstructures. Nature Nanotech. **4**, 824–829 (2009)
11. Erickson, J., Tooker, A., Tai, Y.-C., Pine, J.: Caged neuron MEA: A system for long-term investigation of cultured neural network connectivity. J. Neurosci. Method. **175**, 1–16 (2008)
12. Patolsky, F., Timko, B.P., Yu, G., Fang, Y., Greytak, A.B., Zheng, G., et al.: Detection, stimulation, and inhibition of neuronal signals with high-density nanowire transistor arrays. Science **313**, 1100–1104 (2006)
13. Cohen-Karni, T., Timko, B.P., Weiss, L.E., Lieber, C.M.: Flexible electrical recording from cells using nanowire transistor arrays. Proc. Natl. Acad. Sci. U S A **106**, 7309–7313 (2009)
14. Cohen-Karni, T., Qing, Q., Li, Q., Fang, Y., Lieber, C.M.: Graphene and nanowire transistors for cellular interfaces and electrical recording. Nano Lett. **10**, 1098–1102 (2010)
15. Hu, Y., Xiang, J., Liang, G., Yan, H., Lieber, C.M.: Sub-100 nanometer channel length Ge/Si nanowire transistors with potential for 2 THz switching speed. Nano Lett. **8**, 925–930 (2008)
16. Cohen-Karni, T., Casanova, D., Cahoon, J., Qing, Q., Bell, D., Lieber, C.M.: Synthetically encoded ultrashort-channel nanowire transistors for fast, pointlike cellular signal detection. Nano Lett. **12**, 2639–2644 (2012)
17. Kodambaka, S., Tersoff, J., Reuter, M.C., Ross, F.M.: Germanium nanowire growth below the eutectic temperature. Science **316**, 729–732 (2007)
18. Hille, B.: Ion Channels of Excitable Membranes, 3rd ed., Sinauer Associates Inc., Sunderland (2001)
19. Qing, Q., Pal, S.K., Tian, B., Duan, X., Timko, B.P., Cohen-Karni, T., et al.: Nanowire transistor arrays for mapping neural circuits in acute brain slices. Proc. Natl. Acad. Sci. U S A **107**, 1882–1887 (2010)
20. Timko, B.P., Cohen-Karni, T., Yu, G., Qing, Q., Tian, B., Lieber, C. M.: Electrical recording from hearts with flexible nanowire device arrays. Nano Lett. **9**, 914–918 (2009)
21. Molleman, A.: Patch Clamping: An Introductory Guide to Patch Clamp Electrophysiology. John Wiley & Sons, Ltd., Chichester, West Sussex, England (2003)
22. Tian, B., Cohen-Karni, T., Qing, Q., Duan, X., Xie, P., Lieber, C.M.: Three-dimensional, flexible nanoscale field-effect transistors as localized bioprobes. Science **329**, 830–834 (2010)
23. Jiang, Z., Qing, Q., Xie, P., Gao, R., Lieber, C.M.: Kinked p-n junction nanowire probes for high spatial resolution sensing and intracellular recording. Nano Lett. **12**, 1711–1716 (2012)
24. Weber, L., Gmelin, E.: Transport properties of silicon Appl. Phys. A **53**, 136–140 (1991)
25. Xu, L., Jiang, Z., Qing, Q., Mai, L., Zhang, Q., Lieber, C.M.: Design and synthesis of diverse functional kinked nanowire structures for nanoelectronic bioprobes. Nano Lett., **13**, 746–751 (2013)
26. Duan, X., Gao, R., Xie, P., Cohen-Karni, T., Qing, Q., Choe, H.S., et al.: Intracellular recordings of action potentials by an extracellular nanoscale field-effect transistor. Nature Nanotech. **7**, 174–179 (2012)
27. Gao, R., Strehle, S., Tian, B., Cohen-Karni, T., Xie, P., Duan, X., et al.: Outside looking in: nanotube transistor intracellular sensors, Nano Lett. **12** , 3329–3333 (2012)

28. Goldberger, J., Hochbaum, A.I., Fan, R., Yang, P.: Silicon vertically integrated nanowire field effect transistors. Nano Lett. **6**, 973–977 (2006)
29. Almquist, B.D., Melosh, N.A.: Fusion of biomimetic stealth probes into lipid bilayer cores. Proc. Natl. Acad. Sci. U S A. **107**, 5815–5820 (2010)
30. Alivisatos, P.A., Andrews, A.M., Boyden, E.S., et al.: Nanotools for neuroscience and brain activity mapping. ACS Nano **7**, 1850–1866 (2013)
31. Hai, A., Shappir, J., Spira, M.E.: In-cell recordings by extracellular microelectrodes. Nature Methods **7**, 200–202 (2010)
32. Hai, A., Spira, M.E.: On-chip electroporation, membrane repair dynamics and transient in-cell recordings by arrays of gold mushroom-shaped microelectrodes. Lab Chip **12**, 2865–2873 (2012)
33. Robinson, J.T., Jorgolli, M., Shalek, A.K., Yoon, M-H., Gertner, R.S., Park, H.: Vertical nanowire electrode arrays as a scalable platform for intracellular interfacing to neuronal circuits. Nature Nanotech. **7**, 180–184 (2012)
34. Xie, C., Lin, Z., Hanson, L., Cui, Y., Cui, B.: Intracellular recording of action potentials by nanopillar electroporation. Nature Nanotech. **7**, 185–190 (2012)
35. Spira, M.E., Hai, A.: Multi-electrode array technologies for neuroscience and cardiology. Nature Nanotech. **8**, 83–94 (2013)
36. Tian, B., Liu, J., Dvir, T., Jin, L., Tsui, J.H., Qing, Q., et al.: Macroporous nanowire nanoelectronic scaffolds for synthetic tissues. Nature Mater. **11**, 986–994 (2012)
37. Liu, J., Xie, C., Dai, X., Jin, L., Zhou, W., Lieber, C.M.: Multifunctional three-dimensional macroporous nanoelectronic networks for smart materials. Proc. Natl. Acad. Sci. U S A **110**, 6694–6699 (2013)
38. Duan, X., Lieber, C.M.: Nanoelectronics meets biology: from new nanoscale devices for live-cell recording to 3D innervated tissues. Chem.-Asian J. **8**, 2304–2314 (2013)
39. Duan, X., Fu, T.-M., Liu, J., Lieber, C.M.: Nanoelectronics-biology frontier: from nanoscopic probes for action potential recording in live cells to three-dimensional cyborg tissues. Nano Today **8**, 351–373 (2013)

Chapter 3
In-Cell Recording and Stimulation by Engulfment Mechanisms

Aviad Hai

Abstract Neuroscientists have long been using microelectrodes to record and stimulate neural activity—both in vitro and in vivo. On one end of the spectrum of electrode-based techniques are the sharp-glass and patch micropipette electrodes; on the other end are dense arrays of metal-based microelectrodes. Glass micropipette electrodes enable *intracellular* recording of action potentials and synaptic potentials with excellent signal-to-noise ratio, but because of their bulkiness, they allow for the recording and stimulation of only several neurons at a time. In addition, the injury inflicted on the cell plasma membrane during electrode entry and recording limits the duration of the recording session, usually to a small number of hours at most. By contrast, multielectrode devices are able to record and stimulate much larger populations of neurons for durations of weeks and even months. This is made possible due to fabrication technologies that allow for a scalable design of hundreds or even thousands of electrodes. These devices, however, have been able to provide only *extracellular* recording and stimulation with limited signal-to-noise ratio due to the extracellular positioning of the electrode in respect to the neuron's plasma membrane. The inability to record intracellular signals from many neurons and for long periods of time has thus far prevented neuroscience from answering the most basic and interesting questions regarding learning and memory in large populations of neurons. This is because the vast majority of neurons in complex nervous systems are usually "silent" and will generate an action potential only when their complex synaptic inputs integrate appropriately. We are therefore blind to the rich milieu of synaptic interactions, synaptic plasticity, and subthreshold network oscillations that reflect the state of the studied nervous system. This chapter describes a recently developed technique termed *in-cell recording*. This technique yielded for

A. Hai, Ph.D. (✉)
The Hebrew University of Jerusalem, Jerusalem, Israel

Massachusetts Institute of Technology, Cambridge, MA 02139, USA
e-mail: aviadhai@MIT.EDU

M. De Vittorio et al. (eds.), *Nanotechnology and Neuroscience: Nano-electronic, Photonic and Mechanical Neuronal Interfacing*, DOI 10.1007/978-1-4899-8038-0_3,

the first time simultaneous, multisite, long-term recordings of action potentials and subthreshold synaptic potentials with matching quality and signal-to-noise ratio of conventional intracellular glass electrodes and the scalability of fabricated multi-electrode devices. The in-cell recording and stimulation technique makes use of an array of cell-noninvasive micrometer-size protruding gold mushroom-shaped microelectrodes (gMμEs). The key to the multielectrode in-cell recording approach is the outcome of three converging cell-biological principles: (a) the activation of endocytotic-like mechanisms in which the cultured cells are induced to actively engulf gMμEs that protrude from the substrate, (b) the generation of high seal resistance between the cell's membrane and the engulfed gMμE, and (c) the localization of ionic channels (ohmic conductance) in the plasma membrane that faces the gMμE. We will describe the electrical, ultrastructural, and cell-biological properties of the interface between the cells and the gMμEs and provide the reader with a digest of the published studies carried out for the development of this technique.

3.1 Introduction

The emergence of modern neuroscience began when it was first realized that neurons communicate by generating electrical signals that can be amplified and monitored [1–5]. Understanding the physical, chemical, and biological mechanisms that underlie the generation of the electrical signals, their propagation along axons and dendrites, and transmission between neurons (synaptic transmission) greatly depended on the development of appropriate methods to monitor and generate the electrical activities of neurons. The timeline traverses from initial studies of impulse propagation and generation by extracellular electrodes [6, 7] to studies of membrane biophysics, impulse propagation, and synaptic transmission with sharp intracellular glass microelectrodes [5, 8–10] and to studies of membrane properties, synaptic transmission, and single ion channels by the patch clamp technique [11]. As the field progressed, it was clear that further advancement in the understanding of brain functions would require the simultaneous gathering of functional (electrical) information from large populations of neurons for long periods of time (from hours to days and weeks). As aforementioned, the use of sharp or patch microelectrodes for parallel recording or stimulation from large populations of neurons is technically limited as the micromanipulation of the electrode tips toward target cells requires the use of rather bulky micromanipulators [12, 13]. In addition, the duration of intracellular recording and stimulation sessions by sharp-glass and patch electrodes is limited, because, with time, mechanical instabilities damage the plasma membrane or—in the case of the patch electrodes—perfusion of the cytoplasm into the volume of the glass electrode alters the intracellular composition of the cells [11]. These shortcomings precluded effective analysis of the electrical activity of large neural networks.

A solution to these problems was the impressive development of different forms of imaging technologies ranging from the use of voltage-sensitive dyes [14–16] to multiphoton imaging of activity-related calcium concentration transients [17–19] and to fMRI [20, 21]. The major advantage of these imaging approaches is that they mostly do not harm the cells and can be performed for long periods of time. However, at high enough concentrations, optical imaging agents can be toxic to the cells and also change their phenotype. In addition, most optical neural recording methods require exposure of the tissue to acquire a signal, which falls significantly in deeper planes. It is worth noting that light-based in vivo recording and stimulation of neuronal populations has also recently evolved as a strong field of research by the use of optogenetic techniques [22–25] but currently serves mainly as a population-specific stimulation method, with limited novel complementary methodologies to concurrently record neuronal activity by either microfiber optic probes [26] or optical–electrical hybrid arrays [27, 28]. In the case of using fMRI for the recording of neural activity, the temporal resolution is not sufficient, reaching orders of magnitude below that of single action and synaptic potentials. In addition, fMRI suffers from low spatial resolution and the acquired signal reflects indirect manifestations of neuronal electrical activity such as blood flow in capillaries [29].

Concurrently, large arrays of extracellular metal microelectrodes (MEAs) or semiconductor-based micro- and nano-transistors were developed [30–37]. The use of noninvasive extracellular microelectrode arrays enables to record and stimulate large populations of excitable cells for days and months without inflicting mechanical damages to the cell plasma membrane. The method has a time resolution suitable for the acquisition of spikes and synaptic potentials and allows for the stimulation of single neurons using current or voltage pulses [38–40]. The most severe disadvantage of extracellular recording electrodes is their low signal-to-noise ratio and the ambiguity in defining the origin of the recorded signals. Therefore, the use of extracellular electrodes is limited to recordings of field potentials generated by action potentials [41, 42]. Only in rare cases such as the Schaffer collaterals-pyramidal cells synapses in the CA1 region of the hippocampus can synchronized synaptic potentials in highly ordered neuronal networks be picked up by extracellular electrodes [43]. Single excitatory or inhibitory subthreshold synaptic potentials or membrane oscillations cannot be detected by currently used extracellular electrodes technology [44]. These shortcomings limit the use of extracellular recordings to the analysis of spike patterns and frequencies which relies heavily on spike sorting for the determination of the number of neurons which act as sources of the recorded signals [45, 46].

An ideal multiunit recording system should provide a readout that covers the entire spectrum of membrane potential events from the individually recorded neurons. This includes action potentials (APs), subthreshold excitatory and inhibitory postsynaptic potentials (EPSPs and IPSPs, respectively), and subthreshold membrane oscillations. In addition, it should be possible to modulate the activity of individual neurons within the network by the application of current. The system should provide this readout simultaneously from hundreds of individual neurons and with a stable contact with the neurons for days and weeks.

The in-cell recording and stimulation system described herein [47–52] (Fig. 3.1) makes use of an array of cell-noninvasive micrometer-size protruding gold mushroom-shaped microelectrodes (gMμEs). The gMμE's geometry and chemical functionalization trigger the activation of endocytotic mechanisms in which the

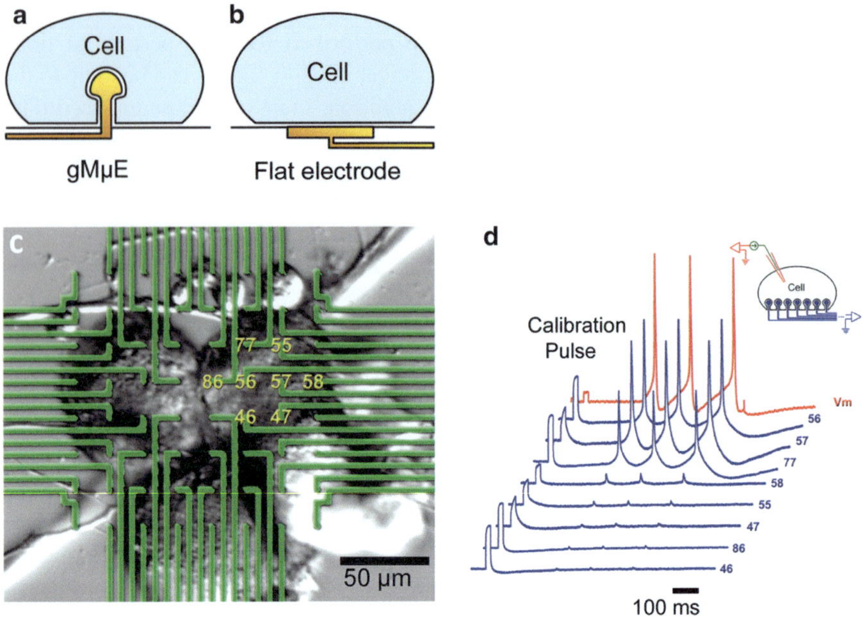

Fig. 3.1 The in-cell recording configuration. (**a**) Schematic representation of a neuron engulfing a gold mushroom-shaped microelectrode (gMμE). (**b**) The geometrical relationships between a neuron and a flat electrode. (**c**) Three *Aplysia* neurons cultured on an array of 62 gMμEs. (**d**) Raw action potential recordings from eight gMμEs (indicated by numbers also in **c**) in response to intracellular stimulation of the neuron by a conventional sharp microelectrode (V_m). Each trace depicts initially a 5 mV, 20 ms calibration pulse and then (following a delay) three action potentials. Reproduced from [51] © The Author(s)

cultured cells are induced to actively engulf the gMμEs protruding from the substrate. These processes generate a high seal resistance (R_{seal}) between the cell's membrane and the engulfed gMμE and also results in increased ohmic conductance, in the form of either voltage-independent ionic channels or membrane pores in the plasma membrane that faces the gMμE.

We begin with a brief technical description of gMμE fabrication and chemical functionalization followed by ultrastructural studies of the cell–gMμE junction. We then examine cytoskeletal and other cell-biological interactions between neurons and gMμEs and finally demonstrate the resulting high electrical coupling that enables long-term stimulation and recording of action potentials and synaptic potentials from multiple cells with signal-to-noise ratio comparable to that of glass-based micropipette electrodes.

3.2 Device Preparation

This section describes some of the technical aspects of device preparation for in-cell recording and stimulation as detailed also in published studies [48–52].

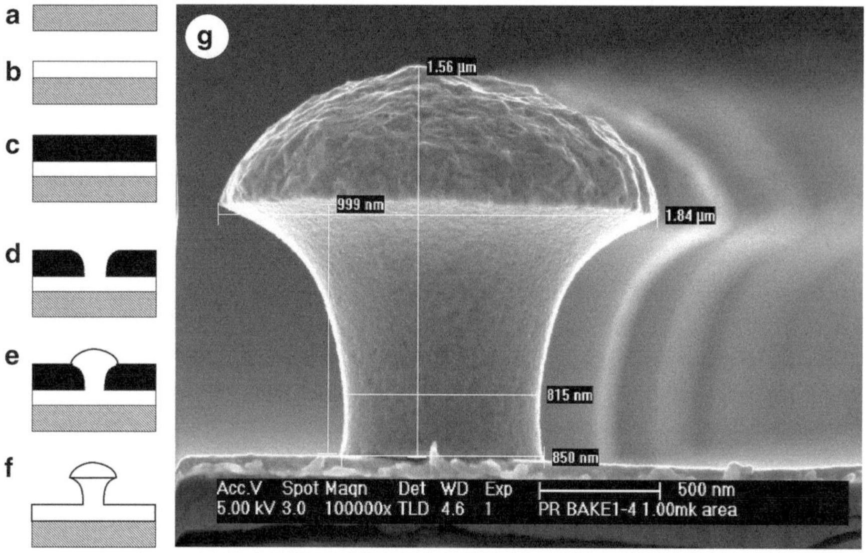

Fig. 3.2 Fabrication process of gMµEs. Silicon or glass samples (**a**) are coated with a Ti/Au layer (**b**) by electron-beam evaporation and are then spin-coated with photoresist material (**c**). Samples undergo photolithography using a photomask to define conducting lines and open holes through the photoresist (**d**). The gMµEs are grown on the surface by way of gold electroplating (**e**) and the photoresist layer is removed (**f**). (**g**) Scanning electron microscopic image of gMµEs fabricated on a glass surface. Reproduced from [48] © The Author(s)

3.2.1 gMµE Fabrication

Device fabrication is based on complementary metal–oxide–semiconductor (CMOS) technology. Arrays of gMµE electrodes are prepared on either glass or silicon wafers (Fig. 3.2 and see also refs. [48–52]). The wafers are first dehydrated at 120 °C for 30 min and are then coated with a Ti (10–15 nm)/Au (45–65 nm) layer by way of evaporation (Fig. 3.2b), spin-coated with photoresist S-1813 (4,000 RPM) and baked for 30 min at 90 °C (Fig. 3.2c). The first photolithographic process is performed followed by Au/Ti wet etch to define the leads of the microelectrode array (Karl Suss MJB UV400 mask aligner was used at $W = 44\ mW\ cm^{-2}$ and exposure time of 4.4–4.6 s). Next, a second photolithographic step with thick photoresist is performed to open holes for the deposition of the gMµE stalks as well as the contact pads (Fig. 3.2d). Gold mushrooms (along with thick metal on the contact pads) are grown by way of electroplating (Fig. 3.2e). The dimensions and shape of the fabricated gold mushroom roughly imitate the shape of naturally occurring dendritic spines [53], with a stem height of approximately 1 µm, diameter of approximately 800 nm, and a mushroom-shaped cap of ~2 µm in diameter. Parameters and electroplating solution are chosen to generate a rough texture of the mushroom head, which increases its effective surface and adhesion to cell membrane (Fig. 3.2g). Next, a layer of silicon oxide (~3,000 Å) is deposited by CVD processing.

Another layer of photoresist is applied for a final lithographic step to expose the contact pads and the heads of the gold mushrooms, followed by wet oxide etch to selectively remove the oxide from the contact pads and the mushroom heads. Wafers are sawed and undergo manual bonding to a 62-pad printed circuit boards to which 21 mm glass rings are attached to create a recording bath chamber for in vitro recording.

3.2.2 Chemical Functionalization

In contrast to the classical approach of having to forcefully push microelectrodes into the cells, we induced engulfment of the gMμE by the cell using a peptide that triggers processes of phagocytosis at the cell–gMμE point of contact. Phagocytosis is a conserved cell-biological mechanism for the internalization of particles [54]. Processes of endocytosis and phagocytosis are known to be mediated by integrins and RGD-based ligands [55] whose participation in processes of cell-matrix adhesion is also well established [56]. We have found that the peptide that induces engulfment of the gMμE by the cell most efficiently is a cysteine (C)-terminated peptide with a number of RGD repeats and a long decalysine (K_{10}) spacer: *CKKKKKKKKKKPRGDMPRGDMPRGDMPRGDM* (MW 3,630 g/mol) [48, 49]. This peptide was referred to as the engulfment promoting peptide (EPP) and was covalently linked to the surface of the device. Functionalization of the gMμE gold surface with the cysteine-terminated peptide makes use of thiol–gold monolayer self-assembly and is done by direct application of the peptide onto the surface (1 mM in phosphate buffer saline at room temperature, overnight). The glass surface in between the gold mushrooms underwent surface functionalization using 3-aminopropyltrimethoxysilane (APTMS, Aldrich, 1 % in MeOH, 10 min in room temperature) to introduce terminal amine groups to the glass surface. Samples were then washed with MeOH to remove uncoupled APTMS. The protein immobilization linker 4-Maleimidobutyric acid sulfo-N-succinimidyl ester (sGMBS, Sigma, 0.5 % in PBS) was then applied to the surface and washed with PBS after 40 min at room temperature. EPP peptide was then applied to the surface and left for 24 h in which the cysteinic thiol residue reacted with the maleimido part of the anchored linker. Samples were then washed with PBS.

3.3 Analysis of Cell–gMμE Interface

3.3.1 Ultrastructural Studies of Cell–gMμE Junction

Using transmission electron microscopy (TEM), we have found that different cell types, including *Aplysia* neurons, rat hippocampal neurons, Chinese hamster ovary cells (CHO), embryonic fibroblast cells (NIH/3T3), rat adrenal medulla cells

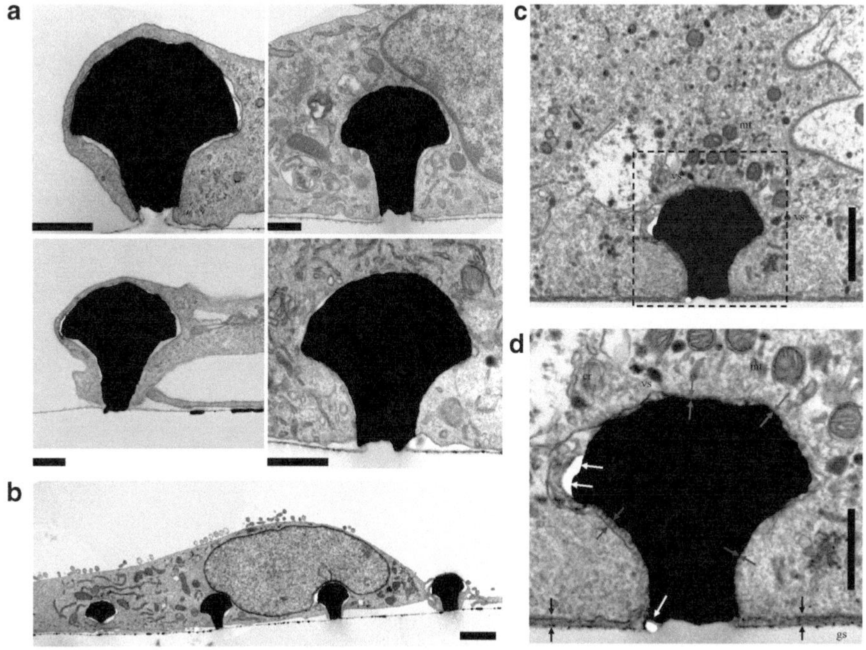

Fig. 3.3 Transmission electron microscopic images of gMμEs engulfed by various cell types. (**a**) CHO, fibroblasts (3T3), cardiomyocytes (H9C2), and rat adrenal medulla cell lines (PC-12). Scale bars, 500 nm. (**b**) Engulfment of multiple gMμEs by a 3T3 cell. Scale bar, 2 μm. (**c**, **d**) *Aplysia* neuron engulfing a gMμE coated with EPP. The gap formed between the plasma membrane and the flat surface is 30–50 nm (*black arrows*). The gap between the gMμE structure and the plasma membrane is less than 30 nm, and in some areas, it cannot be defined and seems to be zero (*gray arrows*). *White arrows* indicate regions where the embedding polymer was torn most probably by post-embedding mechanical tension, as the quality of the fixation is evident by the preservation of the organelles (*mt* mitochondria, *vs* vesicle, *er* endoplasmic reticulum, *gs* area of flat gold substrate). Scale bars, 1 μm and 500 nm. Reproduced from [48] © The Author(s)

(PC-12), and rat myocardium cells (H9C2), all engulf functionalized gMμEs [48, 57, 58].

Cells grown on flat glass cover slides coated by poly-L-lysin (PLL) adhere to the substrate forming a cleft with a thickness of 55.8 ± 28.9 nm, while occasionally the cleft dimension can be larger than 100 nm [48, 59]. When grown on a matrix of gMμEs coated with EPP (Fig. 3.3), cells engulf the protruding gMμE, forming intimate contact with a considerably smaller cleft thickness (Fig. 3.3c, d, and see [48, 58]). For example, the average width of the cleft formed between the plasma membrane of cultured *Aplysia* neurons and the flat substrate between the gMμEs is 56 ± 29 nm, while an average cleft width of 35 ± 21 and 30 ± 17 nm is observed between the gMμE head and stalk, respectively (Table 3.1). Moreover, in many of the electron micrographs, the plasma membrane appears to be in direct contact with the gMμE head and stalk (i.e., without a discernible space between the

Table 3.1 Average width values (nm) of the cleft formed between the plasma membrane of various cell types and the flat gold surface in between the gMµEs, gMµE stalk, and gMµE head, according to the analysis of transmission electron microscopy images

	Flat surface (nm)	gMµE stalk (nm)	gMµE head (nm)	F-value
Aplysia	55.8 ± 28.9	35.2 ± 20.8	30 ± 16.7	$F_{(2,357)} = 43.5$, $p < 0.001$
CHO	89.0 ± 66.4	31.2 ± 27.6	8.5 ± 16.3	$F_{(2,357)} = 89.6$, $p < 0.001$
3T3	80.1 ± 63.1	37.5 ± 35.1	15.7 ± 10.2	$F_{(2,447)} = 90.7$, $p < 0.001$
H9C2	122.8 ± 74.5	49.2 ± 41	49.8 ± 32	$F_{(2,447)} = 97.6$, $p < 0.001$
PC-12	144.8 ± 117.6	31.4 ± 27.6	26.9 ± 24	$F_{(2,447)} = 132.4$, $p < 0.001$

The fifth column depicts the significance of the difference between average cleft width values of the different areas examined (analyzed using one-way ANOVA). Reproduced from [48] © The Author(s)

Table 3.2 Percentage of the areas where the cleft width is in the range of 0–10 nm in various cell types

Cell type	Flat surface (%)	gMµE stalk (%)	gMµE head (%)
Aplysia	0	10	12.5
CHO	0	14.2	24.2
3T3	0.7	18.7	35.3
H9C2	0.6	4	7.3
PC-12	0	10	10.7

plasma membrane and the gold surface; see Fig. 3.3d, gray arrows), suggesting a significantly tighter seal between the plasma membrane and the gMµE compared with the flat surface between the gMµEs. In fact, an estimated 10 % of the contact area between the plasma membrane and the gMµE stalk and 12.5 % of the contact area with the gMµE head form close adhesion with the cleft width in the range of 0–10 nm (Table 3.2). The cleft formed between the flat substrate (in between the gMµEs) and the cell membrane rarely reaches values smaller than 10 nm (Table 3.2) despite the large variability observed (56 ± 29 nm; also see Table 3.1 for other cell types).

The seal resistance (R_{seal}) formed at the cleft between the neuron and the sensing element of the electronic device is a major limiting factor for effective electrical recordings (for review, see [40]). R_{seal} depends on the width of the cleft (d_c), the resistivity of the material within the cleft (ρ_c), and its planar dimensions. R_{seal} can therefore be calculated according to

$$R_{seal} = \frac{\rho_c}{2\pi \cdot r \cdot d_c} \cdot l \tag{3.1}$$

where r is the radius of the gMµE head, taken as 0.95 µm, and l is the length of the resistor across which the transductive extracellular current flows and is taken as the axial circumference of the gMµE (approx. 4 µm). Assuming the specific resistance of the cleft ρ_j to be 100 Ω cm (electrolyte solution), we arrive at $R_{seal} > 67$ MΩ for cleft width $d_j < 10$ nm. The seal resistance formed between cultured *Aplysia* neurons and a flat substrate was found to be 1.2 ± 0.43 MΩ [60]. Thus, our estimation suggests that the neuron–gMµE junction improves R_{seal} by at least 50-fold.

It should be noted that using TEM to assess the nature of the extracellular cleft formed between the plasma membrane of living cells and artificial substrates must be done with great caution as the procedures of chemical fixation, dehydration, embedding, and sectioning might generate alterations in the intracellular osmotic pressure and thereby generate structural artifacts [61]. Nevertheless, as is evident from the well-preserved structure of the mitochondria, vesicles, the endoplasmic reticulum, and the plasma membrane facing the bathing medium (Fig. 3.3d) in the electron micrographs, the fixation, dehydration, and embedding procedures did not produce osmotic pressure artifacts. The presence of electron translucent breaks within the embedding material, mainly (but not exclusively) at the curving junctions between the cells and the gMμE head as well as the stalk (e.g., Fig. 3.3d, white arrows), suggests that mechanical tension generated within the embedding polymer (Agar 100) leads to the detachment of the plasma membrane from the surface of the gMμE head only during sectioning of the agar block or the observations rather than at earlier stages of the procedures.

3.3.2 *Live Confocal Imaging of Actin Cytoskeleton at the gMμE–Neuron Junction*

The difference in the cleft width formed at the neuron–gMμE interface and the neuron–flat-substrate interface as seen in TEM could result from mechanical stretching of the plasma membrane around the gMμE head by cytoskeletal elements that assemble around these structures. In order to assess whether the engulfment of the functionalized gMμE is mediated by actin and other submembrane cytoskeletal elements and molecular motors, we used confocal imaging of fluorescent probes in live *Aplysia* neurons grown on matrices of gMμEs [49].

Neurons cultured on gMμE matrices functionalized by either PLL or EPP were fluorescently labeled by expression of GFP fusion proteins. Initially we expressed GFP alone, thereafter because of the important role of actin in focal adhesion formation [56], and in the generation of the mechanical forces associated with phagocytosis [62], we expressed GFP–actin by microinjection of mRNA encoding for GFP–actin. Expression of GFP–actin in cultured *Aplysia* neurons generates a diffuse signal in the cytosol in association with the homogeneous distribution of globular actin, GFP–actin hotspots in association with focal adhesion, and high-intensity fluorescent GFP–actin stripes in association with F-actin [63, 64]. Confocal microscope scans of 0.2 μm steps on the Z-axis were taken approximately 24 h after injection, and three-dimensional (3D) computer-aided reconstructions were prepared (Fig. 3.4). This procedure revealed that in both EPP- and PLL-functionalized matrices, the cell body and the main axon engulfed many gMμE but to variable degrees: clear differences in engulfment were observed with respect to the coating used as full and tight engulfment of the head and stalk is significantly more frequent on EPP- compared with PLL-coated substrates (Fig. 3.4d–f). Moreover, intense GFP–actin fluorescent signal in the form of a complete or partial "actin ring" was

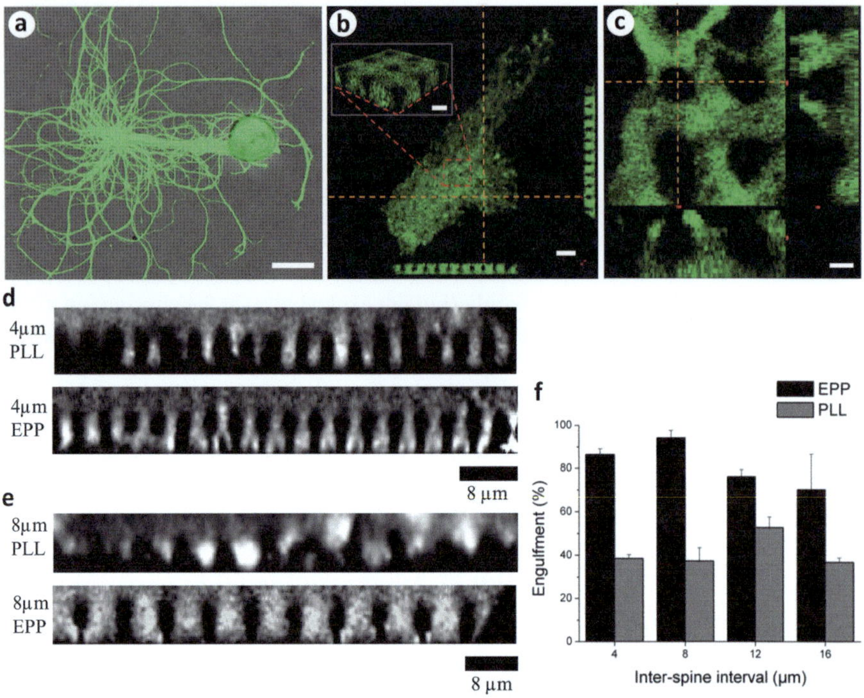

Fig. 3.4 Confocal microscope images of neurons expressing GFP–actin engulfing gMμEs. (**a**) An overview of a neuron cultured on 8 μm spaced gMμE matrix. The image is constructed of superpositioning of transmitted light and confocal GFP–actin images. (**b**) A confocal image taken 0.4 μm above the substrate surface. The equally spaced black *dots* depict single gMμEs. *Inset*: a computer-generated 3-dimensional reconstruction from a series of optical sections. (**c**) Zoom-in of a single PLL-coated gMμE engulfed by the neuron. Scale bars, 100 μm in (**a**), 10 μm for (**b**), 5 μm (**b**, *inset*), and 2 μm (**c**). The *dashed lines* denote plane of scan in both (**b**) and (**c**). (**d**, **e**): comparison of the engulfment of EPP- and PLL-functionalized gMμEs by neurons. Z-axis scans of the cell bodies of neurons cultured on a 4 μm (**d**) and 8 μm (**e**) inter-gMμE intervals. (**f**) Quantitative analysis of the engulfment of EPP-coated (*black*) and PLL-coated gMμE matrices (*gray*) as viewed from z-scans of GFP–actin expressing neurons. Scale bars, 8 μm. Reproduced with permission from [49], © (2009) IOP Publishing

frequently detected surrounding the stem of the EPP-functionalized gMμE (Fig. 3.5), as compared with PLL-coated substrate where hotspots of actin were dispersed in a non-organized fashion throughout the plane of cell–substrate interface in between gMμEs (Fig. 3.5b, d).

We decided to examine the assembly of specialized protein structures around the stalk of the gMμEs using an additional molecular probe, namely, cherry-cortactin. Cortactin is an F-actin-associated protein that regulates actin assembly at multiple binding sites and serves mainly as a stimulator of actin nucleation and branching by activation of Arp2/3 [65]. Cells cultured on EPP-coated matrices and injected with cherry-cortactin displayed ring-shaped cortactin fluorescence surrounding the

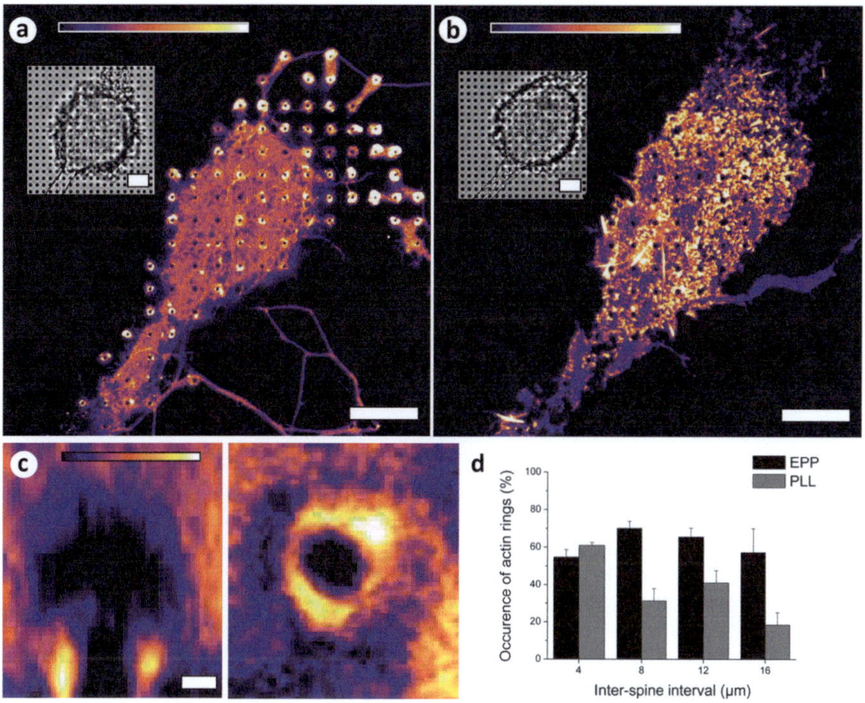

Fig. 3.5 Occurrence of "GFP–actin rings" surrounding the stems of EPP- and PLL-coated gMμEs. (**a**) and (**b**) depict the cell somata of *Aplysia* neurons cultured on EPP- and PLL-coated 8 μm spaced gMμE matrices, respectively. Relative intensity scale of the GFP–actin is shown at the *top* of both (**a**) and (**b**). Images were taken at the plane of the gMμE stem. *Insets*: corresponding transmitted light images of the cell bodies. Scale bars, 25 μm. (**c**) A ring of GFP–actin signal surrounding the stem of a gMμE. *Left-hand side*, ZX-axis scan; *right-hand side*, XY-axis scan taken at the plane of the gMμE stem (scale bar, 1 μm for both scans). (**d**) Comparison of the occurrence of actin rings surrounding the stems of gold spines on 4, 8, 12, and 16 μm matrices, coated with either EPP (*black*) or PLL (*gray*). Reproduced with permission from [49], © (2009) IOP Publishing

gMμE stalks, and in addition, when co-expressed, actin and cortactin fluorescence were generally colocalized [49].

These findings show that the gMμEs functionalized with EPP promote engulfment and actin polymerization around the stalk of the gMμE. Interestingly, in EPP actin rings are dominantly formed around the stalk, whereas the actin density along the substrate and the head of the gMμE is small (Fig. 3.5a, c). This suggests that in addition to the chemical signaling generated by the EPP, the mechanical tension generated by the geometry of the gMμEs also provides meaningful information for actin assembly, as is consistent with other studies [56]. However, as PLL alone does not generate tight engulfment of the gMμEs by the neurons, the geometry of the gMμEs is not sufficient by itself to trigger specific interactions between the cell and substrate.

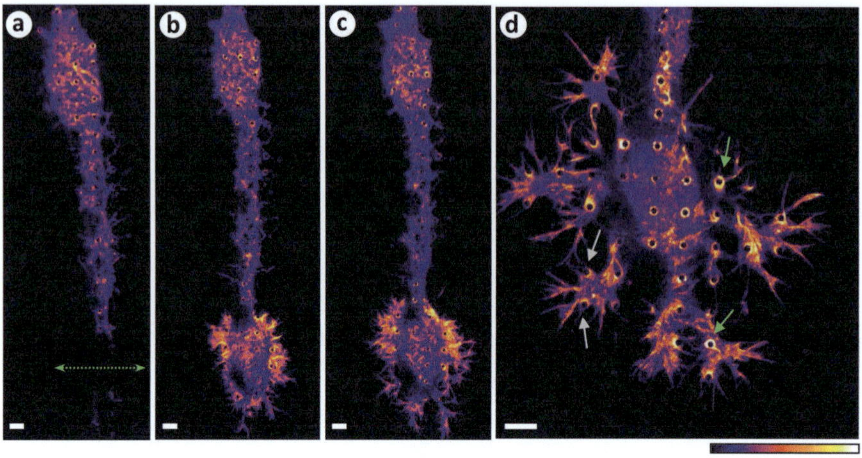

Fig. 3.6 Growth cone extension of transected axon on EPP-coated 8 μm spaced gMμE matrix. (**a**) Shown are the cell body (*upper part*) and its axon. Axotomy was applied along the double-*headed arrow* leading to the formation of a growth cone in the form of an extending flat lamellipodium (*lower part* of the images in **b**–**d**). (**b**) Ten minutes after axotomy, (**c**) twenty-two minutes, and (**d**) thirty-two minutes post axotomy. In (**c**), two *arrows* on the *right* indicate accumulation of GFP–actin in the form of rings around the stem of a gMμE. Two *arrows* on the *left* indicate advancing lamellae in the process of forming actin rings (scale bars, 10 μm). Reproduced with permission from [49], © (2009) IOP Publishing

The observation that the engulfment of gMμEs is associated with the assembly of GFP–actin rings around the stalk raised a number of questions. How fast does an actin ring assemble after the plasma membrane comes in contact with the gMμE? Given that the cytoskeletal elements are generally dynamic structures and that actin skeleton associated with adhesion plaques is highly dynamic [66], how stable are the formed actin rings?

To study the dynamics of actin ring assembly, we took advantage of earlier observations from our laboratory showing that axonal transection of cultured *Aplysia* neurons results in the formation of a growth cone in the form of an extending flat lamellipodium at the tip of the cut axon (for review, see [67]). The extension of a lamellipodium within 10–20 min after axotomy provides an opportunity to image the kinetics of GFP–actin ring formation during the first interaction of the plasma membrane with a functionalized gMμE. We found that within 2–12 min of contact between the flat lamellipodium and a gMμE, distinct actin rings are assembled (Fig. 3.6). Using a criterion of ring integrity (>75 % of the gMμE stalk is surrounded by circular high fluorescent GFP–actin), we found that 48 h after plating, the actin rings are dynamic: a single actin ring disappears and reappears at an average frequency of $4 \cdot 10^{-3}$ Hz $\pm 0.2 \cdot 10^{-3}$. Imaging the actin-ring temporal dynamics once a day over a period of 10 days revealed that even though the number of gMμEs enwrapped by actin rings remains unchanged, the actin-ring temporal dynamics are significantly reduced [49].

The formation of non-static actin rings at the stalk of the gMμEs suggests that the actin cytoskeleton is being constantly remodeled around the protruding structures in response to the extracellular chemical and mechanical signals. This process could be a result of nucleation of new actin filaments and/or consecutive repolymerization and depolymerization of existing actin filaments [68] similarly to adhesion processes occurring at the leading edge of motile or growing cells. We have also shown that the dynamic nature of rings of actin stabilizes around the stalks of the gMμEs approximately 2 weeks after culturing. This might correlate to common behavior of cells in vitro, where they often reach a structural steady state after a certain amount of time [69]. The dynamic nature of actin has also been observed at developing synapses of CA1 pyramidal neurons in rat hippocampal slices [70] where photoactivated GFP–actin was used to assess F-actin turnover rate in plastic dendritic spines. In that study, it was shown that actin exists in distinct pools, with a turnover rate of tens of seconds for the dynamic pool and minutes for the stable pool, not unlike the aforementioned observations. Additionally, previous studies show that monomeric G-actin is constantly being polymerized into ring-shaped networks of F-actin in endocytic hotspots of lamprey synapses [71, 72]. Although in these studies it was demonstrated that the actin rings at areas of vesicle recycling are stable, it was shown nonetheless that there is a constant dynamic turnover as is evident by the application of actin depolymerizing toxin latrunculin B which prevents the incorporation of G-actin into the F-actin rings. Could the dynamic nature of the cytoskeleton at the vicinity of the gMμE reflect the formation of presynaptic structures? In light of the dynamic reorganizing nature of actin and cortactin surrounding the stalks of the gMμEs, it should be no less than exciting to check for the existence of synaptic-related proteins [73] in the area of interaction between the neurons and the gMμE.

3.4 In-Cell Recording and Stimulation

Ultrastructural studies and live confocal imaging demonstrate a tight and active interface between cells and gMμEs. We now turn to the electrical implications of this tight seal. We will explore the ability of gMμEs to provide simultaneous, multisite, long-term stimulation and recording of action potentials and subthreshold synaptic potentials. We will show signal quality and signal-to-noise ratio comparable to that of conventional intracellular glass micropipette electrodes while maintaining scalable design options of multielectrode devices. This section is based on published studies [50–52] in which further technical information can be found.

3.4.1 In-Cell Recording of Action Potentials and Membrane Hyperpolarization

The experiment described in Fig. 3.7 (also described in [50]) depicts the main features of the "in-cell recording" configuration. For the experiment, an *Aplysia*

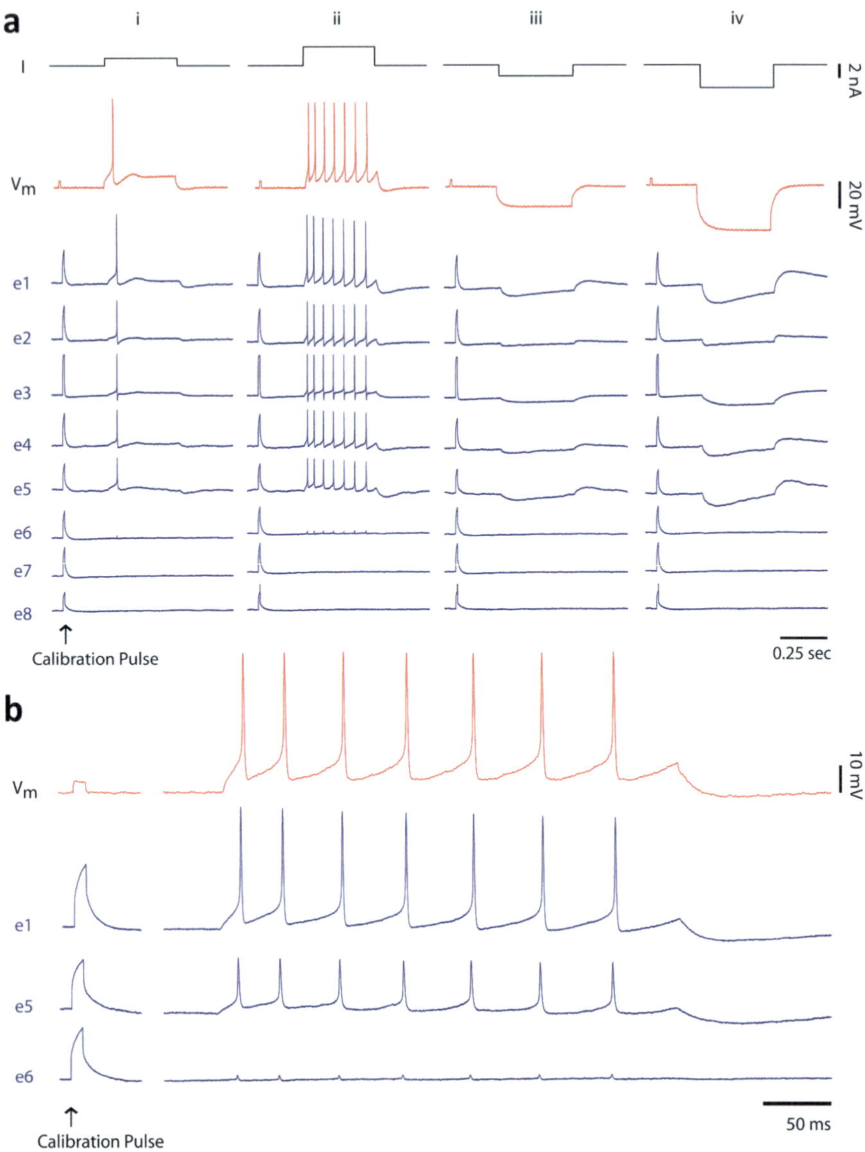

Fig. 3.7 In-cell recording of action potentials and subthreshold hyperpolarizing pulses by gMμEs. A neuron was cultured for 2 days on a gMμEs array device. Recordings of action potentials and hyperpolarizing pulses generated by an intracellular glass microelectrode inserted into the soma were made from eight gMμEs (*e1*–*e8*). Six gMμEs reside under the neuron (*e1*–*e6*) and two away from it (*e7* and *e8*). (**a**) A 5 mV, 10 ms square calibration pulse is delivered at the onset of each voltage trace. Depolarizing current pulses injected through the intracellular electrode (**a**, *i* and *ii*) generated membrane depolarization which reached threshold to fire a single action potential (*i*) and a train of action potentials (*ii*). These action potentials were recorded by gMμEs *e1*–*e6*. The trace showing the trains of action potentials recorded by the intracellular electrode and gMμEs *e1*, *e5*, and *e6* in (*ii*) are enlarged in (**b**). Note the differences in the shapes and amplitude of the 5 mV 10 ms calibration pulses and the action potentials. In columns *iii* and *iv*, hyperpolarizing square pulses were delivered by the intracellular glass microelectrode. The different degrees of electrical coupling between the neuron and the gMμEs and the differences in the filtering of the DC pulse by the different gMμEs are apparent (*ii* and *iv*). Reproduced from [50] © The Author(s)

neuron was cultured for 2 days on an array of 62 gMμEs functionalized with the engulfment promoting peptide (EPP). A sharp-glass microelectrode for both current injection and voltage recording was inserted into the soma of the neurons. Depolarizing current pulses of varying intensities (Fig. 3.7a, columns *i* and *ii*) were delivered by the intracellular electrode and generated a single or a train of action potentials recorded by the same glass microelectrode and the gMμEs. The action potentials recorded by the gMμEs were monophasic, positive potentials that resemble in shape the intracellularly recorded action potentials (APs, Fig. 3.7a, b). In this experiment, APs were recorded by six gMμEs that resided beneath the stimulated neuron's cell body (Fig. 3.7, *e1–e6*, blue). The amplitudes of the raw APs recorded by *e1–e5* were in the range of 5–10 mV, while those recorded by e6 were smaller than 1 mV, reflecting differences in the electrical coupling between the neuron and the individual gMμEs. Interestingly, even the small APs of e6 were positive monophasic potentials. Other gMμEs not residing beneath the neuron's cell body did not record any signals (*e7*, *e8*). This fact demonstrates the lack of cross-talk artifacts from neighboring microelectrodes. Using the same configuration, we recorded in other experiments monophasic action potentials with amplitudes of up to 25 mV [51].

Hyperpolarizing pulses of increased amplitude delivered by the intracellular microelectrode generated membrane hyperpolarization (Fig. 3.7, columns *iii* and *iv*) that were picked up also by gMμEs *e1–e6*. This clearly demonstrates that the interface formed between the neuron and the gMμE generates electrical coupling sufficient to enable parallel multiple site recordings of APs and subthreshold potentials.

3.4.2 In-Cell Recording of Subthreshold Synaptic Events

To further illustrate the quality of the recording provided by the gMμE-based microelectrode array, we cocultured homologous *Aplysia* neurons that form electrical synapses [74]. Figure 3.8 depicts an experiment performed on two neurons cultured on a gMμE array for 2 days. For the experiment, a sharp-glass microelectrode for both current injection and voltage recording was inserted to neuron 1, and recordings were made with a gMμE from neuron 2 (Fig. 3.8a). Hyperpolarizing square current pulse delivered to neuron 1 generated membrane hyperpolarization of neurons 1 and 2 (Fig. 3.8b). The coupling coefficient (the algebraic scaling factor) between neuron 1 (as recorded by an intracellular DC-coupled electrode) and the raw signal recorded in neuron 2 (by the AC-coupled gMμE, Fig. 3.8b*iii*) was estimated to be 0.2–0.3. Consistent with these observations, firing of neuron 1 (Fig. 3.8c*ii*) generated an excitatory postsynaptic potential (EPSP) in neuron 2 (Fig. 3.8c*iii*) which reached threshold to initiate an AP (Fig. 3.8c*iii*). The AP in neuron 2 (Fig. 3.8c*iii*) in turn initiated an EPSP in neuron 1 (Fig. 3.8c*ii*). Trains of action potentials generated by the intracellular electrode in neuron 1 (Fig. 3.8c, right column) generated a train of EPSPs in neuron 2 that summated to fire a train of action potentials (Fig. 3.8c right panel).

It is important to note that the impedance of the gMμEs and the AC amplifiers used for the recording by the gMμE alter the shape of the recorded signals compared

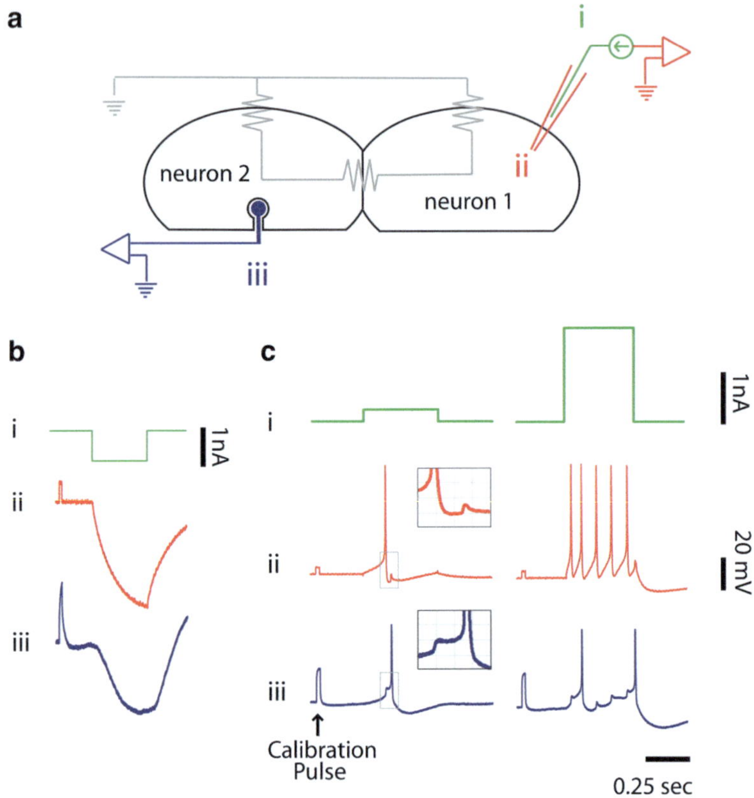

Fig. 3.8 In-cell recording of subthreshold synaptic potentials. (**a**) Experimental setup. (**b**) Hyperpolarizing square current pulse (*i*) delivered to cell 1 (*right-hand side* in **a**), generated membrane hyperpolarization of both cells 1 and 2, and picked up by the glass micropipette electrode and the gMμE (**b** *ii* and *iii*, respectively) demonstrating the electrical coupling between the cells. (**c**) Depolarizing square current pulse (*i*) delivered to cell 1 elicited firing of an action potential (**c** *ii*) and generated an excitatory postsynaptic potential (EPSP) in cell 2 picked up by the gMμE (**c**, *iii*). The AP in cell 2 (**c**, *iii*) in turn initiated an EPSP in cell 1 (**c**, *ii*) (*insets*: close-up of concurrently recorded EPSPs and APs in both cells). Trains of action potentials (*right-hand-side* panels) generated by the intracellular electrode in cell 1 generated a train of EPSPs in cell 2 (*iii*) that summated to fire a train of action potentials. Reproduced from [50] © The Author(s)

with those recorded by the DC-coupled intracellular sharp-glass electrode. The electrical impedance depends on the ionic bilayer (also known as the electrical double layer) [75] formed at the interface between the gMμE and the culture medium and of the AC amplifier used. In addition, the attenuation in the amplitude is further attributed to the quality of the seal resistance formed between the plasma membrane and the gMμE and the conductance of the patch of plasma membrane that faces the gMμE. Since the parameters that represent individual neuron–gMμE junction are not identical, the alterations in the signal shape and attenuation factor differ for individual gMμEs, as is reflected by the differences in the shapes and amplitude of the 5 mV 10 ms calibration pulses and the action potentials (Fig. 3.7). Nevertheless, these individual alterations can be corrected using the calibration pulse as a reference and employing either an offline or a real-time negative capacitance compensation circuit [51, 76, 77]. Thus, in spite of the filtering and attenuation of the signals by the gMμE impedance and AC-coupled amplifier, the quality of recording by the extracellular positioned gMμE is of unprecedented quality.

3.4.3 In-Cell Stimulation

Analysis of neuronal circuits largely relies on repeated use of stimulating electrodes. It is of critical importance that stimuli are delivered without damaging the cells [78]. Whereas current injection (stimulation) by intracellular sharp-glass electrodes or patch electrodes involves no difficulty, stimulation by extracellular high-impedance electrodes is complicated by the relatively limited charge transfer to the plasma membrane, by the risk of damaging the cells by electroporation [7, 79, 80], or by irreversible electrochemical reaction products [81–84]. To overcome these problems, trains of weak capacitive stimuli can be delivered to cultured cells to activate local sodium currents that generate sufficient depolarization to reach the firing threshold [78]. While this approach is safe, it might complicate experimental protocols in which precisely timed consecutive stimuli are to be delivered.

We now examine whether the interface formed between the gMμEs and the neurons as well as the gMμE properties supports effective stimulation of the neurons without inducing irreversible electroporation. To that end, a neuron cultured on a gMμE-based microelectrode array was impaled by a sharp electrode for both current injection and voltage recordings (Fig. 3.9). We next characterized the coupling level between the neuron and the gMμE and selected to concentrate on two gMμEs, one with high coupling (gMμE-1) and the other with low coupling values (gMμE-2) (Fig. 3.9a). Hyperpolarization of the neuron by current injection through the intracellular electrode leads to gMμE-1 and to a much smaller extent gMμE-2 (Fig. 3.9b*i*). Consistently, depolarization of the neuron to fire an action potential leads to the generation of an action potential of ~3 mV recorded by gMμE-1 and ~0.5 mV by gMμE-2 (Fig. 3.9c*i* and d*i*, respectively). Delivering a single depolarization square voltage step of 1–10 ms with an increasing voltage from 1,400 to 1,800 mV by gMμE-1 evoked depolarization of the neuron which reached threshold to fire action potentials as recorded by the glass microelectrode (Fig. 3.9e, upper panel). Applying similar current pulses to gMμE-2 depolarized the neuron but failed to reach threshold (Fig. 3.9e, bottom panel).

To examine whether this series of stimulations (5–10 stimulations per experiment) via the gMμE induced damages to the neuron, we once again delivered hyperpolarizing pulses through the intracellular glass microelectrode and measured the input resistance and the coupling between the neurons and the gMμEs. As seen in Fig. 3.9 (b*ii*, c*ii*, and d*ii*), the stimuli delivered by the gMμE did not induce detectable changes to the neuron's input resistance or altered the coupling coefficient between the neuron and the gMμE.

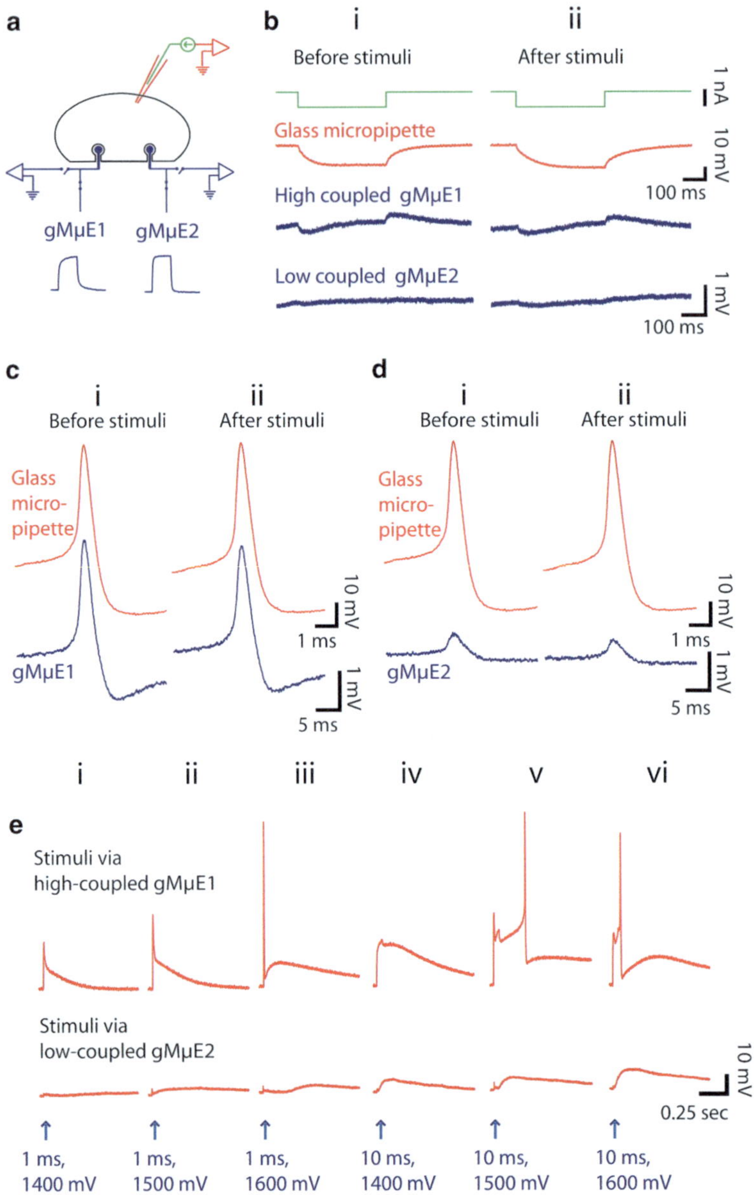

Fig. 3.9 Stimulation by gMμEs without damage to the cell. (**a**) Experimental setup. (**b**) Hyperpolarizing square current pulse delivered to the cell by the glass micropipette intracellular electrode generated membrane hyperpolarization detected by both high- and low-coupled gMμEs (gMμE1 and gMμE2, respectively, in *i*). The input resistance and level of coupling between the neuron and gMμE1 and gMμE2 were not changed before and after the application of stimuli delivered by gMμE1 and gMμE2 stimulation (shown in **e**). (**c**, **d**) Action potential shape, duration, and amplitude as recorded by gMμE1 and gMμE2. These are not changed by the use of gMμE1 and gMμE2 to stimulate the neuron (shown in **e**). (**e**) Stimulation of the neuron by gMμE1 and gMμE2 while recording with the intracellular glass microelectrode. The strength of the applied stimulus and its duration are given below the traces. Note that whereas the high-coupled gMμE1 (*upper* panel) generated an action potential, the low-coupled electrode did not reach threshold. Reproduced from [50] © The Author(s)

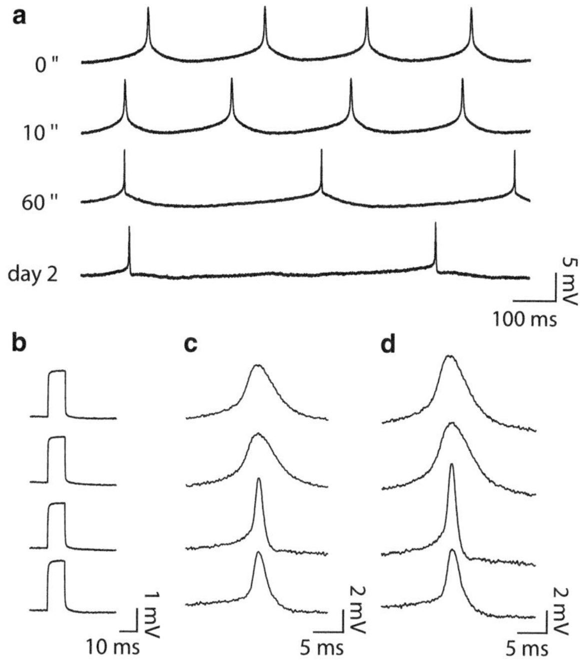

Fig. 3.10 Long-term in-cell recording from a spontaneously active neuron. (**a**) Recording sessions from the same neuron of hours and days are maintained (time is given on the *left-hand side*). (**b**) Calibration pulse along 2 days of recordings. (**c**) The shapes and amplitudes of the raw action potentials recorded by gMμEs along 2 days. (**d**) Deconvolution of the recorded action potentials using the calibration pulses and a capacitance compensation circuit (see [51]). Reproduced from [50] © The Author(s)

We conclude that with the observed high coupling coefficient, the in-cell configuration enables the application of sufficient current to reach firing threshold of the neuron without inflicting damages to the plasma membrane or the junction.

3.4.4 Stability of the gMμE–Neuron Junction

A critical feature of the extensive use of extracellular recording for in vitro and in vivo research purposes as well as for potential future clinical applications is the cell-noninvasive nature of the electrodes which permits long-term recordings. In an earlier study, we established that culturing *Aplysia* neurons on a dense matrix of gold mushroom-shaped protrusions (gMμPs) for over a week does not alter the excitable membrane properties and synaptic physiology of the neurons [48]. We now turn to examining the functional stability of the junctions formed between the neurons and the gMμE. We began to examine this question by culturing spontaneously active *Aplysia* neurons on gMμE-based microelectrode arrays (Fig. 3.10 and see [50]). We found that effective electrical coupling between neurons and gMμEs was maintained for over 48 h. It should be noted that in these experiments, the shape and amplitude of the action potentials over this period were not absolutely stable. We avoided the risk of culture contamination by not taking off the cover from the culture dish for insertion of a conventional sharp intracellular

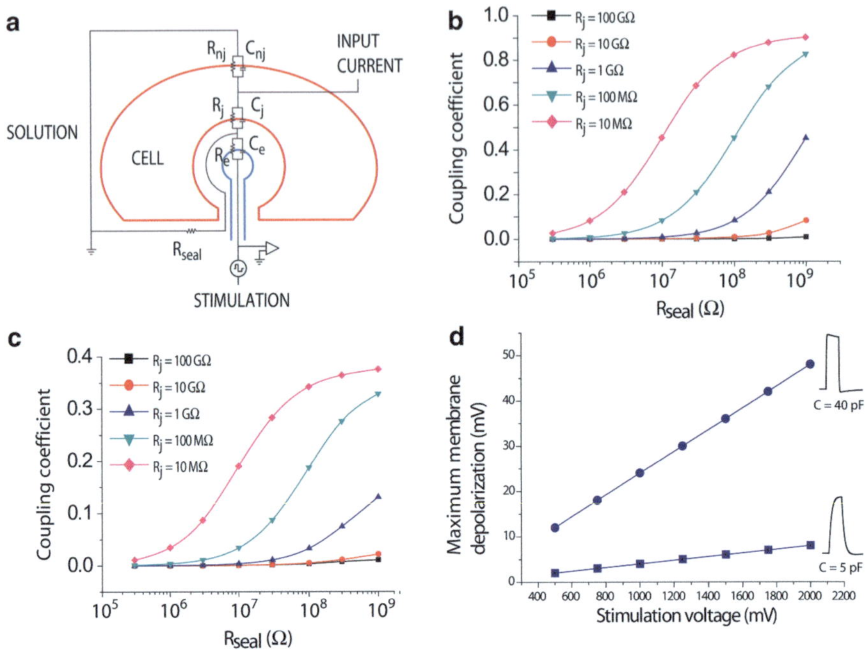

Fig. 3.11 The in-cell configuration: ohmic conductance at the junctional membrane. (**a**) Equivalent circuit of gMμE–neuron junction. (**b**) and (**c**): coupling coefficient as a function of seal resistance and the resistance of the junctional membrane in high-frequency (100Hz) (**b**) and low-frequency (1Hz) (**c**) signals. (**d**) In-cell stimulation: maximum membrane depolarization as a function of amplitude of voltage-based stimulation through the gMμE in high (*circles*) and low (*squares*) capacitance gMμEs. Application of an ~1,500 mV voltage results in a maximum depolarization of ~35 mV, which is sufficient to elicit an action potential. Reproduced from [50] © The Author(s)

microelectrode for comparison. Nonetheless, a number of mechanisms could account for the observed changes. These include small alterations in the neurons' resting potential, changes in the junctional membrane resistance, or seal resistance (see Sect. 3.4.5, equivalent electrical circuit of the in-cell configuration). To the best of our knowledge, such recording sessions of intracellular readout with lengths of over 2 days were never conducted prior to this study.

3.4.5 *In-Cell Configuration: Ohmic Conductance at the gMμE–Neuron Junction*

The unprecedented electrical coupling between neurons and engulfed gMμEs can be explained by the use of the analog electrical *equivalent* circuit shown in Fig. 3.11. The model includes the following: (a) a neuron composed of a non-junctional membrane characterized by a passive RC circuit with parameters (R_{nj}; C_{nj}) and a junctional membrane facing the gMμE (R_j; C_j), (b) the cleft formed between the neuron's plasma membrane and the surface of the gMμE (R_{seal}), and (c) the gMμE itself ($R_{gM\mu E}$ and $C_{gM\mu E}$). The majority of parameters used for the simulation was obtained from direct measurements or by calculations of the physical parameters that fit the specific geometry of the gMμE and the neuron–gMμE interface (for further detail, see [50]). The expected coupling coefficient for action potentials (high frequencies of 100–1,000 Hz) and long (DC) pulses were calculated as a function of R_{seal} using different values for the junctional membrane resistance (ranging from 10 MΩ to 100 GΩ, Fig. 3.11b, c). R_{seal} was taken to be ~100 MΩ (see Sect. 3.3.1 and [48]).

Based on these parameters, we evaluated the range of possible values of the junctional membrane resistance which generate the experimentally observed coupling coefficient (i.e., 0.05–0.5) to be 10–100 MΩ. Assuming that the junctional membrane conductance is increased by recruitment of voltage-independent ionic channels such as potassium channels with single-channel conductance of 10–100 pS, then ~10–100 such channels have to concentrate within the confined area of the junctional membrane. This would imply a density of approximately 0.5–10 channels/μm^2. This density is physiological and was documented in many cell types [85].

It is important to recall that the coupling between the neurons and the gMμEs depends on the value of R_{seal}. Reducing R_{seal} to values below 100 MΩ reduces the coupling coefficient drastically (Fig. 3.11b, c).

Using the same values, it is also possible to simulate the experiments of Fig. 3.9e, modeling the membrane depolarization induced by the delivery of a millisecond long voltage stimulation pulse by a gMμE. Figure 3.11d depicts the membrane depolarization induced by such a pulse assuming that $R_j = 100$ MΩ, $R_{seal} = 100$ MΩ, and gMμE capacitance is either low (5 pF) or high (40 pF). We have found that for high-capacitance gMμEs, an ~1,500 mV voltage results in a maximum depolarization of ~35 mV, which is sufficient to elicit an action potential. This correlates with the empirical findings as shown in Fig. 3.9e.

As discussed in Sect. 3.3, engulfment of gMμEs is much more effective when they are functionalized with EPP than with PLL. This suggests that EPP plays a role in the induction of the engulfment and that binding between receptors displayed on the plasma membrane and the EPP activates a highly conserved cascade leading to phagocytosis-like processes [54]. The neurons then elevate the conductance of their plasma membrane facing the gMμEs in concert with the assembly of a cytoskeletal actin ring around the stalk of the gMμEs. The geometry of the gMμEs plays an important role in these cytoskeletal reorganizations [48] as sensing of surface topography by the cell plays a significant role in the regulation of cell functions (for review, see [86]). Convex surfaces are known to cause BAR-domain protein (Bin, Amphiphysin, Rvs domain) to release Rac (a GTPase), which in turn leads to local cytoskeleton rearrangements. This may result in redistribution of ion channels and the introduction of ohmic conductance at the neuron–gMμE junction. In addition, changes in mechanical tension of the inner and outer faces of the lipid bilayer by the curvature of the gMμE geometry may result in the activation of ion channels or passive increase in local conductance.

3.5 Concluding Remarks

We have seen that the neuron–gMμEs interface forms an unexpectedly tight junction which supports high-quality bidirectional electrical coupling. This configuration enables in-cell recording with quality and signal-to-noise ratio that matches classical sharp- and patch-electrode intracellular recording. The interface also supports "in-cell stimulation" by milliseconds-long single pulses without damaging the cell membrane. Consistent with the extracellular position of the gMμEs with respect to the neurons, the recording sessions could last for over 2 days and most likely for significantly longer periods.

A number of challenges lie in the path toward the translation of in-cell recording and stimulation to in vivo mammalian neural networks [57, 58]. A more sophisticated chemical functionalization of the device, suitable for the diverse population of cell types in the living brain, will most likely be necessary in order to specifically target neurons and prevent professional phagocytotic cells such as microglia from engulfing the electrodes.

Acknowledgments The author was supported by a doctoral scholarship from the Israeli Council for Higher Education, a postdoctoral fellowship from the Edmond & Lily Safra Center for Brain Sciences (ELSC), and a fellowship from the European Molecular Biology Organization (EMBO). The research described in this chapter was originated in the laboratory of Prof. Micha E. Spira in collaboration with Prof. Joseph Shappir of the Hebrew University of Jerusalem, Israel.

References

1. Bernstein, J.: Ueber den zeitlichen Verlauf der negativen Schwankung des Nervenstroms. Pflüger's Arch. ges. Physiol. **1**, 173–207 (1868)
2. Galvani, L.: De viribus electricitatis in motu musculari commentarius. Bon Sci Art Inst Acad Comm **7**, 363–418 (1791)
3. Galvani, L.: Opere edite ed inedite del Professore Luigi Galvani raccolte e pubblicate dall'Accademia delle Science dell'Istituto di Bologna. Dall'Olmo, Bologna (1841)
4. Helmholtz, H.: Note sur la vitesse de propagation de l'agent nerveux dans le nerfs rachidiens. C R Acad Sci (Paris) **30**, 204–206 (1850)
5. Verkhratsky, A., Krishtal, O. A. & Petersen, O. H.: From Galvani to patch clamp: the development of electrophysiology. Pflugers Arch **453**, 233–247 (2006)
6. Hodgkin, A. L.: Evidence for electrical transmission in nerve: Part II. J Physiol **90**, 211–232 (1937)
7. Armstrong, C. M.: Life among the axons. Annu Rev Physiol **69**, 1–18 (2007)
8. Grundfest, H.: The mechanisms of discharge of the electric organs in relation to general and comparative electrophysiology. Prog Biophys Biophys Chem **7**, 1–85 (1957)
9. Eccles, J. C., Eccles, R. M. & Lundberg, A.: Synaptic actions on motoneurones in relation to the two components of the group I muscle afferent volley. J Physiol **136**, 527–546 (1957)
10. Hodgkin, A. H., Huxley, A.F.: Action potentials recorded from inside a nerve fibre. Nature **144**, 710–711 (1939)
11. Sakmann, B. & Neher, E.: Patch clamp techniques for studying ionic channels in excitable membranes. Annu Rev Physiol **46**, 455–472 (1984)

12. Berger, T. K., Perin, R., Silberberg, G. & Markram, H.: Frequency-dependent disynaptic inhibition in the pyramidal network: a ubiquitous pathway in the developing rat neocortex. J Physiol **587**, 5411–5425 (2009)
13. Hochberg, L. R. et al.: Neuronal ensemble control of prosthetic devices by a human with tetraplegia. Nature **442**, 164–171 (2006)
14. Hartline, F. F.: Biological applications for voltage sensitive dyes. Science **203**, 992–994 (1979)
15. Loew, L. M., Cohen, L. B., Salzberg, B. M., Obaid, A. L. & Bezanilla, F.: Charge-shift probes of membrane potential. Characterization of aminostyrylpyridinium dyes on the squid giant axon. Biophys J **47**, 71–77 (1985)
16. Shoham, D. et al.: Imaging cortical dynamics at high spatial and temporal resolution with novel blue voltage-sensitive dyes. Neuron **24**, 791–802 (1999)
17. Stosiek, C., Garaschuk, O., Holthoff, K. & Konnerth, A.: In vivo two-photon calcium imaging of neuronal networks. Proc Natl Acad Sci U S A **100**, 7319–7324 (2003)
18. Higley, M. J. & Sabatini, B. L.: Calcium signaling in dendrites and spines: practical and functional considerations. Neuron **59**, 902–913 (2008)
19. Rothschild, G., Nelken, I. & Mizrahi, A.: Functional organization and population dynamics in the mouse primary auditory cortex. Nat Neurosci **13**, 353–360 (2010)
20. Kwong, K. K. et al.: Dynamic magnetic resonance imaging of human brain activity during primary sensory stimulation. Proc Natl Acad Sci U S A **89**, 5675–5679 (1992)
21. Bandettini, P. A., Jesmanowicz, A., Wong, E. C. & Hyde, J. S.: Processing strategies for time-course data sets in functional MRI of the human brain. Magn Reson Med **30**, 161–173 (1993)
22. Boyden, E. S., Zhang, F., Bamberg, E., Nagel, G. & Deisseroth, K.: Millisecond-timescale, genetically targeted optical control of neural activity. Nat Neurosci **8**, 1263–1268 (2005)
23. Deisseroth, K. et al.: Next-generation optical technologies for illuminating genetically targeted brain circuits. J Neurosci **26**, 10380–10386 (2006)
24. Gradinaru, V. et al.: Targeting and readout strategies for fast optical neural control in vitro and in vivo. J Neurosci **27**, 14231–14238 (2007)
25. Lee, J. H. et al.: Global and local fMRI signals driven by neurons defined optogenetically by type and wiring. Nature **465**, 788–792 (2010)
26. Bradley, P. M., Murphy, D., Kasparov, S., Croker, J. & Paton, J. F.: A micro-optrode for simultaneous extracellular electrical and intracellular optical recording from neurons in an intact oscillatory neuronal network. J Neurosci Methods **168**, 383–395 (2008)
27. Zhang, J. et al.: A microelectrode array incorporating an optical waveguide device for stimulation and spatiotemporal electrical recording of neural activity. Conf Proc IEEE Eng Med Biol Soc **2009**, 2046–2049 (2009)
28. Anikeeva, P. et al.: Optetrode: a multichannel readout for optogenetic control in freely moving mice. Nat Neurosci **15**, 163–170 (2012)
29. Logothetis, N. K.: What we can do and what we cannot do with fMRI. Nature **453**, 869–878 (2008)
30. Berdondini, L. et al.: Active pixel sensor array for high spatio-temporal resolution electrophysiological recordings from single cell to large scale neuronal networks. Lab Chip **9**, 2644–2651 (2009)
31. Bologna, L. L. et al.: Low-frequency stimulation enhances burst activity in cortical cultures during development. Neuroscience **165**, 692–704 (2010)
32. Shahaf, G. et al.: Order-based representation in random networks of cortical neurons. PLoS Comput Biol **4**, e1000228 (2008)
33. Rubehn, B., Bosman, C., Oostenveld, R., Fries, P. & Stieglitz, T.: A MEMS-based flexible multichannel ECoG-electrode array. J Neural Eng **6**, 036003 (2009)
34. Hutzler, M. et al.: High-resolution multitransistor array recording of electrical field potentials in cultured brain slices. J Neurophysiol **96**, 1638–1645 (2006)
35. Wise, K. D. & Najafi, K.: Microfabrication techniques for integrated sensors and microsystems. Science **254**, 1335–1342 (1991)
36. Campbell, P. K., Jones, K. E., Huber, R. J., Horch, K. W. & Normann, R. A.: A silicon-based, three-dimensional neural interface: manufacturing processes for an intracortical electrode array. IEEE Trans Biomed Eng **38**, 758–768 (1991)

37. Shein, M. et al.: Engineered neuronal circuits shaped and interfaced with carbon nanotube microelectrode arrays. Biomedical Microdevices **11**, 495–501 (2009)
38. Eytan, D. & Marom, S.: Dynamics and effective topology underlying synchronization in networks of cortical neurons. J Neurosci **26**, 8465–8476 (2006)
39. McCreery, D., Pikov, V. & Troyk, P. R.: Neuronal loss due to prolonged controlled-current stimulation with chronically implanted microelectrodes in the cat cerebral cortex. J Neural Eng **7**, 036005 (2010)
40. Fromherz, P.:Neuroelectronic Interfacing: Semiconductor Chips with Ion Channels, Nerve Cells and Brain. In: Waser, P. (ed.) Neuroelectronic interfacing: Semiconductor chips with ion channels, nerve cells, and brain. Wiley-VCH, Berlin (2003)
41. Fromherz, P., Offenhausser, A., Vetter, T. & Weis, J.: A neuron-silicon junction: a Retzius cell of the leech on an insulated-gate field-effect transistor. Science **252**, 1290–1293 (1991)
42. Fromherz, P.: Three levels of neuroelectronic interfacing: silicon chips with ion channels, nerve cells, and brain tissue. Ann N Y Acad Sci **1093**, 143–160 (2006)
43. Neves, G., Cooke, S. F. & Bliss, T. V.: Synaptic plasticity, memory and the hippocampus: a neural network approach to causality. Nat Rev Neurosci **9**, 65–75 (2008)
44. Spira, M. E. & Hai, A.: Multi-electrode array technologies for neuroscience and cardiology. Nat Nanotechnol **8**, 83–94 (2013)
45. Shoham, S., Fellows, M. R. & Normann, R. A.: Robust, automatic spike sorting using mixtures of multivariate t-distributions. J Neurosci Methods **127**, 111–122 (2003)
46. Fee, M. S., Mitra, P. P. & Kleinfeld, D.: Automatic sorting of multiple unit neuronal signals in the presence of anisotropic and non-Gaussian variability. J Neurosci Methods **69**, 175–188 (1996)
47. Spira, M.E. et al.: Improved neuronal adhesion to the surface of electronic device by engulfment of protruding micro-nails fabricated on the chip surface. Transducers '07 & Eurosensors Xxi, Digest of Technical Papers, Vols 1 and 2, 1247–1250 (2007)
48. Hai, A. et al.: Spine-shaped gold protrusions improve the adherence and electrical coupling of neurons with the surface of micro-electronic devices. J R Soc Interface **6**, 1153–1165 (2009)
49. Hai, A. et al.: Changing gears from chemical adhesion of cells to flat substrata toward engulfment of micro-protrusions by active mechanisms. J Neural Eng **6**, 066009 (2009)
50. Hai, A., Shappir, J. & Spira, M. E.: Long-term, multisite, parallel, in-cell recording and stimulation by an array of extracellular microelectrodes. J Neurophysiol **104**, 559–568 (2010)
51. Hai, A., Shappir, J. & Spira, M. E.: In-cell recordings by extracellular microelectrodes. Nat Methods **7**, 200–202 (2010)
52. Hai, A. & Spira, M. E.: On-chip electroporation, membrane repair dynamics and transient in-cell recordings by arrays of gold mushroom-shaped microelectrodes. Lab Chip **12**, 2865–2873 (2012)
53. Roelandse, M., Welman, A., Wagner, U., Hagmann, J. & Matus, A.: Focal motility determines the geometry of dendritic spines. Neuroscience **121**, 39–49 (2003)
54. Stuart, L. M. & Ezekowitz, R. A.: Phagocytosis: elegant complexity. Immunity **22**, 539–550 (2005)
55. Dupuy, A. G. & Caron, E.: Integrin-dependent phagocytosis: spreading from microadhesion to new concepts. J Cell Sci **121**, 1773–1783 (2008)
56. Geiger, B., Spatz, J. P. & Bershadsky, A. D.: Environmental sensing through focal adhesions. Nat Rev Mol Cell Biol **10**, 21–33 (2009)
57. Fendyur, A. & Spira, M. E.: Toward on-chip, in-cell recordings from cultured cardiomyocytes by arrays of gold mushroom-shaped microelectrodes. Front Neuroeng **5**, 21 (2012)
58. Fendyur, A., Mazurski, N., Shappir, J. & Spira, M. E.: Formation of Essential Ultrastructural Interface between Cultured Hippocampal Cells and Gold Mushroom-Shaped MEA- Toward "IN-CELL" Recordings from Vertebrate Neurons. Front Neuroeng **4**, 14 (2012)
59. Oren, R. et al.: Electrically conductive 2D-PAN-containing surfaces as a culturing substrate for neurons. J Biomater Sci Polym Ed **15**, 1355–1374 (2004)
60. Cohen, A., Shappir, J., Yitzchaik, S. & Spira, M. E.: Reversible transition of extracellular field potential recordings to intracellular recordings of action potentials generated by neurons grown on transistors. Biosens Bioelectron **23**, 811–819 (2008)

61. Studer, D., Humbel, B. M. & Chiquet, M.: Electron microscopy of high pressure frozen samples: bridging the gap between cellular ultrastructure and atomic resolution. Histochem Cell Biol **130**, 877–889 (2008)
62. Castellano, F., Chavrier, P. & Caron, E.: Actin dynamics during phagocytosis. Semin Immunol **13**, 347–355 (2001)
63. Sahly, I., Erez, H., Khoutorsky, A., Shapira, E. & Spira, M. E.: Effective expression of the green fluorescent fusion proteins in cultured Aplysia neurons. J Neurosci Methods **126**, 111–117 (2003)
64. Sahly, I., Khoutorsky, A., Erez, H., Prager-Khoutorsky, M. & Spira, M. E.: On-line confocal imaging of the events leading to structural dedifferentiation of an axonal segment into a growth cone after axotomy. J Comp Neurol **494**, 705–720 (2006)
65. Decourt, B., Munnamalai, V., Lee, A. C., Sanchez, L. & Suter, D. M.: Cortactin Colocalizes With Filopodial Actin and Accumulates at IgCAM Adhesion Sites in Aplysia Growth Cones. Journal of Neuroscience Research **87**, 1057–1068 (2009)
66. Endlich, N., Otey, C. A., Kriz, W. & Endlich, K.: Movement of stress fibers away from focal adhesions identifies focal adhesions as sites of stress fiber assembly in stationary cells. Cell Motil Cytoskeleton **64**, 966–76 (2007)
67. Bradke, F., Fawcett, J. W. & Spira, M. E.: Assembly of a new growth cone after axotomy: the precursor to axon regeneration. Nat Rev Neurosci **13**, 183–193 (2012)
68. Pollard, T. D. & Cooper, J. A.: Actin and actin-binding proteins. A critical evaluation of mechanisms and functions. Annu Rev Biochem **55**, 987–1035 (1986)
69. Bailey, C. H. & Chen, M.: Time course of structural changes at identified sensory neuron synapses during long-term sensitization in Aplysia. J Neurosci **9**, 1774–1780 (1989)
70. Honkura, N., Matsuzaki, M., Noguchi, J., Ellis-Davies, G. C. & Kasai, H.: The subspine organization of actin fibers regulates the structure and plasticity of dendritic spines. Neuron **57**, 719–729 (2008)
71. Bourne, J., Morgan, J. R. & Pieribone, V. A.: Actin polymerization regulates clathrin coat maturation during early stages of synaptic vesicle recycling at lamprey synapses. J Comp Neurol **497**, 600–609 (2006)
72. Shupliakov, O. et al.: Impaired recycling of synaptic vesicles after acute perturbation of the presynaptic actin cytoskeleton. Proc Natl Acad Sci U S A **99**, 14476–14481 (2002)
73. Malkinson, G. et al.: Calcium-induced exocytosis from actomyosin-driven, motile varicosities formed by dynamic clusters of organelles. Brain Cell Biol **35**, 57–73 (2006)
74. Benbassat, D. & Spira, M. E.: The Survival of Transected Axonal Segments of Cultured Aplysia Neurons Is Prolonged by Contact with Intact Nerve-Cells. European Journal of Neuroscience **6**, 1605–1614 (1994)
75. Mortari, A., Maaroof, A., Martin, D. & Cortie, M. B.: Mesoporous gold electrodes for sensors based on electrochemical double layer capacitance. Sensors and Actuators B-Chemical **123**, 262–268 (2007)
76. McGillivray, R. & Wald, R.: Dual-path capacitance compensation network for microelectrode recordings. Am J Physiol **238**, H930–1 (1980)
77. Wilson, C. J. & Park, M. R.: Capacitance compensation and bridge balance adjustment in intracellular recording from dendritic neurons. J Neurosci Methods **27**, 51–75 (1989)
78. Schoen, I. & Fromherz, P.: Extracellular stimulation of mammalian neurons through repetitive activation of Na+ channels by weak capacitive currents on a silicon chip. Journal of Neurophysiology **100**, 346–357 (2008)
79. Ryttsen, F. et al.: Characterization of single-cell electroporation by using patch-clamp and fluorescence microscopy. Biophys J **79**, 1993–2001 (2000)
80. Rubinsky, B.: Irreversible electroporation in medicine. Technol Cancer Res Treat **6**, 255–260 (2007)
81. Brummer, S. B., Robblee, L. S. & Hambrecht, F. T.: Criteria for selecting electrodes for electrical stimulation: theoretical and practical considerations. Ann N Y Acad Sci **405**, 159–171 (1983)
82. Harnack, D. et al.: The effects of electrode material, charge density and stimulation duration on the safety of high-frequency stimulation of the subthalamic nucleus in rats. J Neurosci Methods **138**, 207–216 (2004)

83. Yao, Y. et al.: Influence of electroporation on the biological activities of primary rat hepatocytes. Zhonghua Gan Zang Bing Za Zhi **9**, 178–180 (2001)
84. Merrill, D. R., Bikson, M. & Jefferys, J. G.: Electrical stimulation of excitable tissue: design of efficacious and safe protocols. J Neurosci Methods **141**, 171–198 (2005)
85. Hille, B.: Ion Channels of Excitable Membranes. Sinauer, Sunderland, MA (2001)
86. Vogel, V. & Sheetz, M.: Local force and geometry sensing regulate cell functions. Nat Rev Mol Cell Biol **7**, 265–275 (2006)

Chapter 4
Nanostructured Coatings for Improved Charge Delivery to Neurons

Takashi D.Y. Kozai, Nicolas A. Alba, Huanan Zhang, Nicolas A. Kotov, Robert A. Gaunt, and Xinyan Tracy Cui

Abstract This chapter explores the variability and limitations of traditional stimulation electrodes by first appreciating how electrical potential differences lead to efficacious activation of nearby neurons and examining the basic electrochemical mechanisms of charge transfer at an electrode/electrolyte interface. It then covers the advantages and current challenges of emerging micro-/nanostructured electrode materials for next-generation neural stimulation microelectrodes.

4.1 Introduction to Electrical Stimulation of Neurons

Stimulation electrodes have many clinical applications. The most common clinically approved stimulator is the artificial cardiac pacemaker, which is implanted into patients with a slow or arrhythmic heartbeat [1]. Artificial pacemakers use electrodes placed directly in contact with the heart muscles to regulate heart rate by delivering a timed series of electrical pulses. Another common stimulation device implanted into many adults and children is the cochlear implant [2]. Cochlear duct electrodes are designed as a series of stimulation contacts arranged along a flexible silicone carrier (Fig. 4.1a). These electrodes are placed into the cochlea where they directly electrically stimulate nerve cells, bypassing damaged hair cells that transduce acoustic vibrations into electrical impulses in the underlying nerve cells.

T.D.Y. Kozai (✉) • N.A. Alba • X.T. Cui
Department of Biomedical Engineering, University of Pittsburgh, Pittsburgh, PA 15260, USA
e-mail: tkozai@umich.edu

H. Zhang • N.A. Kotov
Department of Chemical Engineering, University of Michigan, Ann Arbor, MI 48109, USA

R.A. Gaunt
Department of Physical Medicine and Rehabilitation, University of Pittsburgh,
Pittsburgh, PA 15260, USA

M. De Vittorio et al. (eds.), *Nanotechnology and Neuroscience: Nano-electronic, Photonic and Mechanical Neuronal Interfacing*, DOI 10.1007/978-1-4899-8038-0_4,

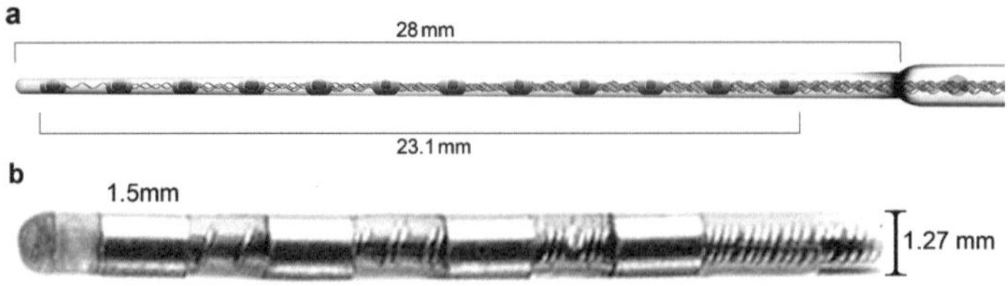

Fig. 4.1 (**a**) The FLEX 28 cochlear duct electrode array. (Image courtesy of MED-EL.) (**b**) The Medtronic 3387 model DBS electrode has 1.5 mm long stimulation electrodes separated 1.5 mm edge-to-edge with a diameter of 1.27 mm. Each stimulation electrode has the surface area of 5.98×10^6 μm^2. Reproduced with permission from [327] © (2009) Springer

Deep brain stimulation (DBS) is another class of clinically available stimulator that employs electrodes implanted into "deep" white matter structures in the brain (Fig. 4.1b). Electrical stimulation in these deep brain structures, such as the basal ganglia, has been used to treat symptoms of Parkinson's disease including tremor and bradykinesia, as well as other movement disorders such as dystonia [3–5]. In addition, DBS is being studied as a treatment for stroke, chronic pain, major depression, and chronic obesity [6–10]. For epilepsy and major depression, vagal nerve stimulation offers an alternative that does not require brain surgery [11, 12]. In the extremities, functional electrical stimulation (FES) is used to deliver electrical current to activate nerves or muscles in disabled patients. FES has been applied to aid in standing, walking, and basic handgrips and for restoring bowel and bladder function [13–15]. Many other electrical stimulation applications exist, including the restoration of somatosensation and vision, the treatment of hypertension and gastroparesis, and the promotion of nerve regeneration [16–19].

While these neurostimulation devices have demonstrated some success in bypassing, replacing, or treating damaged neural circuits, they are far from being able to reliably restore natural function in all patients. In order to understand the variability and limitations of traditional stimulation electrodes, we must first appreciate how electrical potential differences lead to efficacious activation of nearby neurons and examine the basic electrochemical mechanisms of charge transfer at an electrode/electrolyte interface. This chapter will first lay out our current knowledge of these areas and afterwards will cover the advantages and current challenges of emerging micro-/nanostructured electrode materials for next-generation neural stimulation microelectrodes.

4.1.1 Basic Principles Behind Electrical Signals in Neurons

As covered earlier in this book, neurons transmit electrochemical signals throughout the body and brain (Fig. 4.2a). Neurons, muscle, and endocrine cells maintain a small resting membrane potential on the order of −60 to −95 mV, depending on the cell type (Fig. 4.2b) [20–23]. This membrane potential is maintained by the active pumping of positively charged sodium ions out of the cell [24, 25]. Neurons are

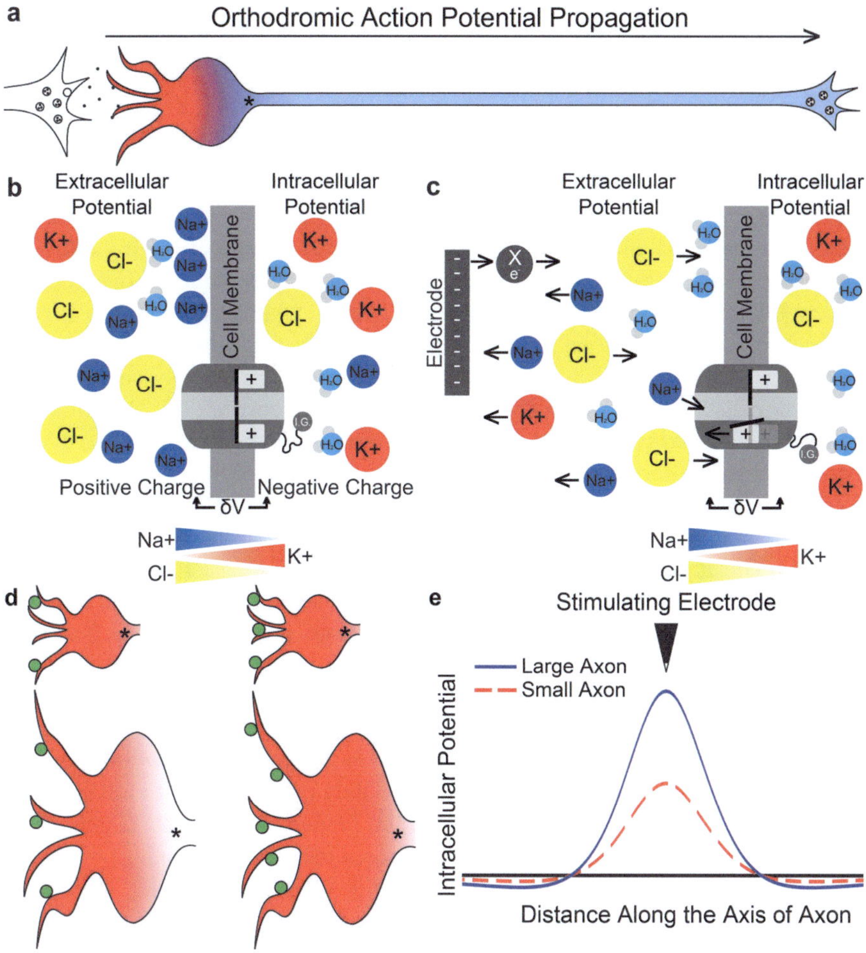

Fig. 4.2 (**a**) Simplified *orthodromic* or natural signal transduction model in neurons. Neurotransmitters bind to the dendrites of a neuron, which cause channels to depolarize or hyperpolarize the membrane. This signal is summed up from the dendrites and cell body at a point just past the axon hillock (*). If an action potential occurs, voltage-gated ion channels propagate the signal down to the axon where it releases neurotransmitters at the presynaptic terminal. (**b**) Model of voltage potential at the membrane of a neuron. The extracellular space is positively charged with high concentrations of sodium ions. (**c**) Cathodic electrical stimulation drives the extracellular space to become more negative, causing depolarization of the membrane potential ($\delta V \rightarrow 0$). (**d**) Two different-sized neurons activated with neurotransmitters (*circles*) in physiological condition. Neurotransmitters bind to ligand-gated ion channels which cause excitatory postsynaptic potentials (EPSPs) (depolarization). *Left*: neurons receive low levels of neurotransmitter inputs from a presynaptic neuron. In smaller neurons, ionic currents generated by EPSP result in higher voltage changes across the membrane which is summed at the axon hillock (*) and reaches the threshold for generating an action potential. On the other hand, larger neurons have decreased EPSP voltage changes which are unable to reach the depolarization threshold by the time it reaches the axon hillock. *Right*: greater synaptic inputs are able to generate action potentials in the larger neuron. (**e**) Activation function diagram of two different-sized axons from electrical stimulation by an electrode along an axon. Cathodic pulse can generate greater depolarization potential in the larger diameter axon (*solid*) than a smaller neuron with half the axon diameter (*dashed*), showing the opposite recruitment order compared to activating neurotransmitters (larger first)

typically activated by specialized receptors that open ion channels in the membrane in response to specific triggers, such as neurotransmitters [26–29]. As sodium ions flow into the cell, the intracellular potential becomes less negative relative to the extracellular space and approaches zero, depolarizing the cell. If enough ions flow into the cell, nearby voltage-gated sodium channels open [30, 31], allowing further influx of sodium that in turn depolarizes other adjacent channels [32]. If a sufficient number of voltage-gated sodium channels have been activated and opened, this action potential propagates. These channels then quickly enter an inactive state (*refractory period*) blocking inward sodium ion flow and preventing continuous activation [32–34]. The cell repolarizes as sodium ions are actively pumped back out of the cell. This process usually lasts on the order of 1 ms [32–34]. The refractory period duration and the types of ions crossing the membrane vary from cell to cell. After inactivation ends and the membrane repolarizes, the sodium channels remain closed until activated again by depolarization.

Action potentials normally travel in one direction from the soma to the axon terminal. This is called either *anterograde* or *orthodromic* signal conduction. Under orthodromic signal conduction, neurotransmitters released from the axon terminal bind to receptors on the target neuron, which are generally on the dendrites (Fig. 4.2a). Once the neurotransmitters bind to the target receptors, ion channels open and contribute to the depolarization of the membrane [23]. Excitatory (depolarizing) and inhibitory (hyperpolarizing) signals travel along the dendrites and soma and are summed at the axon hillock, the point where the soma meets the axon [35]. The action potential is generated immediately beyond the axon hillock prior to the first myelinated segment of the axon and propagates down the length of the axon resulting in the release of neurotransmitter at the axon terminal.

4.1.2 Basic Spatial and Rate Encoding of Neural Information

Action potentials are said to be "all-or-none" since the amplitude of the action potential is independent of processes leading to its generation (including voltage and current) [36]. In other words, the amplitude of an action potential can be considered binary (1 or 0) and does not encode information about function or intensity. Instead, neural information is encoded spatially and temporally [37–39]. For example, the frequencies of sound in the auditory system or spatial information in visual system are encoded topographically by the location of the connected neurons in the brain [40]. Intensity is encoded through the activation order or *recruitment order* of a population of neurons (activated in a specific order, see Sect. 4.1.4) and the rate at which they generate action potentials [41].

4.1.3 Basic Principles of Electrical Stimulation

Electrical stimulation of the neural tissue seeks to bypass normal biological activation processes by introducing an artificially generated electrical impulse that leads to action potential generation in a neuron. In order to artificially generate neural activity using electrical stimulation, charge delivery to neurons is commonly carried out with a cathodic (negative) current pulse. This is accomplished by placing a stimulation electrode near the cells being targeted and placing a larger *counter* electrode (one which *counters* the current generated at the stimulation electrode, i.e., completes the circuit) at a distant site. Normally, the counter electrode should be sufficiently large such that the current density at its surface is low in order to minimize activation of the neural tissue around the counter electrode. During a cathodic (negative) stimulation pulse, positive current flows into the electrode, electrons flow out of the electrode, and nearby positive ions are attracted towards the electrode surface. Cathodic stimulation pulses thereby create a negatively charged microenvironment in the extracellular space around neurons near the electrode. This leads to the depolarization of the cellular membrane, opening of the voltage-gated ion channels, and further depolarization of the cell (Fig. 4.2b, c). On the other hand, an anodic (positive) stimulation pulse results in positively charged microenvironment in the extracellular space around neurons near the electrode, which leads to hyperpolarization of nearby neurons and blockade of action potentials (Fig. 4.2b).

4.1.4 Recruitment Order of Neurons

The natural recruitment order of neurons has been best characterized in the case of spinal motor neurons [42]. Henneman's size principle states that motor units undergo orderly recruitment from smallest to largest, where small motor units consist of small diameter motor neurons innervating a small number of slow-twitch, fatigue-resistant muscle fibers via slowly conducting axons and large motor units consist of large diameter motor neurons innervating a large number of fast-twitch, fatigable muscle fibers via fast conducting axons [42–47]. At any given time, most neurons in the body are quiescent or fire only occasionally until actively recruited [48]. It is generally understood that from this quiescent state, smaller neurons are recruited first because they have greater synaptic input densities, and their smaller surface area leads to a higher input impedance resulting in higher voltage changes from ionic currents generated by postsynaptic potentials (Fig. 4.2d) [42–47, 49–52]. This size-dependent recruitment order phenomenon has also been observed in sensory neurons [53–55].

The recruitment order that occurs during electrical stimulation varies from natural recruitment in important ways. When axons equally distant from an electrode are electrically stimulated with a cathodic pulse, larger axons (which would normally be recruited last under natural conditions [56]) are recruited first due to their larger surface area and lower resistance to electrical currents. For myelinated axons, this is also due to the larger potential difference generated between the nodes of Ranvier (the excitable portion of the axons) [49, 57]. On the other hand, smaller diameter axons have less surface area and therefore higher membrane resistance requiring larger electrical stimulus to reach depolarization thresholds. Additionally, in myelinated axons, smaller diameter axons (which have shorter internodal distance) will see a smaller potential difference between two nodes and therefore require larger stimulus amplitudes to generate depolarizing potentials. As expected, it has been shown that while the voltage membrane thresholds are similar in different-sized neurons, smaller neurons axons require a more negative depolarizing current to generate an action potential [49, 57]. This leads to lower stimulation thresholds for larger axons [49, 58] (Fig. 4.2e). As such, electrical stimulation of neurons is generally recruits axons in reverse physiological order [13]. However, there are situations when this size principle does not hold, as discussed below [59, 60].

4.1.5 Effect of Electrical Stimulation on Neurons

Ideally, the intent of cathodic electrical stimulation, or *cathodic excitation*, is to generate orthodromic action potentials to elicit a specific function in an impaired nervous system. Similarly, the goal of anodic stimulation, or *anodic block*, is to inhibit unwanted neural activity by hyperpolarizing nearby neurons. Unfortunately, neuronal recruitment in vivo in response to electrical stimulation is far more complicated and influenced by many factors. Focal cathodic electrical microstimulation has been shown to activate sparse, distributed, and very distant populations of neurons in the brain and spinal cord [61, 62]. Sparse and discrete populations of neurons in the brain up to 4 mm away from the stimulating electrodes have been shown to become depolarized even when orthodromic synaptic transmission was blocked with a pharmaceutical agent [61]. Similarly, intraspinal microstimulation was shown to elicit activity in dorsal root filaments with entry zones up to 17 mm both anterior and posterior from the stimulation site [62]. In both cases, this is largely due to the fact that electrical stimulation frequently recruits axons before cell bodies themselves [62–66] and that these action potentials can back-propagate towards the soma and dendrites [67, 68], a phenomenon known as *antidromic* activation. This is due, in part, to limitations in both electrode design and the ability to target charge delivery to a specific region of neurons but also reflects basic physiological properties. Overall, these examples demonstrate that electrical stimulation can lead to the activation of neural pathways in a manner that is very different from normal physiological recruitment (Fig. 4.3a) [68].

To further complicate matters, cathodic stimulation does not always depolarize neurons, and anodic stimulation does not always hyperpolarize neurons. During strong cathodic pulses, current flows away from the electrode into the cell, leading to the depolarization of nearby voltage-gated sodium channels. However, depending on the position of the cathode, part of the current that entered the cell simultaneously flows out of the cell towards the electrode at a site further along the axon called the *virtual anode* (Fig. 4.3b) [14, 69, 70]. When virtual anodes flank the depolarized region, positive charge will flow into the neuron and lead to hyperpolarization, which will prevent the propagation of action potentials along the axon. This phenomenon is referred to as a *cathodic block* [71]. Similarly, during strong anodic stimulation pulses, a distant *virtual cathode* can cause positive charge to

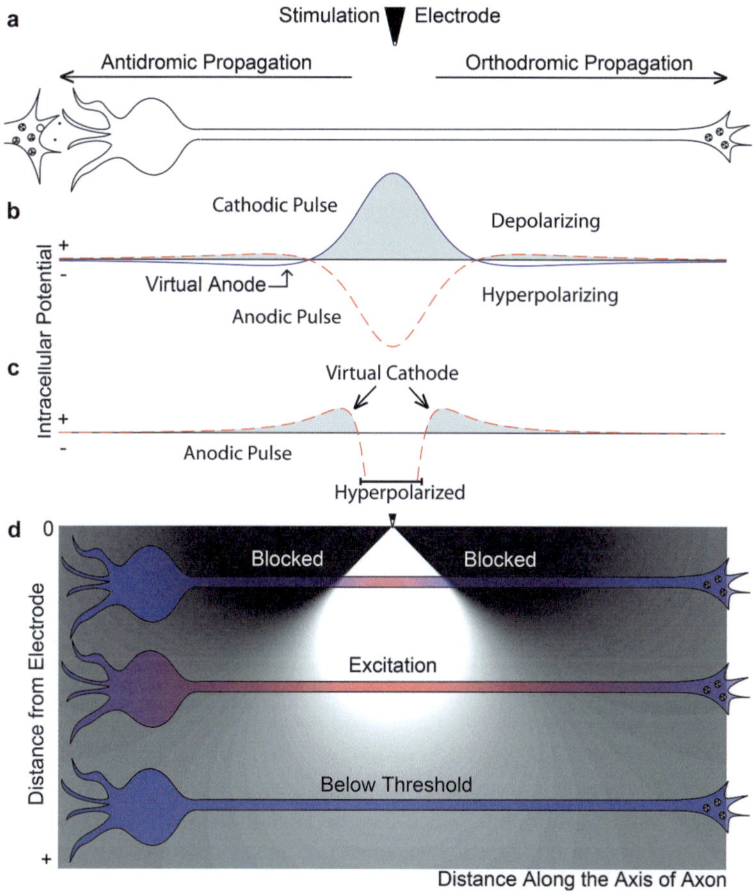

Fig. 4.3 (**a**) Electrical stimulation of a neuron from an electrode near the center of the axon. Action potentials propagating towards the cell body are referred to as *antidromic propagation*. (**b**) Potential experienced along the length of an axon from a cathodic (*solid*) and anodic (*dashed*) stimulation pulse. The *shaded* areas indicate a depolarized of the axon while the *unshaded* areas indicate hyperpolarization. (**c**) Strong anodic currents cause hyperpolarization along a segment of the axon (closest to the electrode) but generate *antidromic* and *orthodromic* action potential through virtual cathodes. (**d**) Visualization of the potential experienced by axons oriented parallel to the x-axis from strong cathodic stimulation pulses as a function of distance from the electrode. *Top*: nearby neurons experience depolarization (*white* region) and neighboring hyperpolarizing (*dark* region; virtual anodes) current. *Middle*: neurons far enough away only experience the depolarizing potentials and full action potential propagation. *Bottom*: neurons far away experience subthreshold potential and do not generate action potentials

flow out of the membrane, leading to action potential propagation away from the anodic electrode, which is generally hyperpolarized from the stimulation pulse (Fig. 4.3b, c) [14, 69, 70]. A phenomenon called *anode break excitation* can also occur when a long anodic pulse is released [23, 71]. While this mechanism is explained in detail elsewhere [23, 71], the hyperpolarization during the anodic pulse leads to the relaxation of the inactivation gate in voltage-gated sodium channels, and the sudden voltage drop at the end of the pulse causes sodium and potassium ions to overshoot the resting membrane potential, causing depolarization and action potential initiation.

While all of these effects may be observed during stimulation, the probability and variability of inducing an effect is highly dependent on the leading pulse polarity, waveform shape, amplitude, type of neuron (excitatory, inhibitory), distance from the neuron, and relative position of the electrode with respect to the axon, soma, and dendrite, as well as the size and geometry of the electrode (Fig. 4.3d). Therefore, it is important to take these issues into account when designing stimulation electrodes and stimulation paradigms.

4.1.6 Subthreshold Electrical Stimulation

In addition to electrical stimulation that directly activates or inhibits neurons, *low-frequency, low-amplitude* stimulation has been studied for modulating the probability of neuronal firing by driving the membrane potential away from the rest without eliciting an action potential. These low-frequency, low-amplitude simulation pulses have been explored for applications in nerve guidance, the promotion of wound healing, the detection of chemical signals, and the promotion of cell differentiation. Following nerve injury, regeneration of axons in the periphery is incomplete and disorganized, leading to functional deficits. Similarly, regeneration of axons in the spinal cord is actively inhibited. Electrical and functional stimulation of nerve targets has been shown to increase the speed and accuracy of axonal regeneration and reinnervation [72–75]. It has also been shown to synergistically modulate neurons to react more strongly in the presence of a neurotrophic gradient [76]. Additionally, electrical stimulation induces specific nerve growth factor (NGF) and neuroprotective molecule upregulation [77, 78] and results in a more accurate reinnervation of target nerve pathways than that observed following NGF injection [79]. Neurons are capable of significant activity-dependent reorganization [80], which can be induced and guided by electrically stimulated activity [19].

For example, neurites have demonstrated preferential growth towards cathodic electrical fields [81]. Experimental tests have determined that these extracellular electrical fields serve as morphological cues and pattern coordinates in the developing embryo [82]. During embryonic development, there is a natural voltage gradient along the neural tube [82]. This voltage gradient appears only once when neurulation (neural tube development) begins and disappears after it ends. The disappearance of this voltage gradient may contribute to poor accuracy of neural regeneration and reinnervation in adulthood. Despite the potential of low-frequency, low-amplitude electrical stimulation in neural interface applications, near-DC electrical stimulation is difficult to achieve with micro-/nanoelectrodes for reasons that will be covered below, and this application will not be extensively covered. However, nanostructured coatings on macroelectrodes have demonstrated that combinations of nanoscale topographical cues and electrical cues can cause divergent stem cell differentiation (a few examples will be covered in Sect. 4.3).

4.2 Introduction to the Electrochemistry of Electrical Stimulation

Electrical stimulation through nanostructured electrodes is influenced by electrochemical principles of charge transfer at the electrode/electrolyte interface. When electrodes are implanted, they come into contact with the body's extracellular *electrolyte* or fluid-containing ions and salts. General chemistry principles dictate that any metal in an electrolytic solution has a tendency to oxidize into metal ions. The electrode's potential, or *half-cell potential*, determines the degree of oxidation of the electrode. Metals with more negative half-cell potentials tend to be more easily oxidized than those with more positive potentials. During oxidation, positively charged metal ions leave the electrode and enter the solution (also called the *dissolution* of metal ions) leaving the electrode with excess electrons accumulated at the surface. This buildup of charge creates an electrical field, which encourages the reverse reaction or *reduction*. The system will therefore reach equilibrium, unless disturbed by an applied external electrical field or current.

At equilibrium, the negative surface charges will attract positively charged ions in the electrolyte that accumulate at high concentration near the electrode. Similarly, positively charged ions will accumulate near a negatively charged surface depending on the half-cell potentials of the working and counter/reference electrodes. A so-called *Helmholtz double layer* or *electrical double layer* is formed by the molecules in the electrolyte at the electrode surface [83–85]. As shown in Fig. 4.4a, the layer immediately adjacent to the charged surface consists of an ordered hydration shell produced as the polar water molecules are aligned by the electrode's electrical field. Beyond this inner layer, a second layer forms which consists of hydrated ions attracted by the electrical field on the electrode surface. This double layer can be modeled as a capacitor that contributes to the charge transfer at the electrode/electrolyte interface.

During electrical stimulation, the electrical charge at the electrode surface (in the form of electrons) must be transferred to the biological electrolyte of the body (in the form of ions) in order to be delivered to nearby neurons. This section discusses how electrical stimulation parameters and electrode designs must be considered to prevent damage to the electrode and/or the nearby tissue. The transfer of charge at the electrode/electrolyte interface generally occurs through two mechanisms: (1) *faradic charge injection* and (2) *non-faradic charge injection.*

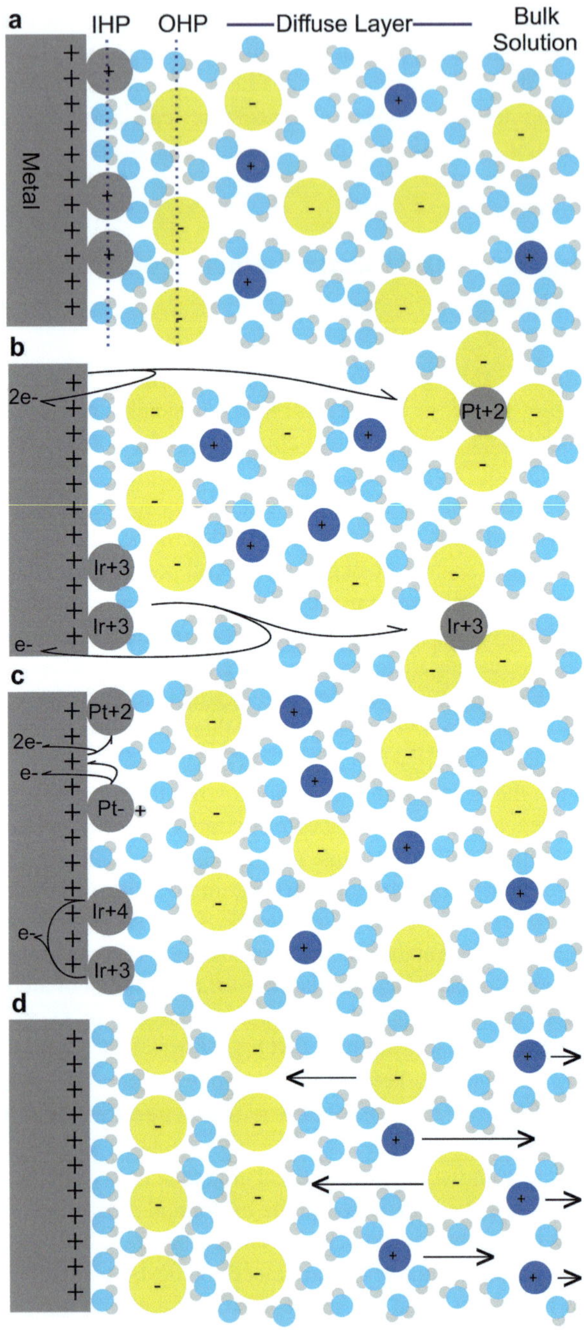

Fig. 4.4 Signal transduction at the electrode interface. (**a**) *Helmholtz double layer*: the inner Helmholtz plane (IHP) consists of ions and water molecules adsorbed onto the electrode material. The outer Helmholtz plane (OHP) consists of hydrated ions and water molecules attracted to the

4.2.1 Faradic Charge Injection

In faradic charge transfer, electrons are transferred between the electrode and the electrolyte via reduction–oxidation (redox) reactions. The simplest such reaction occurs between a solid metal and metal ion. When a positive (anodic) pulse is applied, electrons are removed from the metal electrode, and metal ions are released into the electrolyte. Such oxidation helps current flow across the electrode/electrolyte interface. Conversely, when a negative (cathodic) pulse is applied, electrons are ejected from the electrode but accepted by the metal ions nearby, resulting in them being reduced into metal and redeposited on the electrode surface. The chemical changes at the electrode/electrolyte interface that result from the initial stimulus pulse can be *reversible* or *irreversible* when a stimulation pulse of opposite polarity is quickly applied (Fig. 4.4b). While both reversible and irreversible charge transfer processes can occur, the rate of the reaction can be greatly controlled by strategically selecting the electrode material and electrical stimulation parameters. These factors will be discussed later in this chapter.

Reversible faradic reactions occur when the redox product remains on the electrode surface (such as an oxide) or when the diffusion rate of the product is slow enough to allow a stimulation pulse of the opposite polarity to revert the products before they have sufficient time to diffuse away [86]. Examples of such reactions include the oxide formation of platinum (Pt) or iridium (Ir) and the valence transitions between different forms of iridium oxides. Electrodes with reversible faradic charge transfer capabilities are said to have *pseudocapacitive* properties, as the mode of charge injection is through mass transfer (as in faradic reactions), but the electrode material is preserved as in the case of capacitive charge transfer [87–90] (Fig. 4.4c).

In an *irreversible* faradic reaction, the chemical products resulting from the initial stimulus phase diffuse away before the reverse reaction can take place. Such reactions are to be avoided, as they produce new chemical species that are often harmful to the surrounding tissue. The electrolysis of water into gas during electrical stimulation is also an irreversible reaction and has dire consequences for the underlying tissue and the electrode as gas bubbles are generated [84, 85].

Fig. 4.4 (continued) surface charge on the electrode. Over the diffuse layer, the influence of electrical attraction decreases with distance from the electrode surface until the distribution of free ions is dominated by random thermal motion in bulk solution. (**b**) *Irreversible faradic signal transduction*: the electrode materials (Pt and Ir) are ejected from the electrode during strong electrical stimulation. (**c**) *Reversible faradic signal transduction*: during electrical stimulation, hydrogen atom-plated platinum or platinum metal becomes oxidized, but remains on the surface of the electrode, which can be reversed. Iridium oxide at the surface of the electrode can reversibly change to/from iridium (III) oxide and iridium (IV) oxide. The rapidly reversible properties are called pseudocapacitive and at low charge densities encompass the efficiency and speed of faradic signal transduction and the electrochemical safety of capacitive electrodes (both for the tissue and electrode). (**d**) *Capacitive signal transduction*: charge transfer through the diffusion of ions which is safer, but have limitations in maximum charge injection and charge transfer speed

4.2.2 Non-Faradic or Capacitive Charge Injection

In non-faradic charge transfer, the *Helmholtz double layer* at the electrode/electrolyte interface forms a capacitor, and charge transfer is accomplished by charging and discharging of this capacitor (Fig. 4.4d). Because electrons are not transferred between the electrode and electrolyte, there are no chemical reactions, and the effects of stimulation are completely reversible. The principal limitation of capacitive charge transfer is that it is dependent on the diffusion rates of ions, which are larger and more diffuse and therefore have slower kinetics compared to the direct electron flow produced during faradic reactions. While purely capacitive charge transfer is safer from a tissue perspective, much less charge is transferred than occurs through pseudocapacitive mechanism, limiting its functional effect.

Unfortunately, there are currently no known perfectly capacitive electrodes (*ideally polarizable electrodes*) or perfectly resistive electrodes (*nonpolarizable electrodes*). Stimulation electrodes generally transfer charge through some combination of both faradic and non-faradic processes. This ratio is highly dependent on many parameters including electrode size, material, and geometry, as well as specific stimulation pulse parameters including amplitude, duration, and overall waveform shape. The primary challenge in designing electrodes and stimulation parameters is to eliminate irreversible faradic reactions over the lifetime of the application while still regularly delivering physiologically efficacious electrical stimulation to the target neurons.

4.2.3 Electrical Stimulation Parameters for Safe and Efficacious Charge Injection

From a safety standpoint, stimulation parameters should be designed to keep electrode potentials sufficiently low to avoid the generation of irreversible faradic reactions that may harm the underlying tissue or electrode material. This includes keeping the electrode potential below the threshold where the electrolysis of water into gas occurs. This "water window," or the interfacial potential range within which electrolysis is thermodynamically unfavorable, is typically regarded as –0.6 V and 0.8 V vs. Ag/AgCl, a conservative range for most metals [91]. For this reason, the charge storage capacity (CSC) of an electrode is typically quantified as the charge per geometric surface area transported across the electrode during the application of a single cycle of cyclic voltammetric stimulation (–0.6 to 0.8 V).

In order to accurately characterize the performance of stimulation electrodes in relation to the water window, their *charge injection limit* (or cathodic charge injection capacity (CIC_C) and anodic charge injection capacity (CIC_A)) are frequently reported. This charge injection limit is the maximum charge that can be injected into the tissue without violating the interfacial potential window within which the electrode is considered to operate reversibly. In practice, this limit is determined through the observation of voltage transients or the measured voltage drift during the application of a current-controlled stimulation pulse beyond the access voltage. During controlled-current stimulation, the access voltage (or instantaneous voltage, which includes terms such as the ohmic potential drop within the electrolyte and the concentration overpotential) is the potential drop of the system that does not contribute to the electrolysis of water. Therefore, the charge injection limit for an electrode is defined to be the total charge injected per surface area during a constant-current pulse that results in an interfacial potential drop that crosses either −0.6 V (for cathodic pulses) or 0.8 V (for anodic pulses) *beyond* that of the access voltage [83]. CSC and charge injection capacity are frequently reported in units of mC/cm^2.

Safe stimulation pulses must also be large enough to activate nearby neurons and elicit physiologically relevant responses. However, the charge necessary to evoke functional responses can damage the electrode or local tissue over time [92, 93]. In order to minimize electrochemical damage to the electrode, a commonly used approach has been to follow each stimulation pulse with a pulse of equivalent charge of the opposite polarity [94]. This pulsing strategy attempts to prevent the progressive accumulation of charge at the electrode interface that would lead to irreversible reduction and oxidation reactions. This stimulus pattern is commonly referred to as a *charged-balanced biphasic waveform*. It is important to note that while charge balancing has generally been an effective method to minimize irreversible reduction and oxidation reactions, it cannot prevent damage if the electrode potential within a single pulse exceeds the range for reversible reactions. Charge-balanced waveforms may be current controlled (galvanostatic) or voltage controlled (potentiostatic coulometry).

With current-controlled waveforms, the delivered current during the interpulse period can be set to zero, which prevents reduction and oxidation reactions from occurring at the electrode surface. It is also simpler to control the charge per phase (experimentally the most reliable measure of neural response threshold) by simply setting both the current and duration of each stimulation phase and therefore is more commonly used. However, it may be necessary to monitor the voltage transients associated with the current pulses to check that the maximum electrode polarization does not exceed the water window. One method to assess the electrode polarization during constant-current stimulation is to introduce a delay between the stimulating (typically cathodic) and recovery (typically anodic) phases of the waveform and measure the potential during this period when no current is flowing. At this point, voltage drops associated with electrolyte resistance, and other factors (i.e., the access voltage) do not contribute to the measured voltage, thus allowing measurement of the electrode polarization [83, 91]. Monitoring the polarization is particularly important for small surface area electrodes, as biofouling (adsorption of protein) and scar tissue encapsulation can increase the electrode's impedance and resistivity over the lifetime of the implant.

With voltage-controlled waveforms, *coulometry* (or measurement of total charge injected) must be performed during the leading phase to ensure that the duration of the reverse polarity pulse can be adjusted to maintain charge balance. In addition, during the interpulse periods, the applied voltage should ideally be adjusted to the *equilibrium potential* (instead of simply to 0 V), such that the net current flow between the working electrode and counter electrode is zero (the various stimulating and counter electrodes each possess different half-cell potentials and thus exhibits different combined equilibrium potentials). If the interpulse voltage between the working electrode and counter electrode is set to zero, polarization can occur as a result of the equilibrium potential of the combined electrodes, which may allow continuous irreversible faradic reactions to occur at low levels, especially for nanoscale electrodes [95]. Voltage control is advantageous in that the maximum voltage range may be set within the water window to prevent the system from ever reaching an electrochemical potential that could cause irreversible redox reactions that damage the nearby neurons over time [91]. While safe, this approach can be functionally limiting, as the potential between the working and counter electrodes is a combination of both the electrode polarization and resistive losses in the tissue itself.

One commonly employed technique that increases the population of depolarized neurons during an electrical stimulus without violating the water window is to add a short interphase delay time between the first (usually cathodic) phase and the second (usually anodic) phase of a charge-balanced waveform. This interphase delay time increases the duration that the microenvironment around the electrode remains negative, thereby increasing the probability that nearby neurons depolarize. More recently, there have been efforts examining how different asymmetric charged-balanced waveforms may be able to recruit action potentials in a more physiologically relevant manner than symmetric charged-balanced waveforms [96, 97]. One important note is that as the phase duration and/or interstimulus dwell time increases, the more likely the chemical products of the initial stimulus phase may diffuse away, leading to an irreversible faradic reaction. This is especially true for nanostructured electrodes and requires careful consideration when designing electrodes and stimulus parameters.

While the mechanisms behind the degradation of electrodes during stimulation are well characterized, the exact mechanisms behind the damage of the nearby tissue during stimulation are not completely understood. These mechanisms are discussed in detail elsewhere, but some of the proposed mechanisms implicate combinations of pH change and gas formation, denaturing of proteins, oxidation of biological molecules, depletion of nutrients, metal ion toxicity, generation of free radicals, excitotoxic neural activity, electroporation, thermal toxicity, breakdown of the blood–brain barrier (BBB) from stimulation, thrombosis, and/or generation of toxic electrochemical species depending on the electrode material and stimulus [85, 98–101].

4.2.4 Electrode Design Consideration for Safe and Efficacious Charge Transfer to Neurons

One of the fundamental electrical stimulation parameters for maximum safe charge injection in vivo has been characterized by the following equation [85]:

$$\log\left(\frac{Q}{A}\right) = k - \log(Q) \tag{4.1}$$

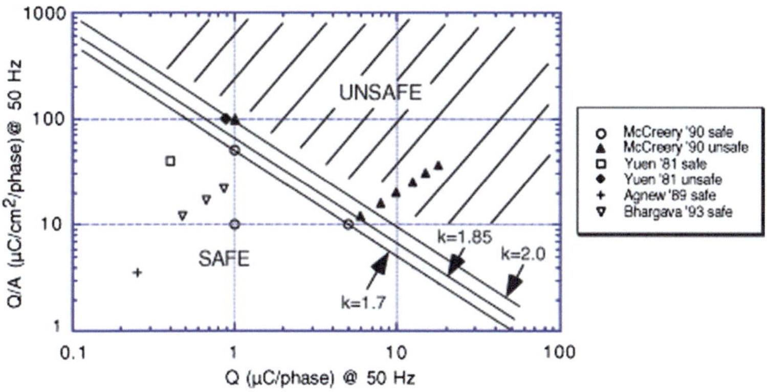

Fig. 4.5 Charge (Q) vs. charge density (Q/A) for safe stimulation. A microelectrode with relatively small total charge per pulse might safely stimulate using a large charge density, whereas a large surface area electrode (with greater total charge per pulse) must use a lower charge density. Reproduced with permission from [85] © (2005) Elsevier. *Note*: "k" and the linearity of this equation may vary for nanostructured electrodes compared to traditional smooth metal electrode due to differences in electrochemical surface area, frequency dispersion, and edge effects. This relationship should be reevaluated with data from micro-/nanoelectrodes

where A is the surface area of the electrode (cm^2), Q is the charge per phase (μC), and k is a constant such that $k \lesssim 1.7$. k has been derived from a series of experimental data examining the functional response of neurons to stimulation and histological analysis of the implanted tissue following electrical stimulation using traditional metal electrodes [85, 102–106] (Fig. 4.5). This may be rewritten as follows:

$$Q \leq \sqrt{A \times 10^k} \tag{4.2}$$

This shows that the maximum charge that can be safely delivered to the tissue is dependent on the surface area of the electrode. In fact, high charge density stimulation (Q/A) using microelectrodes has been used to irreversibly lesion the brain tissue in vivo [107]. This is one reason why current clinically available stimulation electrodes typically have large surface areas. However, with large electrodes, both charge per phase and charge density must be considered as both factors have been implicated in the neural tissue damage [104]. It should be noted, however, that "k" and the linearity of this equation may vary for nanostructured electrodes compared to traditional smooth metal electrode for reasons covered next (Sects. 4.2.5 and 4.2.6). These equations and values should be reevaluated with nanostructured micro-/nanoelectrodes. Nevertheless, it illustrates the importance of charge per surface area.

4.2.5 Advantages and Limitations of High Electrochemical Surface Area

The surface area of electrodes has been discussed up until now in terms of their apparent area, or *geometric surface area*. However, it is possible to introduce

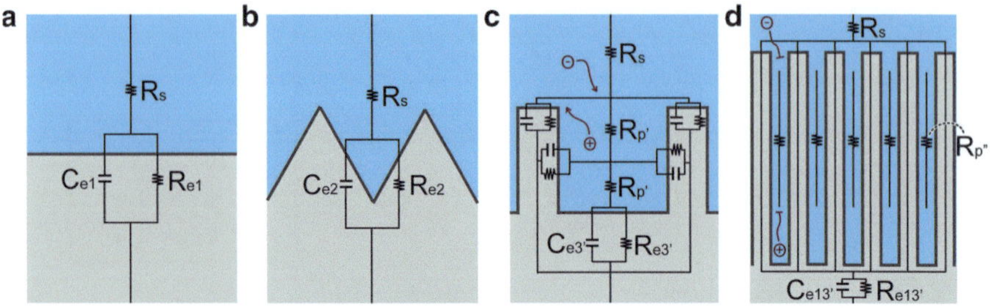

Fig. 4.6 Charge transfer at the electrode interface during electrical stimulation. The electrode is illustrated in *gray* and tissue or extracellular fluid in *blue*. (**a**) Part of the charge is transferred to the solution resistively (R_{el}) and capacitively (C_{el}). The signal must transfer the solution (R_s) before reaching its target neuron. (**b**) *Fractal* electrode surface with twice the surface area has doubled the capacitive charge transfer and half the resistance of (**a**). (**c**) Porous electrode surface with tripled surface area of (**a**). The pore is modeled as a "transmission line" with resistance ($R_{p'}$). $R_{p'}$ is treated as having the A of the cross-sectional area of the pore and l of each "transmission segment." Here, A is relatively large, and Σl is relatively short such that it minimally impacts ion movement. (**d**) While the surface area of the electrode is 13 times (**a**), the A is too small and l is too long that greatly impairs ion movement. As a result, $R_{p''}$ is so large that it is effectively treated as an open circuit element

micro-/nanostructured electrode features, such as roughness and porosity (or *fractal* geometry) onto electrode surfaces, which can increase the effective surface area (or *electrochemical surface area*) of an electrode without increasing its geometric surface area. Increasing the electrochemical surface area allows greater charge to be injected safely into the tissue than comparable smooth electrodes (Fig. 4.6a, b). This is because the capacitance (non-faradic charge injection property) of an electrode is proportional to its surface area, and resistance (faradic charge injection property) is inversely proportional to surface area.

$$Capacitance = \varepsilon_r \varepsilon_0 \frac{A}{d} \tag{4.3}$$

$$Resistance = \rho \frac{l}{A} \tag{4.4}$$

where A is the surface area of the conductor (m^2), ε_r is the relative static permeability or dielectric constant (unitless), ε_0 is the electrical constant ($\sim 8.854 \times 10^{-12}$ F/m), d is the separation of the inner and outer Helmholtz layers (m), ρ is the electrical resistivity (Ω m^2), and l is the length of the conductor (m) which becomes particularly relevant for ionic resistivity in porous surfaces. Additionally, increasing the electrode surface area has the effect of lowering the cutoff frequency of the double-layer capacitor (a frequency below which the impedance magnitude is independent of the frequency), providing the electrode with more uniform impedance over a broader frequency range. This can significantly improve electrode performance in stimulation applications with long pulse widths or improve recording sensitivity in applications such as cardiac pacing, where the electrode serves a dual purpose in sensing biopotentials [108].

Since increasing the electrochemical surface area reduces the electrode impedance, equivalent current amplitudes can be achieved with a lower electrode potential, lessening the chance for electrode and tissue damage. Multiple strategies have been applied to create nanoscale topography to increase the electrochemical surface area of electrodes for improving stimulation outcomes [108–115]. However, there are limitations to the beneficial effects of increasing electrochemical surface area. As surface roughness and porosity (or the presence of surface pores, leading to a spongelike morphology) increases, interfacial charge transfer characteristics will eventually become dominated by pore ionic resistance instead of by surface area. This is due to the fact that in order to access surface area deep within pores, ionic current is forced to penetrate an increasing thickness of restrictive channels which impede ion motion [116].

The penetration depth of a sinusoidal AC waveform into a surface pore of known size can be theoretically expressed as

$$\lambda = \frac{1}{2}\sqrt{\frac{\kappa r}{c_d \omega}} \tag{4.5}$$

where r is the pore radius (cm), κ is the electrolyte conductivity (1/Ω cm), C_d is the surface capacitance (F/cm^2), and ω is the angular frequency of the waveform (rad/s) [116]. Note that as this function is dependent on the frequency content of the stimulation waveform, there will be an increasing degree of *frequency dispersion* within the electrode or a nonuniformity of interfacial electrical characteristics with changing frequency [116]. In essence, this phenomenon leads to a reduction of theoretically available surface area, especially for higher frequency stimuli. It is important to note that in practice, actual stimulation waveforms often consist of square-wave pulses and thus possess multiple frequency components, which substantially complicates theoretical assessments of waveform penetration and frequency dispersion.

In order to better understand the influence of pore length and ionic diffusion resistance on electrode properties, the interface may be characterized as a "transmission line" equivalent circuit model (see Fig. 4.6c), where the interface is broken down into a simple and idealized set of circuit elements which approximate its real-world behavior. In this model, ionic diffusion resistance is represented by a series of resistors that increase in number with the depth of the pore [116, 117]. As expressed in (4.5) above, for a particular stimulation frequency, exposed surface area in this transmission line beyond the depth λ cannot contribute to charge transfer, suggesting that further increases to film surface area will provide no additional benefit [116]. In practice, this means that attempts to create better stimulating electrodes simply by increasing surface roughness and porosity to finer degrees will eventually negatively impact charge transfer properties (electrochemical surface area will begin to decrease even though actual surface area continues to increase) (Fig. 4.6d).

This reduced effectiveness of increasing electrochemical surface area has the potential to be magnified when the electrode is implanted in vivo, as the reactive tissue encapsulation may clog pores along the electrode surface, further restricting ionic diffusion [118]. In fact, the encapsulating influence of the inflammatory tissue that chronically develops around an implanted electrode has the potential to overwhelm the performance improvement provided by surface area enhancements. This was observed in cardiac pacing electrodes, where one study demonstrated that stimulation thresholds increase by 193 % between the third and eighth week of implantation for platinized electrodes, compared to nearly unchanged (18 %) threshold increase observed from steroid-releasing electrodes designed to dampen local inflammation [119]. Similar impedance peaking has been observed in cortical recording electrodes during the initial 2 weeks of implantation in both cortical recording [120–122] and DBS electrodes [123]. Therefore, in implanted systems, benefit of nanostructuring electrode surfaces may be found not only for optimizing electrical characteristics but also for controlling the local tissue response.

4.2.6 Limitations of Traditional Stimulation Electrodes

Current limitations for implantable electrical stimulation microelectrodes include the need for surface areas that are sufficiently large to maintain a low charge density per phase and low electrode potentials, especially for the viability of long-term electrical stimulation. Microelectrodes with low geometrical surface area in theory have more specificity in accessing small populations of neurons, but are more likely to reach high charge densities (Q/A) that damage the electrode (Sect. 4.2.1) or damage the tissue (Sect. 4.2.4) prior to reaching thresholds necessary to elicit physiological responses. This is one reason that current clinical electrical stimulation systems use electrodes with high geometric surface area electrodes for neural stimulation.

As a result of the large geometric size of conventional stimulation electrodes, activating specific neuronal populations is difficult. For example, in cochlear implants, 8–22 large stimulation electrodes are implanted to replace 16,000 hair cells to activate underlying auditory nerve cells. This drastically reduces the amount of information that can be extracted from a perceived sound (16,000 channels of information are reduced to 8 channels). Despite this limitation, the cochlear implant remains amongst the most successful neural stimulation systems and can restore speech perception in most subjects [124].

In addition to challenges of activating specific and small neuronal populations, it is more difficult to position a large stimulation electrode to target specific regions of an individual neuron to achieve a specific effect (i.e., cathodic excitation, anodic excitation, anodic block, cathodic block, anodic break excitation, high-frequency block). Cell somas of neurons in the body are relatively large, but many axons from distant neurons pass between the cell bodies, forming a complex and dense mesh. Electrical stimulation using large microelectrodes tends to activate both nearby neuronal axon hillocks (somas) (orthodromic activation) and distant neurons with axons that pass through the stimulated region (antidromic and orthodromic activation) [62, 125]. This leads to a physiologically unnatural form of activation. Similarly in the peripheral nervous system, bundles of nerves contain autonomic and somatic nerves; broad activation of both of these pathways with a large electrode results in physiologically unnatural activation. It is important to note that functionally distinct

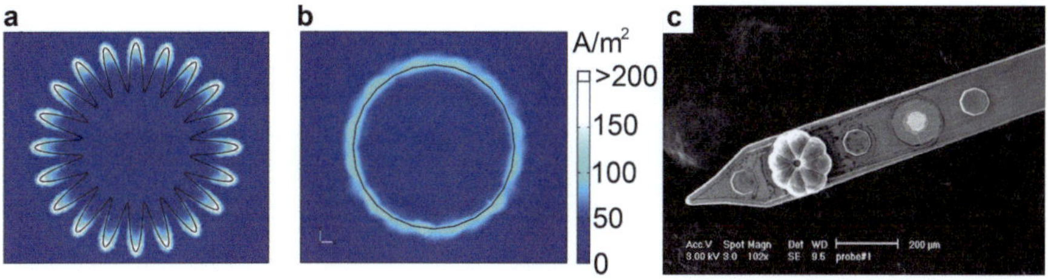

Fig. 4.7 Edge effect. Charge density distribution during electrical stimulation around (**a**) a planar high-perimeter electrode and (**b**) a planar circular electrode. The charge densities are preferentially distributed towards the edges of the electrode [126]. (**c**) Polypyrrole (PPy) electrochemically deposited on an octagonal recording site. Eight teardrop segments of PPy preferentially grew from the corners of the octagon, and an absence of coating can be found at the very center of the electrode

neural networks that reside in the same region (such as inhibitory pathways) can be unnaturally activated during electrical stimulation.

This broad combination of activation and block of excitatory and inhibitory neurons is further complicated by the fact that the charge density is not evenly distributed across the surface of the electrode during stimulation. Following Coulomb's law, like charges on the electrode during each phase of the electrical stimulation repel each other. Charges are then more concentrated at the edges of the stimulation electrode in what is referred to as the "edge effect" [126, 127] (Fig. 4.7). Because of this, even when relatively safe stimulation parameters are followed, irreversible redox reaction may still occur at the edges which may lead to corrosion of the electrode and damage of the tissue, as well as facilitate the delamination of the electrode sites from around its edges.

4.2.7 Challenges for Nanostructured Stimulation Electrodes

Micro-/nanostructured stimulation electrodes increase the effective or electrochemical surface area of the electrode. In addition, nanostructured materials may be employed to design surfaces that more evenly distribute charges by essentially making the entire electrode's geometric surface an "edge." Arrays of micro- and nano-sized stimulation electrodes would also allow "current steering" to pattern the electrical field generated by simultaneously stimulating a combination of nearby electrodes with cathodic and anodic pulses.

As the stimulation electrode features become nanometer/angstrom scale, the nanoscale ultrastructures may become more susceptible to corrosion and degradation due to the "edge effect," impacting long-term in vivo applications. Additionally, when nanostructured electrodes are implanted into the tissue, proteins and cells can seal off porous structures, reducing the benefit provided by increased electrochemical surface area. However, if the nanostructured electrodes can be positioned

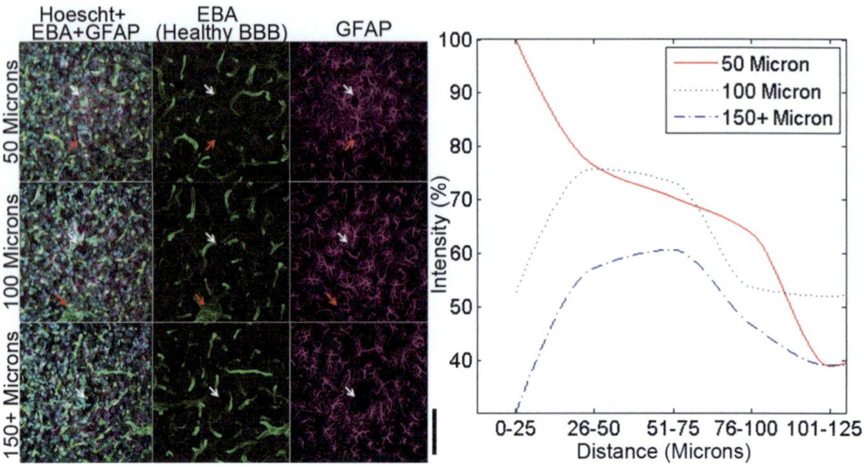

Fig. 4.8 In vivo challenges. Astroglial reactivity (GFAP) to chronically implanted probes (*white arrow*) increases with respect to implantation proximity to major vasculature (*red arrow*; EBA with >25 μm diameters). Scale bar indicates 100 μm. This suggests that microstructures in the brain may be associated with chronic performance. Decreasing the probe's profile as well as avoiding key vascular microstructures during insertion may be critical in targeting charge transfer coupling with specific neuron(s) or subcellular structures [134]. Adapted with permission from [134] © (2012) Nature Publishing Group

optimally (i.e., adjacent to the axon hillock or have nanostructures that bind to neurons), lower currents may be necessary for efficacious neural stimulation, which could also lead to improved tissue safety and extended battery life [67]. Specific targeting of a neuron or a region of neurons in vivo has not been explored extensively, in part due to the lack of opportunity: (1) safe and efficacious subcellular-sized stimulation electrodes for long-term implantation have yet to be demonstrated; (2) knowledge of the complex brain network is limited, leaving investigators to speculate on the precise subcellular neural targets; (3) technologies for cellular and subcellular level mapping of the nervous system are still maturing to enable targeting of specific neuronal regions for in vivo and clinical applications; and (4) methods and techniques for minimally damaging delivery (for both the electrode and implanted tissue) of micro-/nanostructured electrode in vivo are still to be developed for many of the emerging technologies demonstrated in vitro (Fig. 4.8). Parallel advancement in these technologies will create new opportunities and applications for stimulation electrodes in basic neuroscience and clinical applications.

4.3 Micro- and Nanostructures for Electrical Stimulation

The successful application of micro-/nanostructured stimulation electrodes must meet several key design parameters to minimize the negative outcomes discussed previously: (1) electrodes must be able to deliver sufficient charge to nearby neurons to evoke an efficacious response and (2) irreversible faradic charge transfer must be minimized to limit corrosion, dissolution of the electrode, electrolysis of water into gas, and degradation of electrical characteristics. This is particularly important at the sharpest and thinnest edges of electrode materials due to the "edge effect" charge distribution. (3) Materials need to be nontoxic to the cells and tissue. This need eliminates materials that undergo spontaneous faradic reactions in the body such as cobalt, copper, iron, silver, and germanium, while mildly reactive materials such as molybdenum, nichrome, and tantalum require careful consideration [128–132]. For chronic in vivo applications, additional parameters require consideration: (1) nanostructures must be delivered to the target tissue region in a minimally damaging manner for both the nanostructures and the tissue [133, 134] and (2) ejected nanoparticles and ions from the electrode must be nontoxic to the tissue. The remainder of this chapter explores conventional and leading-edge nanostructured electrodes for electrical stimulation of the neural tissue.

It should be noted that the fabrication of micro electrodes often utilizes multiple layers of metal, usually as adhesion layers for the electrode site or the insulation material. If the multiple layers of the metal are exposed to the electrolyte, either by design, corrosion of electrode edge, or cracking of the insulation, it creates a *galvanic cell* in which metal surfaces corrode irreversibly over time. In addition, during electrical stimulation, multilayered materials may result in differential expansion, leading to mechanical delamination over many stimulation cycles.

It should also be noted that diameters of individual atoms are on the angstrom scale (10^{10} m), and therefore nanostructures cannot be made smaller due to this limitation.

4.3.1 Nanotextured Metal and Metal Oxide Electrodes

Nanotextured electrodes aim to increase the electrochemical surface area of smooth metal electrodes. Often, this is done by growing a metal oxide film. Perhaps the most common, micro-/nanostructured stimulation electrodes are activated iridium oxide electrodes [104]. Activated iridium oxide film (AIROF) is formed by applying electrical pulses to iridium electrodes in the presence of a buffer solution [135] (Fig 4.9a, b). As a result, a thin film of iridium on the surface of the electrode is oxidized to iridium oxide. AIROF is pseudocapacitive with a particularly unique ability to transition between Ir^{+4} and Ir^{+3} and has a rough surface texture which improves its charge injection capacity [88, 136]. The roughness and surface texture can be somewhat controlled by the type of buffer, pulse parameter, and number of pulses during activation [137].

Iridium oxide can also be electrodeposited (EIROF) onto the surface of the electrode to improve charge injection capacity [138]. However, because AIROF and EIROF are generated in a solvent (usually water), the iridium oxide on the surface is hydrated, which reduces the density and stability of the oxide layer drastically compared to pure iridium oxide crystals [139, 140]. Additionally, activation and electrochemical deposition often result in fragile nanostructured oxide growth,

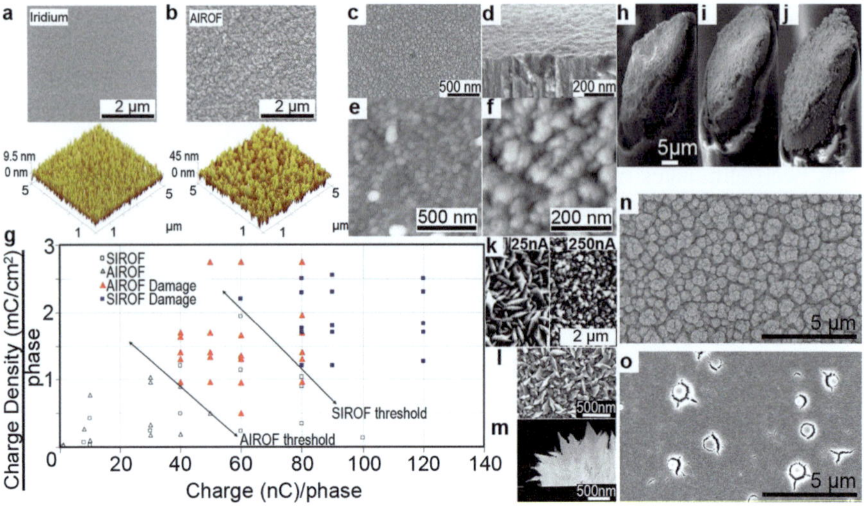

Fig. 4.9 Metal. (**a**) Iridium film imaged with SEM (*top*) and AFM (*bottom*). (**b**) AIROF film imaged with SEM (*top*) and AFM (*bottom*). (**c**) SEM of SIROF surface and the edge of a fractured SIROF sample (**d**) showing columnar microstructures [329]. (**e, f**) AFM of SIROF surface with topographical features >150 nm. The contrast scale with 256 gray scale correspond to 100 nm (**e**) and 25 nm (**f**) [329]. (**g**) The effect of charge injection and charge density per phase of the electrode on the electrode damage after 1.26 million stimulation pulses was passed at 50 Hz with varying current amplitude and pulse width. (**h**) SEM of uncoated tungsten microwire [144]. (**i, j**) SEM of wires electrodeposited with Pt black with ultrasonication (**i**) and without (**j**) [144]. (**k**) SEM showing electrochemical deposition current densities (*left*, 25 nA; *right*, 250 nA) and gold plating solution concentration impact gold nanoflake structures. (**l**) SEM of nanoflake stimulation electrode [146]. (**m**) TEM of (**l**) from a cross-sectional view. (**n**) SEM of platinized Pt black electrode surface. (**o**) SEM of nanovoid Pt electrode surface electrochemically deposited in reverse micelle solution. (**a, b**) Reproduced with permission from [328], © (2013) Elsevier. (**g**) Reproduced with permission from [142] © (2010) Elsevier. (**k**) Adapted with permission from [145] © (2003) Elsevier. (**l, m**) Reproduced with permission from [146], © (2010) IOP Publishing. All rights reserved. (**n, o**) Adapted with permission from [147], © (2010) American Chemical Society

particularly on the edges, due to high charge density caused by the edge effect. In contrast, sputtered iridium oxide films (SIROF) are produced by sputtering deposition of iridium in an oxygen plasma [141] (Fig. 4.9c–f). While the ejection of iridium ions during electrical stimulation can still be observed, SIROF-coated microelectrodes have been demonstrated to have improved charge injection properties (charge injection limit of 0.7–4 mC/cm^2) and stability from delamination compared to AIROF after prolonged stimulation cycles [142] (Fig. 4.9g). Moreover, SIROF properties can largely depend on deposition parameters such as deposition power and oxygen concentration, which may be further explored [83, 143].

Like iridium, platinum is a pseudocapacitive material. *Platinum black*, or finely powdered platinum so-named because of its dark color, is a material that is often electrochemically deposited onto smooth platinum surfaces to improve surface characteristics of neural microelectrodes [136]. While the surface morphology and electrical properties are enhanced, platinum black microelectrodes are typically unstable for reasons similar to those discussed with EIROFs [139, 144]. This has limited their popularity for chronic electrical stimulation applications (and electrochemical sensing). One method applied to improve the platinum black nanotexture is to electrochemically deposit the platinum within an ultrasonicating bath [144] (Fig. 4.9h–j). The ultrasonication process tends to break off unstable nanostructures generated due to the edge effect and thus maximizes the growth of stable nanostructures on the microelectrode. This method can be adapted to the electrochemical deposition of any nanostructured electrode, but it should be noted that some microelectrode substrates, such as thin crystalline silicon, will shatter during ultrasonication.

Other emerging techniques aim to deposit metals using conditions that generate nanostructures. For example, gold nanoflakes have been electrochemically deposited onto thin-film recording sites to generate three-dimensional multi-crystallized structures under specific deposition currents and solution concentrations [145, 146] (Fig. 4.9k–m). Electrochemical deposition of platinum in a reverse micelle (with micelles consisting of a hydrophilic core and hydrophobic shell) surfactant solution also generates nanostructured microelectrode surfaces which were observed to increase charge injection limit from 0.05–0.15 to 3 mC/cm^2 [83, 91, 147, 148] (Fig. 4.9n, o). As expected, these structures reduce impedance and allow for depolarization of neurons using a lower voltage. Due to the edge effect, ultrathin and ultrasharp nanostructures may be more susceptible to mechanical and electrochemical degradation in vivo; however, lower electrical currents and more selective efficacious neural stimulation may be achieved with these nanostructured microelectrodes. Their stability over prolonged stimulation in vivo will need to be extensively characterized to understand these trade-offs and limitations.

As described in an earlier chapter, nanostructured coatings on metals like platinum and iridium can be used to detect neurotransmitters and pH in the brain [137, 149, 150]. These detection methods utilize a small electrical potential to detect redox reactions of target molecules at the surface of the microelectrode. While these devices can be effective in acute experiments, there are many challenges in designing long-term implantable nanobiosensors, including but not limited to faradic degradation and biofouling. Nevertheless, these micro-/nanotextured metals have demonstrated feasibility in favorably extending electrical stimulation properties and in some cases extending functionality to chemical detection.

4.3.2 Capacitive Nanostructured Electrodes

Capacitive electrodes are attractive for electrical stimulation [151–153], particularly because they are capable of preventing or limiting faradic reactions. Several examples of capacitive electrodes include tantalum/tantalum pentoxide and titanium nitride [83, 85, 151–154]. One challenge with capacitive microelectrodes (with a reported charge injection limit of 0.02–0.9 mC/cm^2) is delivering enough current for

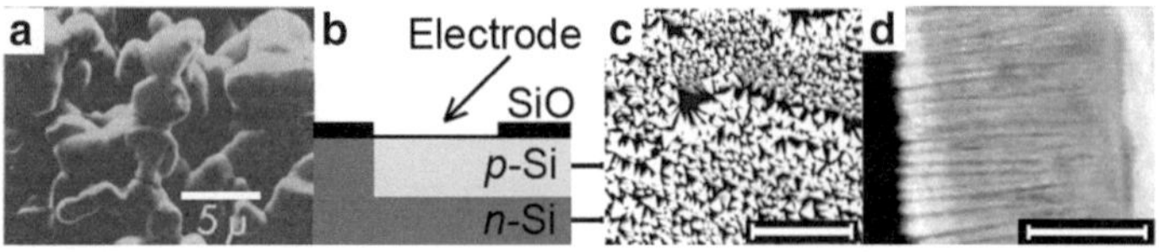

Fig. 4.10 Capacitive stimulation electrodes. (**a**) Ultrastructured tantalum/tantalum oxide electrode. (**b**) Schematic of planar silicon capacitive electrodes using a thin 10 nm oxide dielectric layer. A bias voltage is applied to the *n*-typed silicon giving rise to a blocked *p–n* junction. Electrical stimulation applied to the *p*-type silicon transfers charge into the electrolyte [163]. (**c**) Highly fractal titanium nitride electrodes surface with porous features imaged from the top [116]. (**d**) Cross-sectional scanning electron microscopy image of titanium nitride electrodes from the side showing the long porous structure. (**a**) [153] (**c**, **d**) Reproduced with permission from [116] © (2005) The Electrochemical Society

efficacious stimulation. Another challenge is materials with a high dielectric constant can compromise band gaps during high current stimulation which leads to larger faradic leakage currents[1] [152, 155].

One of the earliest capacitive electrode materials explored was tantalum/tantalum pentoxide (Ta/Ta_2O_5). As mentioned earlier, tantalum is known to be reactive in the brain tissue. This reactivity is prevented by *polarizing* the electrode or applying high positive voltage bias, particularly during the interpulse period to prevent toxic faradic reactions [156]. Despite the fact that Ta_2O_5 has both a high band gap and dielectric constant [157, 158], additional obstacles limit the extensive adoption of Ta/Ta_2O_5 electrodes. Because the charge injection is completely capacitive, it relies entirely on ion diffusion for charge transfer. In order to reach efficacious charge injection values, the electrode must have a *fractal* surface geometry (Fig. 4.10a) [152, 153]. Improving the electrochemical surface area of Ta/Ta_2O_5 electrodes has been explored by increasing its porosity. However, the limiting factor was observed to be the ion flow rate through the nanopores [159, 160].

Similarly, titanium/titanium dioxide (Ti/TiO_2) electrodes have faced challenges in adoption for neural stimulation electrodes. TiO_2 semiconductor microelectrode sites separated by <10 nm from electrode traces to prevent leakage currents have been demonstrated in vitro to depolarize neurons that are in contact with the electrode surface through weak stimulation currents [161]. One concern is the possibility that if TiO_2 nanoparticles are expelled from the electrode site, they could cause a deleterious reactive response in the brain [162].

Alternatives to Ta_2O_5 and TiO_2 surfaces include capacitive silicon field-effect transistor electrodes with conductive *p-doped* silicon (silicon which has been *positively* charged through the addition of *dopants* such as boron) separated from the

---

[1] Band gaps in solid materials describe an energy range where electrons cannot exist and refers to the energy gap between the material's valence band and conduction band in insulators (large band gap) and semiconductors (smaller band gaps). Materials with very small or no band gaps (because the valence band and conduction band overlap) are conductors. At high stimulation currents, band gaps can be compromised, turning good capacitive electrodes (with high charge injection limit) into faradic electrodes.

tissue electrolyte with a thin silicon oxide layer that is nanometers in thickness (Fig. 4.10b) [163–165]. As is the case with Ta_2O_5 and TiO_2, the currents these devices are capable of delivering are small and require the cell to be in contact with the electrode surface, which is difficult to accomplish in vivo due to the presence of other cell types, extracellular matrix, and protein adsorption.

Sputtered titanium nitride (TiN) microelectrodes have demonstrated good biocompatibility and relatively high charge injection capacities, which has made them the preferred electrode material for many stimulation electrode applications [117, 156, 166–168]. While their charge injection capacity is lower than other pseudocapacitive metals, TiN demonstrates a wider safe voltage range to avoid electrolysis of water into gas [83]. As with other electrodes, highly porous surface geometries increase the capacitive surface area and charge injection capacity (Fig. 4.10c, d).

In general, capacitive electrode materials have been favorably viewed, as charge injection does not occur through irreversible faradic process and their generally broader range of safe stimulation voltages. However, their use in microelectrodes has been limited due to their lower charge injection limits.

4.3.3 Three-Dimensional Nanostructured Electrodes

Advances in nanofabrication technology have enabled the development of solid-state nanostructured electrodes. Advanced three-dimensional nanostructures allow interfacing and charge delivery to neurons in new ways. For example, gold nanoparticles (20 nm) have been adsorbed onto positively charged surfaces pretreated with polyethyleneimine [168]. Electrical stimulation of the resulting gold nanoparticle surface led to a five- to tenfold increase in neurite outgrowth of PC12 neuronal cells in culture, demonstrating the synergistic effect of electrical and topographical signal [168].

Gold mushroom-shaped microelectrodes (gMμE) or functionalized gold-spine electrodes are cell culture electrodes that protrude from the cell culture substrate [169]. The 1–1.5 μm tall mushroom-shaped protrusions are functionalized and coated with an engulfment-promoting peptide [170] (Fig. 4.11a–b). The peptide initializes phagocytosis of the protrusion by the neuron. An actin ring then forms around the stalk of the mushroom-shaped protrusion thereby creating a tight seal. When neurons or cardiac muscle cells lay on gMμE or ~150 nm diameter platinum nanowires (1.5 μm tall), small electrical pulses through the electrode can cause *electroporation* of the cells (using high-density electrical pulses to generate ionic pores in the cell membrane) [171–174]. During electroporation, sharp and sparsely spaced nanopillars can penetrate the cells before the membrane repairs itself allowing for intracellular recording and stimulation.

Titanium (IV) nitride (Ti_3N_4) and tungsten micro-nail electrodes (1.2 μm diameter) also enable high-density arrays for extracellular electrical stimulation [171, 175] (Fig. 4.11c–e). While the high density may not allow for intracellular penetration, it does enable electrical stimulation of individual cells or even different regions of cells.

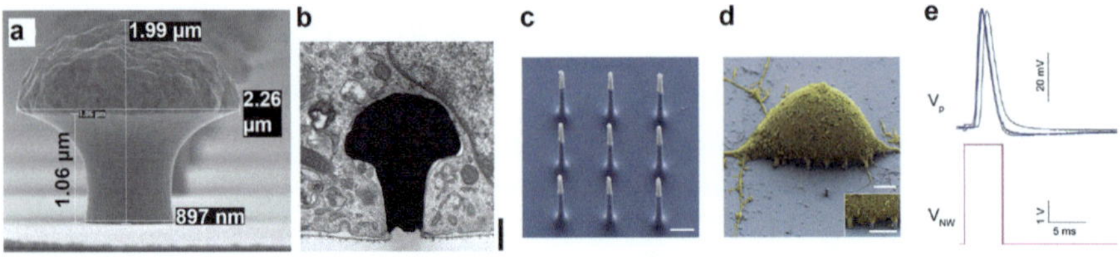

Fig. 4.11 (**a**) SEM of a gMµE. (**b**) TEM of a PC12 cell engulfing a gMµE. Scale bar is 500 nm. (**c**) SEM image of VNEA shows metal-coated tips (*white*) and insulating silicon oxide (*blue*). (**d**) SEM image of a rat cortical cell on top of a VNEA pad (scale bar, 2.5 µm), showing nanowires interfacing with the cellular membrane (*inset*; scale bar, 2.5 µm). (**e**) Neuronal action potential measured by a patch pipette showing reliably evoked action potential requires large voltage pulse (≫1 V) to the VNEA. (**a**) Reproduced with permission from [170] © (2009) IOP Publishing. All rights reserved. (**b**) [174]. (**c**–**e**) [176]

Other vertical nanowire electrode arrays with the dimensions of ~150 nm diameter, 3 µm length, and 2 µm pitch have also been developed for cell culture applications [176]. These nanowires are made from a degenerately doped silicon core encapsulated by silicon dioxide insulation and capped with a titanium/gold electrode tip. The membrane potential of these nanowires can change immediately upon current injection (200–400 pA), suggesting that the current electroporated the membrane enough to penetrate the cell [176]. Neurons with vertical nanowire electrode arrays (VNEA) penetrating into the cell body were visually confirmed with confocal microscopy. VNEA have demonstrated the ability to control membrane potential through the injection of a 200 pA current as well as the capacity to generate action potentials through a ~4.5 V, 5 ms monophasic pulse.

One important note is that as recording sites reach the nanoscale, impedance increases and higher voltages may be necessary to elicit a neural response. This may be less important for the lifetime of cell cultures, but it generates an additional design challenge for long-term in vivo implantation, particularly in terms of minimizing faradic reactions. The careful consideration of charge injection, charge injection density, and surface area will need to be balanced to prevent damage to the cells and electrode over time.

At the nanoscale extreme, boron nitride nanotubes (BNNT) have been used to electrically stimulate cells from within [177]. Once the cells engulf and internalize the BNNT, the piezoelectric property of BNNT allows for the transduction of external ultrasound stimulation into intracellular electrical stimulation. For in vivo applications, care must be taken to titrate the toxicity level of the nanotubes to effective dosage for efficacious stimulation.

Three-dimensional nanostructures have enabled interfacing with cells in new ways, such as tight cellular coupling, intracellular penetration, and wireless excitation of neurons. However, additional research, evaluation, and engineering are necessary for in vivo and long-term applications, such as the stability of the cell after electroporation, degradation of electrode characteristics after stimulation, corrosion and dissolution of electrode material on tightly coupled cells, insertion method, and mechanical stability of the nanostructures after implantation. Nevertheless, these nanostructured stimulation electrodes demonstrate early feasibility and versatility over traditional electrodes as well as new opportunities for basic neuroscience research.

4.3.4 Nanostructured Intrinsically Conductive Polymers

Electrodes modified with conducting polymers (CPs) have been shown to exhibit stimulation performance characteristics that are greatly improved over unmodified probes, including increases to both CSC and charge injection limit. CP coatings accomplish this by extending the effective surface area of the electrode as well as by introducing additional faradic charge transfer mechanisms. The stimulation performance of these coatings can be further improved through nanostructuring techniques which greatly extend the effective surface area of the films. With careful consideration of electrode design and stimulation paradigms, these mechanisms will simply involve the transport of ions into or out of the polymer coating without degrading the polymer or damaging the surrounding tissue.

CPs are polymeric macromolecules that feature an uninterrupted π-conjugated backbone that becomes electrically conductive upon oxidation [178, 179]. The oxidized polymer is coupled with one or several counterion dopants during synthesis, which end up distributed throughout the polymer film and ionically bound to cationic centers along the oxidized backbone, maintaining charge balance [179, 180]. Several examples are shown in Fig. 4.12a, with the most commonly studied being polypyrrole (PPy) [181, 182], poly(3,4-ethylenedioxythiophene) (or PEDOT), and their derivatives [183–187]. Polypyrrole has seen frequent use due to its ease of growth and low toxicity [188, 189]. PEDOT, despite possessing a monomer with lower but still acceptable aqueous solubility, exhibits greatly improved chemical and electrochemical stability over polypyrrole [183–185, 190]. Due to these advantages, PEDOT has been the subject of much recent development. The stability of PEDOT is critical for in vivo stimulation applications as it imbues the coating with a resistance to biological reducing agents, electrochemical breakdown, and conductivity loss from chronic stimulation [183, 184, 191, 192].

Conducting polymers may be either chemically or electrochemically synthesized [179, 180], often deposited onto surfaces as a film but also more recently into three-dimensional structures such as hydrogels [193] or tissue scaffolds [194]. Electrochemical deposition, the most common method, is generally performed by immersing the electrode in an aqueous mixture of monomer and dopant and applying a voltage which surpasses the oxidation potential of the monomer. As the monomer oxidizes, it polymerizes through addition and precipitates onto the electrode while incorporating the dopant as a counterion [179, 182]. Electrodeposition can be performed galvanostatically or potentiostatically, with each method producing varying morphology and properties depending on electrode substrate, geometry, and dopant [182, 183, 195]. Figure 4.12b depicts an example of the contrast in

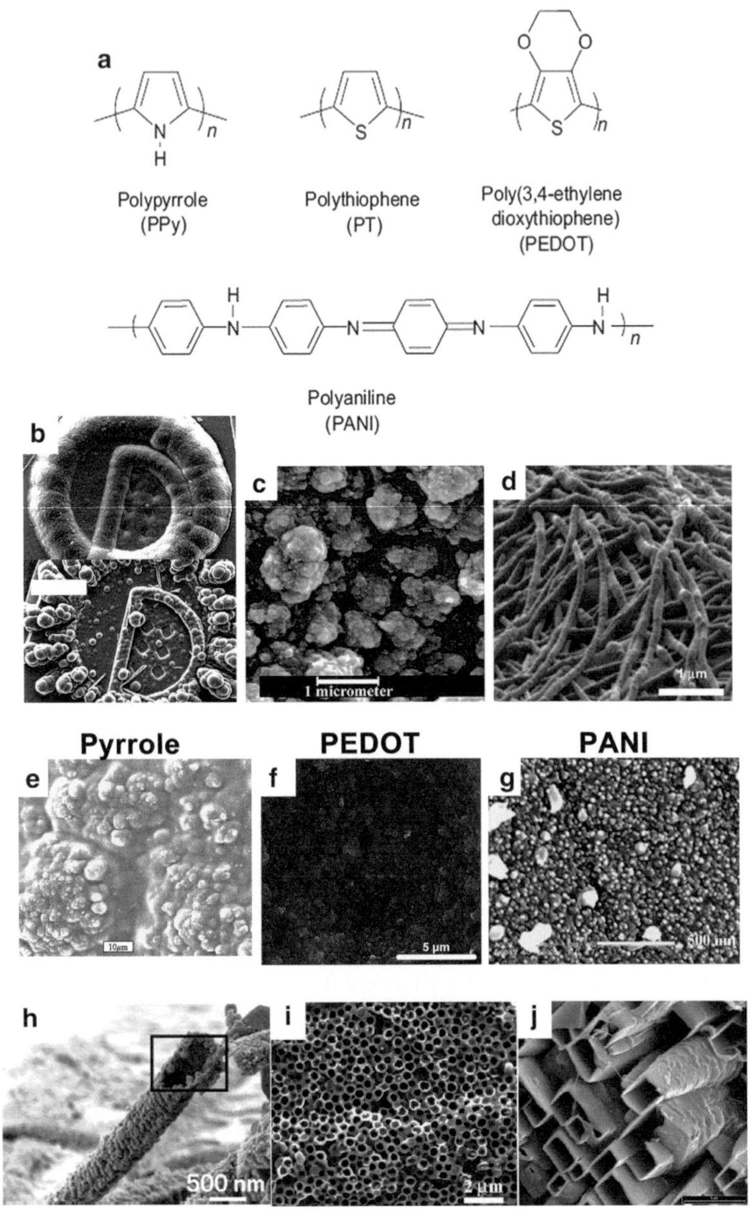

Fig. 4.12 Intrinsically conducting polymers in neural stimulation. (**a**) Chemical structures of the reduced forms of the most commonly used CPs in neural engineering. (**b**) Surface morphology of PEDOT following galvanostatic (*top*) or potentiostatic (*bottom*) electropolymerization. Note that in this study, galvanostatic deposition produced a smooth, uniform coating, while potentiostatic produced a discontinuous, nodular coating (SB = 20 μm). (**c**) "Cauliflower-like" surface morphology typically observed in electrodeposited polypyrrole or PEDOT doped with polystyrene sulfonate (SB = 1 μm). (**d**) Nanofibrous morphology of PEDOT nanotubes prepared by electrodepositing PEDOT over sacrificial PLLA nanofibers (SB = 1 μm). (**e–g**) Representative surface textures of

microstructural surface morphology achieved by coating PEDOT using either galvanostatic (top) or potentiostatic (bottom) methodology, with the former producing a smoother and more robust coating. It should be noted that these results are unique to this particular combination of electrode and coating conditions and that different conditions may produce very different final morphologies.

Altering the deposition parameters and dopants, or coating the surface of templates or nanotextured substrates, enables fine control over the nanostructure of coatings and their resulting electrical properties and stimulation performance. The polyanionic dopant polystyrene sulfonate (PSS) has seen popular use due to excellent coating characteristics [182, 196]. Large dopants such as PSS are typically trapped and immobilized within the bulk of the CP; in vitro electrochemical quartz crystal microbalance (EQCM) studies have verified that polyanionic dopants such as PSS (MW: 100,000) [197] remain within CP films as the CP is reduced and that the ionic flux into and out of the film in such systems consists of small cations which associate with the polyanionic species. It is important to consider the size of the dopant molecules, as smaller dopants tend to leach out of the CP matrix over time, leading to loss of conductivity. CPs doped with smaller ions usually perform well in static, diffusion-driven in vitro environments; however, they can rapidly fail in dynamic systems such as in vivo, where ion concentrations are actively driven by many types of cells.

When applied to metal-based biopotential electrodes, CP coatings have produced substantial changes to their electrical characteristics, including reduced impedance, increased CSC, and increased charge injection limit [133, 181, 183–185, 198–202]. The reduced impedance (often by one or two orders of magnitude at 1 kHz) is in part due to the greatly increased effective surface area of the electrode, with the rough and porous CP coating acting as an electronic extension of the metal electrode underneath. CP coatings have been observed to form a variety of surface morphologies, from cauliflower like (Fig. 4.12c) to nanofibrous (Fig. 4.12d), with the morphology being a product of many factors including monomer chemical structure, dopant and electrolyte selection, reagent concentration, and polymerization method [181–184, 203]. Figure 4.12 exhibits typical surface morphologies of electrodeposited polypyrrole (e), PEDOT (f), and polyaniline (g). Note the fine granular surface morphology of polyaniline compared to the rough, cauliflower-like surface of typical polypyrrole and PEDOT surfaces.

Fig. 4.12 (continued) polypyrrole (SB = 20 μm), PEDOT (SB = 5 μm), and polyaniline (PANI) (SB = 500 nm). (**h**) Detail image of PEDOT nanotube, showing hollow interior left by sacrificial PLLA nanofiber (SB = 500 nm). (**i**) Porous nanostructure generated by electropolymerization of PEDOT around polystyrene nanobeads, which were later dissolved (SB = 2 μm). (**j**) Self-assembled rectangular tube nanostructure in PEDOT following electropolymerization in β-naphthalenesulfonic acid (SB = 5 μm). (**a**) Reproduced with permission from [330], © (2007) Elsevier. (**b**, **c**) Adapted with permission from [183] © (2003) Elsevier. (**d**) Adapted with permission from [234] © (2008) Elsevier. (**e**) Adapted with permission from [181] © (2001) Elsevier. (**f**) Adapted from [185]. (**g**) Adapted with permission from [331] © (2010) Elsevier. (**h**) Adapted with permission from [305] © (2006) John Wiley and Sons. (**i**) Adapted with permission from [201] © (2004) Elsevier. (**j**) Adapted with permission from [235] © (2011) Elsevier

The reduced impedance is also due to the increased electrode CSC, which also leads to improved stimulation performance. The increase in CSC is generated by the inherent redox activity of the conducting polymer film in addition to the high capacitive surface area. It should be noted that these enhancements to electrode impedance and charge storage characteristics come despite oxidized CPs possessing much lower electrical conductivity than typical metal substrates; a typical example is PEDOT doped with PSS, which in the solvent dimethyl sulfoxide exhibits a conductivity of $8*10^2$ S/m [204] compared to platinum which possesses a bulk conductivity of $9.4*10^6$ S/m. This indicates that the changes to electrode performance following CP coating are not due to bulk CP electrical properties but are instead the result of changing interfacial characteristics. Interestingly, thin films of PEDOT doped with PSS on thin film Pt electrodes exhibited linear increases of CSC with increasing deposition charge (and thus coating thickness) until a plateau was reached [184]. This suggests that improving film performance only by way of increasing coating thickness will eventually yield diminishing returns [184]. One potential explanation for this phenomenon is that thicker films possess a lower doping level and a greater number of defects.

Different values of charge injection limit have been observed in PEDOT-coated electrodes depending on the electrode substrate material, including 2.3 mC/cm^2 (on platinum microelectrodes) [184], 3.6 mC/cm^2 (on tin-doped indium oxide electrodes) [205], and 15 mC/cm^2 (on PtIr or Au microelectrodes) [206]. These are comparable to values obtained from IrOx and Ti_3N_4 (0.9–4 mC/cm^2) [135, 155, 207] and much higher than thin-film Pt or PtIr alloy (0.05–0.15 mC/cm^2) [91, 148]. The higher charge injection limit of PEDOT-coated electrodes compared to Pt is partly due to the lower impedance but is also due to the redox activity of PEDOT, which may contribute to the charge injection capacity by providing additional mechanisms for charge transfer across the electrode/electrolyte interface. However, the extent of this contribution is limited depending on the stimulation parameters. In the case of pulses containing very high-frequency components, the redox reactions may not proceed at a fast enough rate to play a significant role in facilitating charge transfer. Polypyrrole-coated electrodes exhibit poor charge injection properties compared to PEDOT, with injection limit being dependent on the dopant employed [208].

Although CP coatings such as PEDOT have showed substantially improved electrode stimulation properties, a number of limitations need to be addressed before clinical translation can occur [180]. PEDOT coatings doped with PSS were observed to exhibit cracking or delamination in over 20 % of the electrodes following 2 weeks of in vitro stimulation with incidence increasing as coating thickness increased [184]. This could be the result of substantial volume changes and mechanical strain during electrical stimulation due to the electrochemical conversion of the polymer and the resulting mass transport of dopants and water [179, 180]. Electrochemical characteristics have also been observed to degrade following chronic use, particularly after periods of long-term stimulation. Polypyrrole films doped with PSS lose up to 95 % of their conductive character following 16 h of constant polarization (400 mV), while PEDOT films subjected to similar stimulation suffered an 11 % reduction [192]. Despite this observed mechanical breakdown and electrochemical breakdown, CPs have exhibited increased resistance to electrochemical breakdown compared to IrOx. Galvanostatically deposited (6 nA) PEDOT coatings doped with PSS were observed to remain stable at current densities beyond those known to cause extensive delamination in IrOx coatings [185]. PEDOT coatings subjected to chronic pulsing stimulation (20 nC 25 Hz charge-balanced square wave) in vitro exhibited a mean time-to-failure of 127 million pulses before voltage excursion exceeded water window limitations [209]. However, while PEDOT coatings are substantially more electrochemically stable than polypyrrole, they have still been observed to exhibit degradation (and resulting conductivity loss) following long periods of exposure to air, with the degradation rate increasing with elevated temperature [210] or the presence of a platinum catalyst [211]. While this degradation will only occur over long, multi-month periods of time in typical storage conditions, PEDOT degradation in vivo has not been well studied and should thus be a subject of consideration. Also, while the biocompatibility and neurotoxicity of many CP films have been well characterized in cell culture [183, 189, 212, 213], the chronic tissue response to coated surfaces in vivo has not been extensively evaluated, particularly in the CNS [202]. CP degradation rates due to stimulation or long-term exposure in vivo may in practice be substantially shorter than values determined in vitro above.

A significant advantage of CP coatings in neural electrode applications is their capacity for modification with bioactive agents, either through doping or surface binding. Such agents can provide intriguing possibilities for modulating cellular interaction and tissue response. Bioactive molecules that have been incorporated into CPs include laminin peptides [214–217], silk fibronectin [214], hyaluronic acid [218, 219], heparin [219], adenosine triphosphate (ATP) [220], collagen [221], dexamethasone [222], and neurotrophic proteins such as neurotrophin 3 (NT-3) [223–226], brain-derived neurotrophic factor (BDNF) [227], and nerve growth factor (NGF) [216, 221]. Dopant chemistry has been shown to have an impact on polypyrrole biocompatibility and surface properties and is known to play a large role in film nanotexture, hydrophobicity, conductivity, and stability [196]. As an example, PSS was found to produce films with a high degree of roughness, hydrophilicity, stability, and in vitro biocompatibility, but yielded conductivity an order of magnitude lower than tosylate, an alternative dopant [196]. Also, the incorporation of large dopant macromolecules (i.e., proteins) results in substantial consequences in terms of coating nanostructure and mechanical properties, with protein-doped PEDOT coatings reported to be brittle and powdery in nature with reduced mechanical stability [216].

A number of alternative CPs have been explored in neural stimulation applications, with encouraging nanostructural possibilities. Polyaniline (PANI) has generated interest following the determination of the biocompatibility of the emeraldine salt form [228]. In addition, ex vivo testing of PANI-coated stimulation electrodes in rat retina revealed reduced protein adsorption, as well as stable coating morphology following 6 months of in vitro stimulation compared to bare platinum controls [229]. Within this study, PANI was polymerized using an oxidative addition reaction, yielding a rough nanoparticulate morphology. It was noted that PANI imparts a high degree of corrosion resistance to the underlying metal through the formation of a passivating compound at the interface that is capable of reducing oxygen through catalysis [229]. Polydopamine, a biomimetic material inspired by the versatile adhesion capability of mussels [230], is a self-assembling polymer capable of adhering to a very broad assortment of organic and inorganic materials and nanotextures through the expression of both catechol- and amine-binding groups. These groups provide both versatile surface adhesion and a mechanism for grafting a variety of substances to the polydopamine surface, including metals and bioactive macromolecules [230]. While not intrinsically conducting, polydopamine has been explored as a coating for neural devices as a simple means to functionalize commonly employed materials [231]. Coated samples were functionalized with polylysine and hippocampal neurons were cultured on the surfaces for both biocompatibility and neural recording and stimulation. The neurons attached to these surfaces and extended robust networks, and both recording and stimulation were successful [231].

Significant development of novel nanostructures of CP coatings has been carried out in order to optimize electrical performance through the maximization of effective surface area. Polypyrrole and PEDOT nanotube coatings have been synthesized through the encapsulation of sacrificial poly(L-lactic acid) nanofibers within electrochemically deposited CP [232, 234]. Following CP polymerization, the poly(L-LACTIC acid) core is dissolved using dichloromethane, leaving behind a hollow tube. After coating, a 95.8 % 1 kHz impedance decrease in polypyrrole nanotube-coated electrodes and a remarkable 99.5 % impedance decrease in PEDOT nanotube-coated electrodes were observed compared to bare iridium electrodes. This constituted a 400 % greater reduction than conventional non-nanotube PEDOT films [232]. The electrode CSC also increased from 0.1 mC/cm^2 for bare iridium to 240 mC/cm^2 for conventional PEDOT film and to 392 mC/cm^2 for the PEDOT nanotube coating. PEDOT nanotube-coated electrodes were additionally implanted into rat cortex for chronic in vivo impedance and recording performance assessment and performed similarly or better than uncoated sites for each measure [233]. A different morphology may be produced by electrodepositing PEDOT or polypyrrole onto a template of sediment-deposited polystyrene latex nanobeads, which can be dissolved with toluene following CP polymerization [201] (Fig. 4.12i). The resulting coating features a network of nanoscale interconnected spherical voids with greatly enhanced surface area [201]. Alternatively, methods have been found to produce self-assembled CP nanostructures, such as the generation of rectangular polypyrrole nanotubes on electrodes following electrodeposition within β-naphthalenesulfonic acid [235] (Fig. 4.12j). CPs have also been successfully incorporated into nanofibrous neural tissue regeneration scaffolds, typically as a copolymer with a mechanical reinforcing or biodegradable material such as poly(lactide-co-glycolide) (PLGA) [236] or polycaprolactone [237], though many combinations have been explored [189, 227, 238–246]. The CP provides the scaffold with a degree of electrical conductivity through which electrical stimuli may be applied [247].

In summary, conducting polymers, through their unique combination of conductivity, electrochemical activity, and processability, have proven to provide substantial benefits when applied to stimulation electrodes. However, while these coatings have been well characterized in vitro as reviewed above, their utility within chronic in vivo applications has only been evaluated to a limited degree. In practice, their chronic performance will likely be strongly dependent on the degree that the inflammatory tissue encapsulation counteracts or negates the benefits supplied by their increased electrochemical surface area. Future work in CP coating development should focus on both improving electrochemical properties and modulating chronic tissue response and protein adsorption.

4.3.5 Carbon

Carbon material's unique properties at nanoscale dimensions have renewed interest for neural stimulation electrodes. As discussed in previous chapters, graphene is composed of strong carbon–carbon sp^2 bonds, which are also potential sites for chemical functionalization. Popularity for graphene as an electrode material has increased due to its low sheet resistivity, good mechanical characteristics, and capacitive charge transfer.

In cell culture, low-voltage electrical stimulation through graphene electrodes has promoted cell–cell interactions [248]. Here, two large sheets of graphene were placed on polyethylene terephthalate substrates. Neuroblastoma cells were cultured between the graphene electrodes, and subthreshold voltage stimulation pulses (4.5 mV/mm biphasic) were applied. The subthreshold stimulation delivered through these graphene sheets resulted in morphological changes in the neurons and promoted neuron–neuron coupling demonstrating graphene electrode's charge injection ability to induce changes in neurons [248].

Carbon nanotubes (CNTs) are sheets of graphene rolled into a small tube. This makes CNT one of the mechanically strongest and the best conductive materials currently available [249–251], with charge injection limits reported up to 3 mC/cm^2 [252, 253]. Impedance spectroscopy analysis has shown that CNT electrodes have nearly an order of magnitude higher capacitive charge transfer capacity, as well as over eight orders of magnitude greater resistance than traditional metals, indicating minimized faradic reactions [254]. Additionally, CNTs can be closed-ended or hollow, which can allow ions to flow into the tubes [255]. Hollow CNTs (diameter>0.5 nm) can allow the passage of ions which drastically increases the surface area for the electrode's double-layer capacitance [256–258]. For these reasons, CNTs demonstrate far less faradic reaction during stimulation and improved capacitive charge transfer over traditional metals including titanium nitride [254, 259]. Careful consideration is needed when selecting the type of CNT and/or the underlying substrate material as it is possible to cause faradic reactions on the underlying substrates. The hollow tubes may facilitate the reversibility of the faradic reaction by confining its diffusion, but it is possible that with large diameter hollow CNT coatings enough irreversible faradic reaction on the underlying substrate can eject the carbon nanotubes.

It is also important to note that the electrical properties of carbon nanotubes depend largely on diameter (n,m) indices (the orientation of the graphene when it is

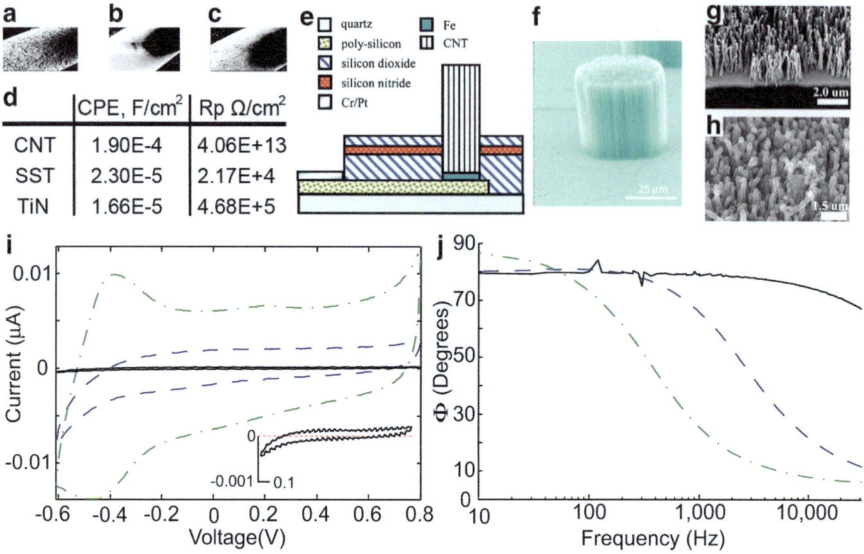

Fig. 4.13 CNT nanostructured microelectrodes. SEM images of 1/8 diameter electrodes with electrode site made from (**a**) carbon nanotube (CNT), (**b**) stainless steel (SST), and (**c**) titanium nitride (TiN). (**d**) Table showing capacitive (constant-phase element; CPE) and electrolyte resistance (R_p) of (**a**)–(**c**) calculated from circuit modeling with impedance spectroscopy data [254]. (**e**) CNT pillar electrode schematic of the cross section (not to scale); (**f**) a 50 μm diameter CNT pillar electrode [253]. SEM micrograph of VACNF electrode arrays at 45° perspective view: (**g**) as-grown VACNFs on electrodes and (**h**) a similar sample after PPy coating in solution and then dried in the air [266]. (**i**) CV (1 V/s) and (**j**) phase angle (Φ) of 38.5 μm² electrode sites with blunt carbon fiber (*solid line*: CSC = 0.6 mC/cm², CSC_C = 0.24 mC/cm²), 50 nC deposition of PEDOT/PSS (*dashed line*: CSC = 11 mC/cm², CSC_C = 5.5 mC/cm²), and 400 nC deposition of PEDOT/PSS (*dash-dot line*: CSC = 46 mC/cm², CSC_C = 23 mC/cm²). (**i**) CV *insert* shows carbon has very flat profile indicating very limited faradic charge transfer. (**j**) Large Φ also supports the fact that carbon is more dominated by capacitive charge transfer [133]. (**a–d**) Adapted with permission from [254], © (2007) American Chemical Society. (**e, f**) Adapted with permission from [253] © (2006) American Chemical Society. (**g, h**) Reproduced from [266] © (2009) Springer. (**i, j**) Adapted from [133], © (2012) The Author(s)

rolled into a tube) and whether it is single walled (SWNT), double walled (DWNT), or multiwalled (MWNT) [257, 260, 261] [For (n,m); CNTs are metallic ($n = m$), quasi-metallic (n-m = 3I, where I is an integer), semiconducting (all others)]. Various forms of carbon nanotube electrodes have demonstrated feasibility in cell cultures and tissue slices with better charge injection capacity and capacitive charge transfer properties than other electrode materials, such as TiN [254, 262, 263] (Fig. 4.13a–d) or even PEDOT and IrO_2 [264].

Dense CNT meshwork electrodes have been deposited onto glass surfaces by first functionalizing CNTs with a 1, 3-dipolar cycloaddition reaction followed by annealing at 350 °C under N_2 after extensive washes [265]. These electrodes demonstrated the ability to stimulate neurons grown on top of the SWNT mesh and raised the possibility that parts of the SWNT may be intimately contacting the cell membrane, though further investigation is necessary.

Another type of CNT pillar microarrays composed of 40 μm tall CNTs have been grown vertically on 50 μm × 50 μm electrode sites using a catalytic thermal chemical vapor deposition system have also been investigated [253] (Fig 4.13e, f). These CNT pillars showed charge injection limits comparable to traditional metals and elicited neural activity through electrical stimulation.

More recently, vertically aligned carbon nanofiber (VACNF) has been grown on a Pt recording site insulated with 400 nm-thick silicon nitride [266]. VACNF (2–4 μm tall, 100–250 nm diameter) was grown using nickel island catalysts that were later removed. The resulting microbrush arrays (MBA) had better electrical characteristics than comparable tungsten or platinum electrodes as evidenced by a lower voltage response to stimulation currents (Fig. 4.13g). However, VACNF MBAs were not able to evoke field potential waves without first being coated with PPy/KCl at 1.5 V for 120 s (Fig 4.13h) [266].

Some CNTs result in stronger cellular adhesion and more extensive neurite growth [252, 267, 268]. However, difficulties related to implanting these micro-/nanostructured electrodes into the body in a minimally damaging way as well as addressing the health concerns of toxicity from loose CNTs have been encountered [269]. To this end, chronically implantable 4–8 μm carbon fiber electrodes have been developed [133, 270]. While the electrical properties of these electrodes are mostly capacitive and do not show any redox peaks between −0.6 and 0.8 V, their CSC and charge injection capacity are relatively limited compared to traditional metals and conductive polymers, demonstrating again that pseudocapacitive charge transfer is much more efficient than capacitive charge transfer (Fig. 4.13i, j).

4.3.6 Layer-by-Layer-Assembled Composites

Layer-by-layer (LbL)-assembled thin-film composites are tightly alternating monolayers of oppositely charged materials. The assembly is based on the principle of charge interaction, which was originally introduced for polyelectrolytes by Decher et al. in 1992 [271] independently of the earlier version of the same technique introduced by Iler [272]. All material surfaces have net charge once immersed in solution due to oxidation, hydrolysis, and preferential adsorption of ions. When a negatively charged surface is immersed in a solution of positive polyelectrolyte, such as poly(dimethyldiallylammonium chloride) (PDDA), the positive electrolyte absorbs onto the surface. After rinsing with water to remove all of the loosely bound polyelectrolyte, the net charge on the surface would be positively charged from the remaining monolayer of polyelectrolyte. This surface could then be immersed in a solution with a negative polyelectrolyte and the net charge on the surface would reverse to negative. With a cyclic process, a multilayer film of alternating polyelectrolyte monolayers can be deposited on the surface with controlled structure and thickness (Fig. 4.14a). This process is not limited to polyelectrolytes. Many nanomaterials are dispersed in solution with a net charge and could be used as a component in the LbL assembly process creating textures and nanostructures based on the nanomaterial [273]. Because of the multilayer nature of the LbL process, it allows

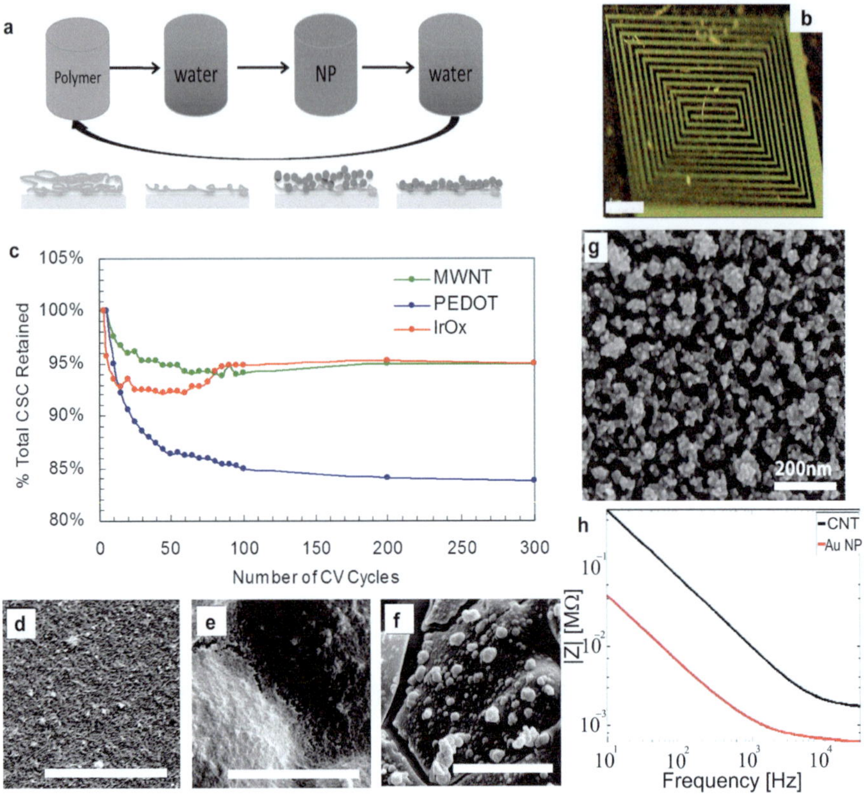

Fig. 4.14 (**a**) Schematics of the LbL assembly process. (**b**) Optical image of direct-write lithography patterned CNT LbL nanocomposite. Scale bar 10 μm [186]. (**c**) CSC of the LbL, IrOx, and PEDOT electrodes coatings over 300 cyclic voltammetry scanning cycles [281]. (**d**–**f**) SEM images of the different electrodes coatings after 300 cyclic voltammetry scanning cycles [281]. SEM images reveal a nearly intact surface on the CNT-coated electrode (**d**) and a slightly and extensively cracked surface for the (**e**) PEDOT- and (**f**) IrOx-coated electrodes, respectively. Scale bars are 5 μm. (**g**) SEM image of LbL-assembled Au NP/PDDA film [283]. (**h**) Impedance measurement showing improved electrical characteristics of Au NP LbL film over CNT LbL films [283]

homogenous high loading of nanoscale components in the material matrix. With fairly simple instruments, LbL assembly can also easily control coating thickness with nanoscale accuracy, incorporate different material components, and adjust other relevant properties to achieve multiple functionalities. It is especially relevant for biological applications where tailored mechanical, electrical, and biological properties are required. Therefore, LbL assembly is one of the most versatile methods for preparation of nanostructures on various surfaces.

Moreover, LbL-assembled nanostructures have many electrochemical advantages for neural stimulation by significantly increasing the electrochemical surface area of the stimulating electrode, thus enhancing the charge–transport capability [274]. Since LbL assembly is a bottom-up fabrication method, it can be applied to a variety of electrode surfaces. The versatility of the LbL system offers opportunities to fabricate nanostructures with various nanocomposites of most diverse composition and nanoscale organization incorporating metal nanoparticles (NPs), semiconductor NPs also known as quantum dots (QDs), and carbon nanotubes (CNTs). Selecting nanomaterials based on unique electrical and optical properties allow for flexibility in different modes of neural stimulation (direct current or photostimulation). Freestanding film from LbL assembly cannot only retain excellent electrochemical properties but also offer mechanical flexibility and unusually high strength [275]. Flexible neural electrodes are believed to play an important role in improving chronic in vivo tissue integration [133, 276, 277]. Recent work indicates that LbL-assembled film can be successfully patterned with lithography techniques [278] opening new opportunities for designing neural electrodes (Fig. 4.14b).

CNTs have electrical properties, chemical stability, and desirable dimensions to create nanostructures suitable for neural modulation. They have drawn much attention for neural interface applications for the past decade. LbL-assembled CNT nanostructures combine excellent electrical conductivity, biocompatibility, and strength to advance neural interface applications. A single-walled carbon nanotube (SWNT)/polyelectrolyte LbL film successfully supported and stimulated neuroblastoma/glioma hybrid cells (NG108) [279, 280]. In a systemic comparison study, LbL-assembled CNTs showed lower impedance and higher CSC compared to electrochemically deposited PEDOT and iridium oxide. After 300 fast-charging cycles, the CNT film retained 95 % of its CSC without any surface cracks compared to PEDOT and iridium oxide, which showed minor or major cracks, respectively [281] (Fig. 4.14c, d). The structural stability of the LbL CNT nanostructures is related to the homogenous high loading of CNT in the polymer matrix and the close interaction between each nanoscale layer. The homogenous loading of CNT in the polymer matrix enables the development of materials that are mechanically strong and flexible. Flexibility of the implantable neural electrode is believed to be a critical parameter for in vivo implantation and prevents glial encapsulation of the nanostructure electrode site [277]. A recent study showed that LbL CNT nanostructure could be successfully fabricated into mechanically flexible neural electrodes by optical lithography techniques [282].

Similar to CNTs, metallic NPs may also be used to create high electrochemical surface area nanostructures with excellent electrical properties (Fig. 4.14g). Gold NPs have been investigated in many other areas of biomedical research including drug delivery, photothermal therapy, and biomedical imaging. The advantages of gold NPs include lower cytotoxicity and facile synthesis. In a recent study [283], densely packed gold nanoparticle LbL nanostructures were fabricated with PDDA. The electrochemical properties were measured with specially designed electrodes. The results indicated that densely packed gold NP nanostructures have better electrochemical performance than SWNT nanostructures with respect to impedance and CSC. The gold NP nanostructures improved the electrochemical properties by enhancing the electrical conductivity of the nanostructures and increasing interfacial capacitance. In terms of structural stability, gold NP nanostructures retained the same percentage of the CSC as CNT nanostructures after 500 fast-charge cycles.

LBL-assembled nanostructures have demonstrated superior electrochemical performance over other materials. Nevertheless, the next-generation devices for neural stimulation require not only outstanding electrical performance but also the ability to facilitate and promote cell growth. One of the advantages of LbL-assembled nanostructures is their ability to control composition and surface properties to promote cell growth and enhance biocompatibility, while maintaining electrical performance.

To illustrate the biocompatibility of the LBL-assembled nanostructure, previous work indicates that LbL-assembled SWNT nanostructures support the growth and differentiation of neural stem cells [279]. Unlike common model cell lines, neural stem cells are very sensitive to their culture environment and other physical and chemical signals. Here, the SWNT nanostructures did not alter the progression of neural stem cell differentiation. To further demonstrate the versatility of the LbL method, extracellular matrix protein (laminin) was incorporated as a complementary layer into the LbL film with SWNTs to create biocompatible nanostructures [284]. By creating a laminin/SWNT composite, cell adhesion was enhanced through the adhesion-promoting nature of laminin and the nanoscale features of the SWNTs. This example demonstrated the successful differentiation of neural stem cells and the ability to stimulate the neural cells. It also illustrates how LbL-assembled nanostructures could be an excellent tool for multifunctional engineering of nanostructures.

Other than improving cell adhesion and biocompatibility, one can further incorporate other functional components into the LBL-assembled nanostructures. With this approach, it is possible to prepare biologically active neural modulation devices. To demonstrate this concept, a plasmid DNA layer was deposited on top of the electrically active LbL CNT nanostructure [285]. The plasmid DNA on the CNT nanostructure transfected neural cells with greater efficiency compared to traditional solution-based methods. Collectively, these examples showcase that the LBL assembly method can create multifunctional nanostructures to neural applications.

4.3.7 Carbon Nanotube Reinforced Conductive Polymer Blends

Recent advancements in CP coatings for neural stimulation electrodes have focused on overcoming various limitations, including poor mechanical resilience, adhesion strength, and electrochemical stability. One approach is the addition of CNTs as a dopant, with the goal of providing mechanical reinforcement, increased surface area, and improved electrical characteristics. Both single-walled and multiwalled CNTs have been incorporated, generally following an acid bath, which cleaves and functionalizes the CNTs to create negatively charged functional groups for ease of doping. CNTs have been successfully incorporated into both polypyrrole [286–288] and PEDOT [289–293] using electrochemical deposition, where the film is deposited using a solution of monomer and functionalized CNTs. As the coating is deposited, CNTs become trapped and enveloped within the polymer, creating an open, nanofibrous architecture that contrasts greatly with the typical cauliflower-shaped morphology of non-CNT coatings [286].

Fig. 4.15a and b show typical nanofibrous surface morphologies of polypyrrole and PEDOT co-doped with MWNTs and the anti-inflammatory corticosteroid dexamethasone; note the similarity of textures despite differing synthesis methods (polypyrrole coated galvanostatically at 500 μA, PEDOT coated potentiostatically at 1.2 V). Both MWNT-doped polypyrrole and PEDOT coatings exhibit a very high degree of porosity which greatly increases surface area. Despite the morphological similarity between the two coatings, PEDOT has demonstrated substantially increased electrochemical and chemical stability compared to polypyrrole as described in Sect. 4.3.4, making it the more suitable of the two in vivo [183, 192, 209]. Deposition time influences nanostructure as well, with Fig. 4.15c demonstrating the cross-sectional morphologies of PEDOT/SWNT coatings of identical composition but different deposition charge (left, 200 nC; right, 800 nC). Note that increased deposition charge produces both increased bulk coating thickness and increased size of surface features. It is also important to note that different final coating nanostructures can be attained by changing the size of the CNTs incorporated. Smaller CNT-doped PEDOT results in a typical cauliflower-like morphology following electrodeposition [292] (OD: <8 nm, length: 0.5–2 μm, Fig. 4.15d), compared to the open nanofibrous morphology using larger CNTs [294] (OD: 20–30 nm; length, 10–30 μm, Fig. 4.15e).

CNT-doped coatings have demonstrated improved performance in terms of both electrochemical properties and mechanical resilience. The charge injection limit of SWNT-doped polypyrrole was shown to increase 50 % over PSS-doped polypyrrole from 8 to 12 mC/cm^2 and retained 92 % of its capacity following stimulation testing [288]. MWNT-doped PEDOT demonstrated an over 200 % increase in charge injection limit compared to PSS-doped PEDOT, from 2.6 to 8.4 mC/cm^2, and maintained over 97 % of its capacity poststimulation [292]. Each case demonstrated a substantial increase in electrochemical stability compared to non-CNT-doped-conducting polymer. The increased charge injection performance observed following CNT doping is a consequence of several mechanisms, including increased coating surface area, CNT electrical conductivity, and the additional charge transfer mechanism of small cationic molecules entering the coating to serve as counterions against the trapped anionic CNTs when the CP is reduced and neutral. These attributes provide the coating with substantial capacitive and faradic character in both oxidized and reduced states [286]. Mechanical stability tests following both acute and chronic stimulation, as well as chronic soaking, revealed MWNT-doped PEDOT (PEDOT/MWNT) exhibited none of the delamination or cracking that are typical of PEDOT coatings [291].

4.3.8 Photoelectric Stimulation Electrodes

The photocurrent effect or photoelectric effect is a well-known phenomenon for semiconductor materials and is widely utilized in photovoltaic technology.

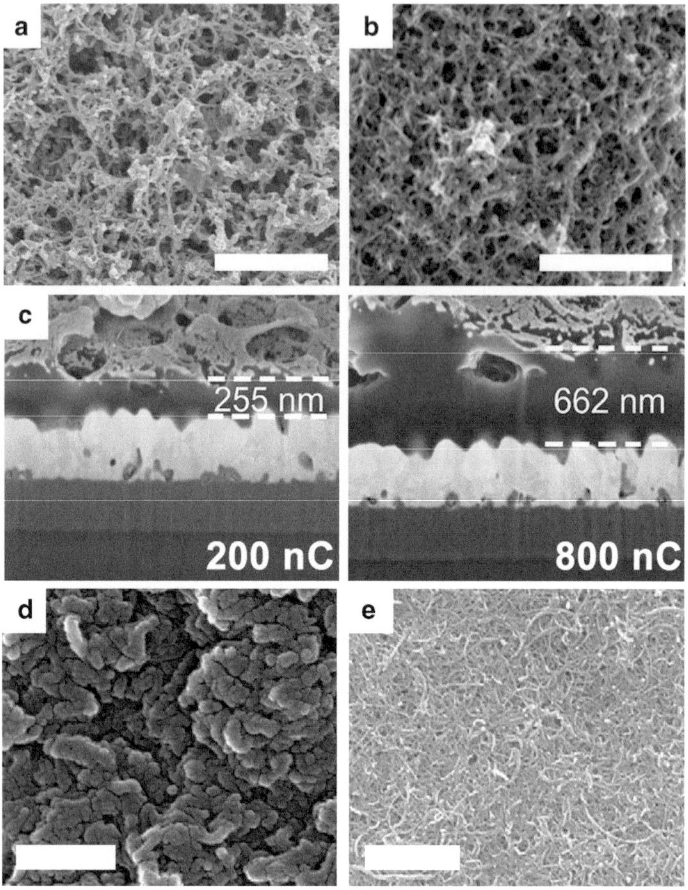

Fig. 4.15 Surface morphology of conducting polymer films doped with carbon nanotubes. (**a**) Nanofibrous latticelike nanostructure of polypyrrole co-doped with acid-functionalized CNTs and the drug dexamethasone. Note the large degree of porosity, which greatly enhances coating surface area (SB = 3 μm). (**b**) Similar nanofibrous nanostructure of PEDOT co-doped with CNTs and dexamethasone (SB = 2 μm). (**c**) Contrast in cross-sectional coating thickness and morphology of CNT-doped PEDOT electropolymerized to a total charge of 200 nC (*right*) and 800 nC (*left*), demonstrating the increased coating thickness and surface pore size with greater deposition time [290]. (**d**) "Cauliflower-like" surface created by small (OD, <8 nm; length, 0.5–2 μm) MWNT dimensions (SB = 0.5 μm). (**e**) Open, nanofibrous surface created by large (OD, 20–30 nm; length, 10–30 μm) MWNTs (SB = 2 μm). (**a**–**c**) adapted from [290]. (**d**) adapted with permission from [293] © (2013) Elsevier. (**e**) Adapted with permission from [292], © (2011) Elsevier

Light energy excites an electron in the valence band to conduction band and generates an electron–hole pair. The electron transfer happens through electrons in the conduction band flowing to an electron acceptor. Other electrons combine with the holes in the semiconductor to generate a current flow. This electrical current generated from a wireless light source can in turn electrically stimulate nearby neurons to elicit a physiological response.

(100)-oriented boron-doped (*p*-doped) silicon wafers have been used to electrically stimulate nearby neurons with a laser [295]. When a negative voltage is applied to *p*-type silicon, major carriers become depleted from the surface. Upon illumination with a laser pulse, photogenerated currents are driven to the surface. The voltage bias (−0.4 to −1.4 V), laser intensity, and laser pulse duration can be tuned to control the amount of photocurrent generated. Similarly, *p-i-n* and *p–n* thin-film photodiode[2] nanostructures have been applied to artificial retina [296].

On the other hand, a high-density cell culture array of TiN microelectrodes can be activated by a 150 nm-thick amorphous silicon photoconductor switch [297]. The switch is activated by a laser passing through the underlying transparent indium tin oxide (ITO) electrode trace. While the switch is activated by the laser, electrical current can be conducted from the underlying ITO trace to the TiN electrode site. Thin-film ITO are mostly transparent conductors that have been used in cell culture, although they are attracting attention due to emerging optical applications covered in other chapters [298, 299]. A challenge with in vivo electrical stimulation with ITO is the oxidization of the material, which results in darkening of the film and a dramatic rise in impedance [300].

Semiconductor nanoparticles or quantum dots (QDs) are a unique class of nanomaterials that are both optically and electrically active. Thus, photocurrent can be generated and utilized for neural stimulation by QDs. Moreover, due to the quantum confinement effect, QDs can be tailored to respond to specific wavelengths of light as compared to bulk semiconductor materials. By combining both the photoelectric and quantum confinement effect, nanostructures can be engineered to stimulate neurons with specific spectral and electrical responses. By applying LbL assembly of mercury telluride (HgTe) nanoparticles and PDDA, polyelectrolyte-nanostructured films have been created with photostimulation capabilities [301] (Fig. 4.16a–c). The nanoscale features can be clearly observed under atomic force microscopy (Fig. 4.16b). Neuroblastoma/glioma hybrid cells (NG108) were successfully cultured and differentiated on the HgTe/PDDA film (Fig. 4.16d, e). While stimulating with a 532 nm laser, depolarization of the differentiated cells on HgTe/PDDA surface was observed, as measured with a standard patch clamp setup (Fig. 4.16f).

Although initial work utilized HgTe, PbSe photoelectric nanoparticles have been coated on the inside of glass pipettes more recently [302]. These NPs adhere to the glass surface, partially due to their low solubility. When the pipette is stimulated with an infrared light, it generates a current that is funneled through the small pipette opening to depolarize nearby neurons.

Photoelectric stimulation has demonstrated proof of concept; however, the miniaturization of implantable lasers, stability of the electrode, and/or toxicity of the materials have limited its adoption. Addressing these concerns will be critical in engineering next-generation chronically implantable photoelectric stimulation electrodes.

---

[2] *p–n diode is two adjoining p-doped and n-doped semiconductor. p-i-n diode has a lightly doped intrinsic semiconductor layer separating the p-doped and n-doped layer.*

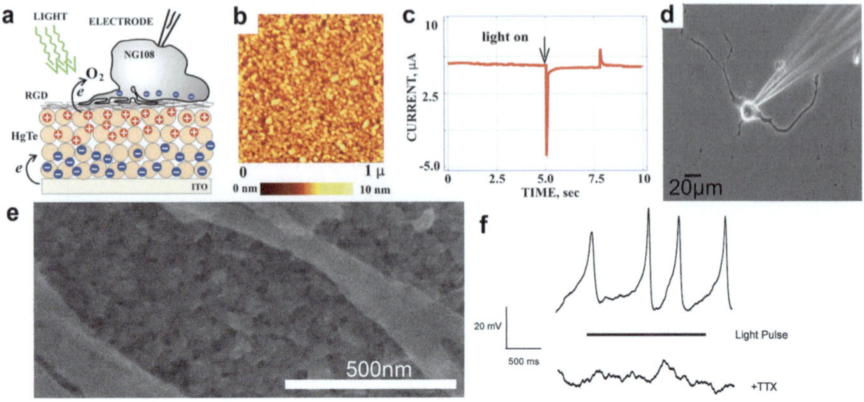

Fig. 4.16 (**a**) Schematics of the photoinduced stimulation of neurons by the layer of nanoparticles. (**b**) AFM image of the nanoparticle layers before adhesion of neurons. (**c**) Kinetics of the photoinduced current in the NP layer. (**d**) Optical microscopy of the patch clamp contacting the neuron under investigation. (**e**) SEM of one of the cell's neuritis on the LbL film. (**f**) Train of the transmembrane potential signals of neuron adhering to the NP film excited by the train of light pulses. *Lower curve*: the same signals after addition of tetrodotoxin (TTX). Adapted with permission from [301], © (2007) American Chemical Society

4.3.9 Drug-Eluting Nanostructures

4.3.9.1 Mechanism

CPs, in addition to reducing electrode impedance and increasing CSC and safe charge injection limit, also feature the ability to release bioactive substances on demand. These substances can include anti-inflammatory drugs, neurotrophic or neuromodulatory agents, and neurotransmitters. CPs are able to reversibly "switch" from an oxidized state to a reduced state following electrical stimulation. This change results in a number of immediate physical and electrical transitions within the polymer due to changes to the electronic configuration of the backbone and the mass transport of ions into or out of the film [179, 303, 304]. These transitions, which manifest in the form of changes to polymer conductivity, volume, and elasticity, may be harnessed to create controlled drug-releasing mechanisms [179, 304] including through (1) the drug incorporation into the coating as the dopant itself, leading to release through ionic repulsion as the polymer reduces [179, 303, 304]; (2) the volume change of the film releasing loosely bound macromolecules such as proteins from the coating surface [223, 224]; and (3) the volume change creating a "squeezing" action as the coating shrinks, expelling drug trapped within nano-reservoirs in the coating [305].

The most extensively studied form of drug release from CPs involves the incorporation of the drug as a dopant itself [179, 303, 304]. As discussed previously in this chapter, oxidized CPs incorporate an anionic dopant to maintain charge balance [179]

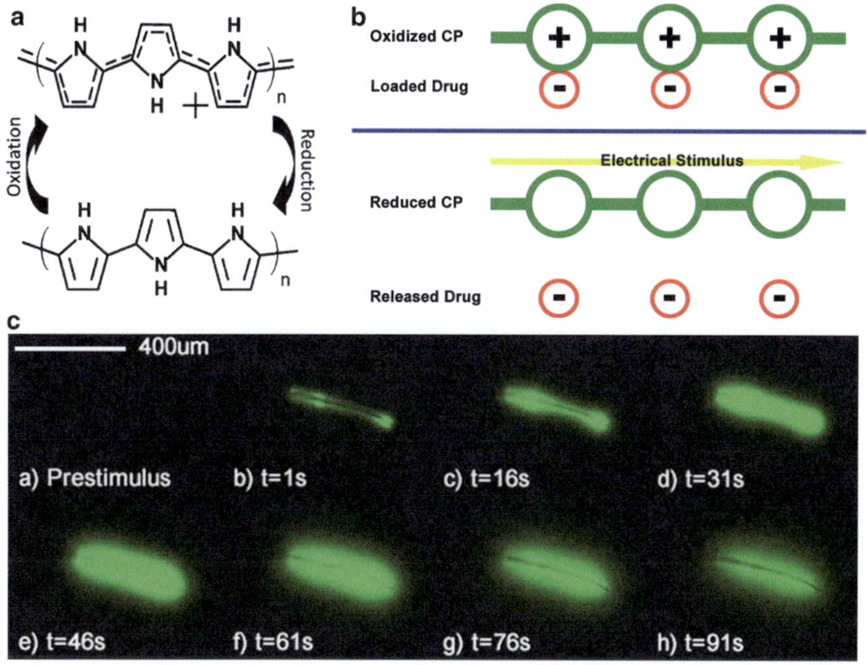

Fig. 4.17 Drug-releasing capability of conducting polymer coatings. (**a**) CPs are switchable between an oxidized state and a reduced state upon application of an electrical stimulus which surpasses the polymer's redox potential (diagram reproduced with permission from [179] © (2010) Elsevier). (**b**) When the polymer is reduced, the dopant which served as the counterion of the oxidized polymer is released and diffuses out of the film. (**c**) Fluorescein dye release from a polypyrrole-coated microwire upon application of a −2 V constant stimulus

and are capable of undergoing a reversible conversion to a reduced state when a voltage that crosses the redox potential of the polymer is applied (Fig. 4.17a). As the CP reduces, the backbone neutralizes, releasing the bound dopant counterions (Fig. 4.17b). If small enough to diffuse through the bulk polymer, the dopant will be driven away through electrostatic repulsion, and a portion will escape the polymer into the surrounding electrolyte, producing an increased drug concentration around the electrode (Fig. 4.17c). However, larger dopants such as organic macromolecules or nanoparticles typically remain trapped and immobilized within the polymer [306, 307].

This method has been used to release a wide assortment of bioactive agents including dexamethasone [222], risperidone [308, 309], progesterone [310], dopamine [129, 311], chlorpromazine [306], ATP [312], (2R)-amino-5-phosphonopentanoate (AP5) [313], 6-cyano-7-nitroquinoxaline-2,3-dione (CNQX) [313], salicylate [314], naproxen [314], N-methylphenothiazine [315], penicillin [316], and glutamate [317]. In addition to molecular size and steric considerations, requirements for drug selection include redox stability within the potential ranges employed for release, as well as a low effective dose for bioactivity due to the limited release yield of most films [179]. While the method is most suited for anionically charged drugs, methods have been developed to release both cationic [129, 306, 311] and electrically neutral [315] molecules, typically through the addition of large anionic intermediary molecules such as melanin, polystyrene sulfonate, or cyclodextrin. Similar release dynamics were observed upon doping CP with proteins such as NT-3 [223–226], BDNF [227], and NGF [216] together with an anionic co-dopant. As these proteins are much too large to escape the polymer film if encapsulated, their release is likely the product of partially entrapped protein molecules detaching from the polymer surface during the actuation process. Recent work has shown the selection of the co-dopant to be crucial in determining the properties and performance of the final polymer [226].

4.3.9.2 Nanostructured Drug-Releasing CP Coatings

Recent advances in CP-based drug release have focused on expanding the variety of bioactive agents that may be incorporated as well as increasing the load and release capacity. One method is to introduce nanodomains into the CP [318]. This can be accomplished by incorporating drug (NGF) into mesoporous silica nanoparticle reservoirs, which are then embedded within the CP along with the co-dopant PSS during electrochemical deposition resulting in a CP-based composite [318] (Fig. 4.18a). Cell culture testing using neuronal PC12 cells verified that NGF released from the films was active after electrical stimulation. However, the films produced a substantial amount of drug "leakage" in the absence of stimulation, although stimulation nearly doubled the final release quantity [318].

An alternative method is to produce a coating composed of PEDOT nanotubes, synthesized through the encapsulation of sacrificial PLGA nanofibers within electrochemically deposited CP [305]. These coatings are morphologically identical to those described in Sect. 4.3.4 above (Fig. 4.12h). The PLGA can be preloaded with a variety of bioactive agents, which remain trapped within the PEDOT tubes after the PLGA is dissolved. Electrical actuation of the PEDOT produces a "squeezing" action that forces the drug into the environment through pores and cracks in the tube walls. As PLGA is capable of holding a very wide assortment of drug types, this method has the potential to be a very flexible basis for drug release technology [305].

A similar approach is to produce a dense network of self-assembled polypyrrole nanowires through electrochemical deposition (Fig. 4.18b) and load drug into the

Fig. 4.18 (continued) pore structure produced by electropolymerizing polypyrrole over colloidal crystalline poly(methyl methacrylate), which is then dissolved (SB = 2 μm). (**h**) Self-assembling nanowires of polypyrrole produced through electrochemical deposition with the dopant adenosine triphosphate (SB = 3 μm, detail *inset* SB = 500 nm). (**a**) Adapted from [318] ©, (2009) IOP Publishing. All rights reserved. (**b**) Adapted with permission from [319], © (2013) Elsevier. (**c–f**) Adapted with permission from [320] © (2009) Elsevier. (**g**) Adapted with permission from [309], © (2013) Elsevier. (**h**) Adapted with permission from [220] © (2011) Elsevier

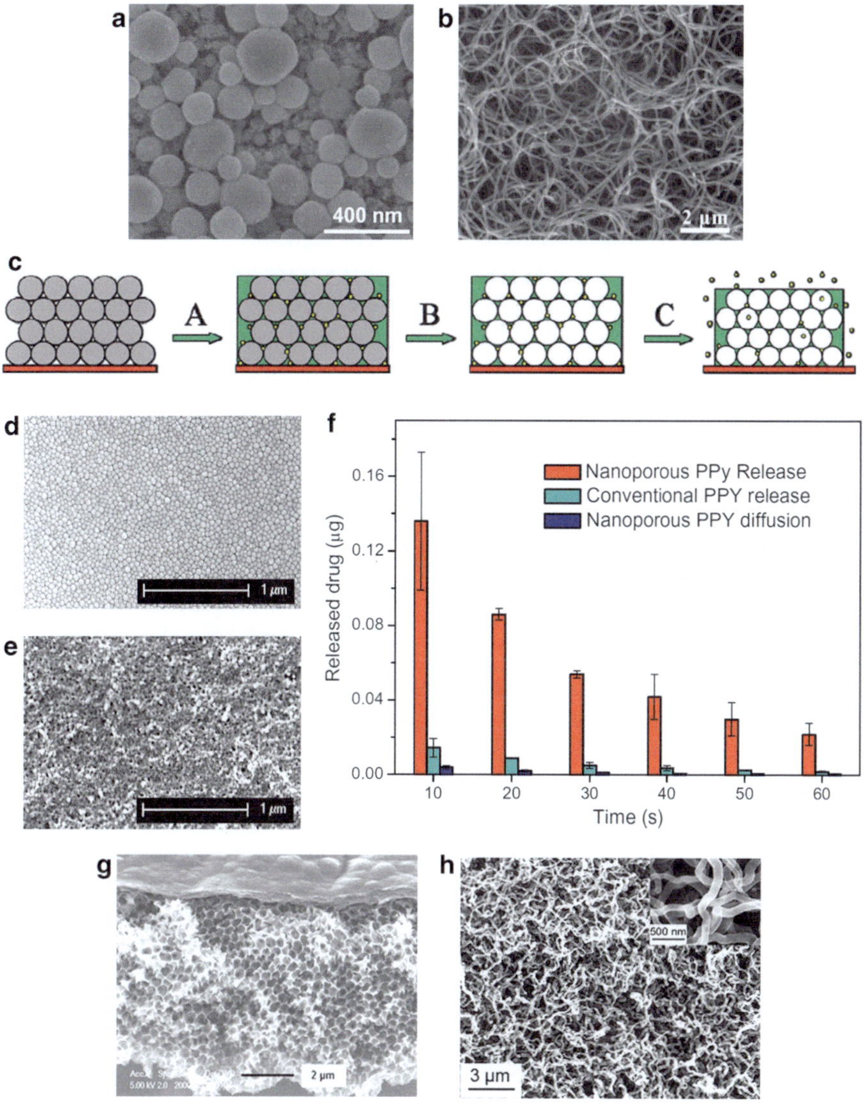

Fig. 4.18 Nanostructures of drug-eluting coatings. (**a**) Polypyrrole-coated silica mesoporous nanoparticles. The nanoscaled pores within the particles act as internal reservoirs of drug (SB = 400 nm). (**b**) Electrochemically deposited self-assembled polypyrrole nanowires. The coating is immersed in a drug solution that is drawn up into the gaps between the nanowires, which act as a reservoir (SB = 2 µm). (**c**) Representation of the synthesis steps of a nanoporous film produced using sacrificial polystyrene nanobeads. First, a nanobead-coated electrode is immersed into a pyrrole solution containing drug, and polypyrrole is coated over the layered beads using electrochemical deposition (*A*). The polystyrene beads are then dissolved in toluene (*B*). Finally, an electrical stimulus is applied to release the contained drug (*C*). (**d**) An SEM image of the polystyrene nanobead template (SB = 1 µm). (**e**) An SEM image of the nanoporous polypyrrole film after template removal (SB = 1 µm). (**f**) Drug release quantities from the nanoporous film with and without release stimulation, as well as from a comparable conventionally prepared film. (**g**) "Inverse opal"

micro- and nano-sized gaps between the wires by immersing them in a droplet of drug solution [319]. This method was reportedly capable of holding and releasing a wide variety of bioactive agents (both lipophilic and hydrophilic). Passive release could be prevented by electrochemically overcoating the wires with a protective polypyrrole layer [319].

An effective way to increase the drug yield of CP coatings is to incorporate synthesis techniques that maximize the surface area of the resulting film. For example, a nanostructured CP drug-releasing coating can be made using a polystyrene nanobead template [320]. After drug-loaded polypyrrole is electrochemically deposited over the template, the beads are selectively dissolved in toluene to leave behind an open, spongy architecture, which through its porosity drastically increases the exposed film surface area (Fig. 4.18c–e). This increased surface area results in a significantly increased amount of drug release compared to non-nanoporous coatings [320] (Fig. 4.18f). Later work demonstrated that these pores may be filled with additional drug and capped using a final layer of electrodeposited CP, producing a network of nanostructured internal drug reservoirs [321]. Such a system is capable of releasing multiple types of drugs simultaneously, useful in cases where it may be advantageous to release two complementary agents together, such as an enzyme with its cofactor, a drug with an adjuvant, or a synergistic neuromodulators [321]. A similar nanostructure may be produced by electrodepositing polypyrrole over a self-assembled colloidal crystal poly(methyl methacrylate) (PMMA) [309]. Following CP polymerization, the PMMA is selectively etched away to reveal an "inverse opal" three-dimensionally ordered film (Fig. 4.18g).

An alternative concept is to incorporate acid-treated multiwalled CNTs into the CP as a co-dopant during electrochemical deposition, after soaking the tubes in dexamethasone solution to load drug within [322]. The CNTs serve multiple roles within the film, including acting as an internal drug reservoir, reinforcing mechanical stability, and providing the structural basis for an open, nanofibrous architecture to maximize surface area [322]. Similar films have also been produced through the layer-by-layer deposition of single-walled CNTs and dexamethasone-doped PEDOT onto a gold substrate through cysteamine binding [323]. Similarly, drug-doped CP can be synthesized through electrodeposition onto a nanotubular titanium substrate [324] or onto platinum nanopillars deposited on a gold-coated polycarbonate template [325]. An analogous nanofibrous morphology can be achieved with polypyrrole coatings upon the incorporation of much smaller dopants such as ATP during electrochemical deposition (Fig. 4.18h), likely through the self-alignment of growing polypyrrole due to hydrogen bonding with encapsulated ATP [220]. Each method has the result of significantly increasing both the coating surface area and drug release yield.

An alternative mechanism was developed which takes advantage of electrophoretic transport through CP film to create an ion pump that drives bioactive molecules from an integrated reservoir into the surrounding tissue on electrical command [326]. These ion pumps are fabricated using multilayer photolithography whereby PEDOT/PSS and the photoresist SU-8 are patterned at a micron-scale resolution to define the functional regions of the device [213]. A key element is the overoxidation of a strip of PEDOT between the source reservoir and the target, rendering the

PEDOT electrically insulative yet ionically conductive. As a potential is applied across the device, ionic drug molecules are drawn electrophoretically through the overoxidized PEDOT, which remains porous to ionic flow due to the open nano-architecture of polymeric material [231]. Nanopores and channels between the polymeric macromolecules allow for the passage of small molecules to the target.

Despite these advances, a number of critical challenges remain that limit the incorporation of this technology into clinical devices and applications. While the timing of release is controllable to a fine degree, the drug load of even the more advanced high-surface area CP films is limited due to lack of reservoir for refill. In vivo application has been limited to a small number of studies, and the chronic stability and release performance of the coatings within living tissue has not yet been extensively assessed. Also, attention must be applied to ensure that the coating selected for use is compatible both mechanically and electrochemically with the electrical stimulation paradigm employed, as the coating materials are electroactive by nature.

4.4 Summary

Electrical stimulation can be used to activate or modulate neuronal activity and, to a lesser extent, to promote neurite growth, neural regeneration, and cell differentiation. While many fields and clinical applications have developed around the application of electrical stimulation, a major current challenge is to understand the physiology of high spatial resolution neural stimulation within a dense, highly networked, and complex nervous system. In order to achieve high spatial resolution neural stimulation, subcellular-sized stimulation electrodes with the ability to target specific neurons will be needed.

The development of subcellular neural stimulation microelectrodes have been limited by irreversible faradic reactions and increased impedance as the electrodes become increasingly small. This can be compensated by using stable nanostructured electrodes to increase electrochemical surface area. However, there are limitations as ion diffusion may become the confining factor in nanostructured electrodes. For chronic in vivo applications, additional considerations must be made for delivery to target tissue regions in a minimally damaging manner for both the nanostructures and the tissue as well as preventing biofouling and glial scar formation. Preventing biofouling and glial scarring will be necessary for maintaining the advantages of high electrochemical surface area, particularly for the high-frequency content of the electrical stimulus. Advances in nanostructured stimulation electrode technologies will create new opportunities and applications for electrical stimulation in vitro and in vivo basic neuroscience and clinical applications.

Acknowledgments The authors would like to acknowledge Kip A Ludwig and Jim Hokanson for critical insight and scientific discussion, as well as valuable discussions from Christi L Kolarcik, Kasey Catt, and Cassandra Weaver. The authors would also like to thank technical contribution by Zhannetta Gugel, Alberto L Vazquez, Noah Snyder, James R Eles, Zhanhong Du, Linn Zhang, Ali Aneizi, and Paras R Patel.

References

1. McWilliam, J.A.: Electrical stimulation of the heart in man. BMJ 1, 348–350 (1889). doi:10.1136/bmj.1.1468.348
2. House, W.F.: Cochlear implants. Ann Otol Rhinol Laryngol **85 suppl 27**, 1–93 (1976)
3. Vidailhet, M. et al.: Bilateral Deep-Brain Stimulation of the Globus Pallidus in Primary Generalized Dystonia. New England Journal of Medicine 352, 459–467 (2005). doi:10.1056/NEJMoa042187
4. Benabid, A.L. et al.: Long-term suppression of tremor by chronic stimulation of the ventral intermediate thalamic nucleus. The Lancet 337, 403–406 (1991). http://dx.doi.org/10.1016/0140-6736(91)91175-T
5. Deuschl, G. et al.: A Randomized Trial of Deep-Brain Stimulation for Parkinson's Disease. New England Journal of Medicine **355**, 896–908 (2006). doi:10.1056/NEJMoa060281
6. Halpern, C.H. et al.: Deep brain stimulation in the treatment of obesity. J. Neurosurg. **109**, 625–634 (2008). doi:10.3171/JNS/2008/109/10/0625
7. Sani, S., Jobe, K., Smith, A., Kordower, J.H. and Bakay, R.A. Deep brain stimulation for treatment of obesity in rats. J. Neurosurg. **107**, 809–813 (2007). doi:10.3171/JNS-07/10/0809
8. Mayberg, H.S., et al.: Deep Brain Stimulation for Treatment-Resistant Depression. Neuron **45**, 651–660 (2005). http://dx.doi.org/10.1016/j.neuron.2005.02.014
9. Richardson, D.E. and Akil, H.: Pain reduction by electrical brain stimulation in man. Part 1: Acute administration in periaqueductal and periventricular sites. J. Neurosurg. **47**, 178–183 (1977). doi:10.3171/jns.1977.47.2.0178
10. Richardson, D.E. and Akil, H.: Pain reduction by electrical brain stimulation in man. Part 2: Chronic self-administration in the periventricular gray matter. J. Neurosurg. **47**, 184–194 (1977). doi:10.3171/jns.1977.47.2.0184
11. Penry, J. K. and Dean, J.C.: Prevention of intractable partial seizures by intermittent vagal stimulation in humans: preliminary results. Epilepsia 31 Suppl 2, S40–43 (1990).
12. Rutecki, P.: Anatomical, physiological, and theoretical basis for the antiepileptic effect of vagus nerve stimulation. Epilepsia **31 Suppl 2**, S1–6 (1990).
13. Peckham, P.H. and Knutson, J.S.: Functional electrical stimulation for neuromuscular applications. Annu. Rev. Biomed. Eng. **7**, 327–360 (2005). doi:10.1146/annurev.bioeng.6.040803.140103
14. Durand, D.M.: Electric Stimulation of Excitable Tissue, Ch. 17. In: Bronzino, J.D. (ed.), The Biomedical Engineering Handbook. 2nd edn., CRC Press LLC, Boca Raton, FL (2000).
15. Gaunt, R.A. and Prochazka, A.: Control of urinary bladder function with devices: successes and failures. Prog. Brain. Res. **152**, 163–194 (2006) doi:10.1016/S0079-6123(05)52011-9
16. Dobelle, W.H., Mladejovsky, M.G. and Girvin, J.P.: Artificial vision for the blind: electrical stimulation of visual cortex offers hope for a functional prosthesis. Science **183**, 440–444 (1974)
17. Abell, T. et al. Gastric electrical stimulation for medically refractory gastroparesis. Gastroenterology 125, 421–428 (2003). http://dx.doi.org/10.1016/S0016-5085(03)00878-3
18. O'Doherty, J. E. et al. Active tactile exploration using a brain-machine-brain interface. Nature **479**, 228–231 (2011). doi:10.1038/nature10489, nature10489 [pii]
19. Van den Brand, R., et al.: Restoring voluntary control of locomotion after paralyzing spinal cord injury. Science **336**, 1182–1185 (2012) doi:10.1126/science.1217416, 336/6085/1182 [pii]
20. Hirst, G.D. and Edwards, F.R.: Sympathetic neuroeffector transmission in arteries and arterioles. Physiol. Rev. **69**, 546–604 (1989)
21. Hodgkin, A.L., Huxley, A.F. and Katz, B.: Measurement of current-voltage relations in the membrane of the giant axon of Loligo. J. Physiol. **116**, 424–448 (1952)
22. Kuffler, S.W. and Vaughan Williams, E.M.: Small-nerve junctional potentials; the distribution of small motor nerves to frog skeletal muscle, and the membrane characteristics of the fibres they innervate. J. Physiol. **121**, 289–317 (1953)
23. Hodgkin, A.L. and Huxley, A.F.: A quantitative description of membrane current and its application to conduction and excitation in nerve. J. Physiol. **117**, 500–544 (1952)

24. Skou, J.C.: Enzymatic Basis for Active Transport of Na+ and K+ across Cell Membrane. Physiol. Rev. **45**, 596–617 (1965)
25. Skou, J.C.: The influence of some cations on an adenosine triphosphatase from peripheral nerves. Biochim. Biophys. Acta **23**, 394–401 (1957). http://dx.doi.org/10.1016/0006-3002(57)90343-8
26. Fonnum, F.: Glutamate: a neurotransmitter in mammalian brain. J. Neurochem. **42**, 1–11 (1984)
27. Spehlmann, R.: Acetylcholine and the synaptic transmission of non-specific impulses to the visual cortex. Brain **94**, 139–150 (1971).
28. Foote, S.L., Freedman, R. and Oliver, A.P.: Effects of putative neurotransmitters on neuronal activity in monkey auditory cortex. Brain. Res. **86,** 229–242 (1975). 0006-8993(75)90699-X [pii]
29. Loewi, O.: Über humorale Übertragbarkeit der Herznervenwirkung. Pflügers Arch. **204**, 629–640 (1924). doi:10.1007/bf01731235
30. Agnew, W.S., Moore, A.C., Levinson, S.R. and Raftery, M.A.: Identification of a large molecular weight peptide associated with a tetrodotoxin binding protein from the electroplax of Electrophorus electricus. Biochem. Biophys. Res. Commun. **92**, 860–866 (1980). 0006-291X(80)90782-2 [pii]
31. Beneski, D.A. and Catterall, W.A.: Covalent labeling of protein components of the sodium channel with a photoactivable derivative of scorpion toxin. Proc Natl Acad Sci U S A **77**, 639–643 (1980)
32. Bendahhou, S., Cummins, T.R., Tawil, R., Waxman, S.G. and Ptacek, L.J.: Activation and inactivation of the voltage-gated sodium channel: role of segment S5 revealed by a novel hyperkalaemic periodic paralysis mutation. The Journal of neuroscience: the official journal of the Society for Neuroscience **19**, 4762–4771 (1999)
33. Lin, W.H., Wright, D.E., Muraro, N.I. and Baines, R.A.: Alternative splicing in the voltage-gated sodium channel DmNav regulates activation, inactivation, and persistent current. J. Neurophysiol. **102**, 1994–2006 (2009).
34. Keynes, R.D. and Elinder, F.: Modelling the activation, opening, inactivation and reopening of the voltage-gated sodium channel. Proc. Biol. Sci. **265**, 263–270 (1998). doi:10.1098/rspb.1998.0291
35. Eyzaguirre, C. and Kuffler, S.W.: Further study of soma, dendrite, and axon excitation in single neurons. The Journal of General Physiology **39**, 121–153 (1955). doi:10.1085/jgp.39.1.121
36. Cannon, W.: Biographical Memoir: Henry Pickering Bowdich. National Academy of Sciences **17**, 181–196 (1922)
37. Theunissen, F. and Miller, J.P.: Temporal encoding in nervous systems: a rigorous definition. J Comput. Neurosci. **2**, 149–162 (1995)
38. Optican, L.M. and Richmond, B.J.: Temporal encoding of two-dimensional patterns by single units in primate inferior temporal cortex. III. Information theoretic analysis. J. Neurophysiol. **57**, 162–178 (1987)
39. Buonomano, D.V. and Merzenich, M.M.: Temporal Information Transformed into Spatial Code by Neural Network with Realistic Properties. Science **267**, 17 (1995)
40. Andersen, R.A., Essick, G.K. and Siegel, R.M.: Encoding of spatial location by posterior parietal neurons. Science **230**, 456–458 (1985)
41. Hubel, D.H. and Wiesel, T.N.: Receptive fields of single neurones in the cat's striate cortex. The Journal of Physiology **148**, 574–591 (1959)
42. Henneman, E., Somjen, G. and Carpenter, D.O.: Functional Significance of Cell Size in Spinal Motoneurons. J. Neurophysiol. **28**, 560–580 (1965)
43. Cope, T.C., Sokoloff, A.J., Dacko, S.M., Huot, R. and Feingold, E.: Stability of motor-unit force thresholds in the decerebrate cat. J. Neurophysiol. **78**, 3077–3082 (1997)
44. Burke, R.E., et al: A HRP study of the relation between cell size and motor unit type in cat ankle extensor motoneurons. The Journal of comparative neurology **209**, 17–28 (1982) doi:10.1002/cne.902090103
45. Ulfhake, B. and Kellerth, J.O.: Does alpha-motoneurone size correlate with motor unit type in cat triceps surae? Brain Res. **251**, 201–209 (1982). 0006-8993(82)90738-7 [pii]

46. Milner-Brown, H.S., Stein, R.B. and Yemm, R.: The orderly recruitment of human motor units during voluntary isometric contractions. J. Physiol. **230**, 359–370 (1973)
47. Binder, M.C., Bawa, P., Ruenzel, P. and Henneman, E. Does orderly recruitment of motoneurons depend on the existence of different types of motor units? Neurosci. Lett. **36**, 55–58 (1983)
48. Shoham, S., O'Connor, D.H. and Segev, R. How silent is the brain: is there a "dark matter" problem in neuroscience? Journal of Comparative Physiology a-Neuroethology Sensory Neural and Behavioral Physiology 192, 777–784 (2006). doi:10.1007/s00359-006-0117-6
49. Carrascal, L., Nieto-Gonzalez, J.L., Torres, B. and Nunez-Abades, P.: Diminution of voltage threshold plays a key role in determining recruitment of oculomotor nucleus motoneurons during postnatal development. PLoS ONE **6**, e28748 (2011) doi:10.1371/journal.pone.0028748. PONE-D-11-20770 [pii]
50. Henneman, E., Somjen, G. and Carpenter, D.O.: Excitability and inhibitability of motoneurons of different sizes. J. Neurophysiol. **28**, 599–620 (1965)
51. Burke, R.E., Walmsley, B. and Hodgson, J.A.: HRP anatomy of group Ia afferent contacts on alpha motoneurones. Brain Res **160**, 347–352 (1979)
52. Burke, R.E. and Rymer, W.Z.: Relative strength of synaptic input from short-latency pathways to motor units of defined type in cat medial gastrocnemius. J. Neurophysiol. **39**, 447–458 (1976)
53. Calancie, B. and Bawa, P.: Voluntary and reflexive recruitment of flexor carpi radialis motor units in humans. J. Neurophysiol. **53**, 1194–1200 (1985)
54. Schmied, A., Morin, D., Vedel, J.P. and Pagni, S.: The "size principle" and synaptic effectiveness of muscle afferent projections to human extensor carpi radialis motoneurones during wrist extension. Experimental brain research. Experimentelle Hirnforschung. Experimentation cerebrale **113**, 214–229 (1997)
55. Muniak, M.A., Ray, S., Hsiao, S.S., Dammann, J.F. and Bensmaia, S.J.: The neural coding of stimulus intensity: linking the population response of mechanoreceptive afferents with psychophysical behavior. The Journal of neuroscience: the official journal of the Society for Neuroscience **27**, 11687–11699 (2007). doi:10.1523/JNEUROSCI.1486-07.2007. 27/43/11687 [pii]
56. Henneman, E.: The size-principle: a deterministic output emerges from a set of probabilistic connections. The Journal of experimental biology **115**, 105–112 (1985)
57. McNeal, D. R.: Analysis of a model for excitation of myelinated nerve. IEEE Trans. Biomed. Eng. **23**, 329–337 (1976)
58. Llewellyn, M.E., Thompson, K.R., Deisseroth, K. and Delp, S.L.: Orderly recruitment of motor units under optical control in vivo. Nat. Med. **16**, 1161–1165 (2010). doi:10.1038/nm.2228. nm.2228 [pii]
59. Gregory, C.M. and Bickel, C.S.: Recruitment patterns in human skeletal muscle during electrical stimulation. Phys. Ther. **85**, 358–364 (2005)
60. Bourbeau, D.J., Hokanson, J.A., Rubin, J.E. and Weber, D.J.: A computational model for estimating recruitment of primary afferent fibers by intraneural stimulation in the dorsal root ganglia. J. Neural. Eng. **8**, 056009 (2011). doi:10.1088/1741-2560/8/5/056009. S1741-2560(11)80621-X [pii]
61. Histed, M.H., Bonin, V. and Reid, R.C.: Direct activation of sparse, distributed populations of cortical neurons by electrical microstimulation. Neuron **63**, 508–522 (2009). doi:10.1016/j.neuron.2009.07.016
62. Gaunt, R.A., Prochazka, A., Mushahwar, V.K., Guevremont, L. and Ellaway, P.H.: Intraspinal microstimulation excites multisegmental sensory afferents at lower stimulus levels than local alpha-motoneuron responses. Journal of Neurophysiology **96**, 2995–3005 (2006) doi:10.1152/jn.00061.2006
63. Jankowska, E. and Roberts, W.J.: An electrophysiological demonstration of the axonal projections of single spinal interneurones in the cat. The Journal of Physiology **222**, 597–622 (1972)
64. Roberts, W.J. and Smith, D.O.: Analysis of threshold currents during microstimulation of fibres in the spinal cord. Acta Physiol Scand **89**, 384–394 (1973). doi:10.1111/j.1748-716.1973.tb05533.x

65. Nowak, L.G. and Bullier, J.: Axons, but not cell bodies, are activated by electrical stimulation in cortical gray matter. II. Evidence from selective inactivation of cell bodies and axon initial segments. Experimental brain research. Experimentelle Hirnforschung. Experimentation cerebrale **118**, 489–500 (1998)
66. Nowak, L.G. and Bullier, J.: Axons, but not cell bodies, are activated by electrical stimulation in cortical gray matter. I. Evidence from chronaxie measurements. Experimental brain research. Experimentelle Hirnforschung. Experimentation cerebrale **118**, 477–488 (1998)
67. Tehovnik, E. J. and Slocum, W. M. Two-photon imaging and the activation of cortical neurons. Neuroscience (2013). doi:10.1016/j.neuroscience.2013.04.022
68. Eccles, J.C.: The central action of antidromic impulses in motor nerve fibres. Pflugers Arch **260**, 385–415 (1955)
69. Malmivuo, J. and Plonsey, R.: Bioelectromagnetism: Principles and Applications of Bioelectric and Biomagnetic Fields. Oxford University Press, New York, NY (1995)
70. Basser, P.J. and Roth, B.J.: New currents in electrical stimulation of excitable tissues. Annu. Rev. Biomed. Eng. **2**, 377–397 (2000). doi:10.1146/annurev.bioeng.2.1.377
71. Mortimer, J. and Bhadra, N.: Peripheral nerve and muscle stimulation. In: Horch, K., Dhillon, G.S. (eds.) Neuroprosthetics: theory and practice. Singapore: World Scientific Publishing Co. Pte. Ltd. (2004)
72. Al-Majed, A.A., Neumann, C.M., Brushart, T.M. and Gordon, T.: Brief electrical stimulation promotes the speed and accuracy of motor axonal regeneration. The Journal of neuroscience: the official journal of the Society for Neuroscience **20**, 2602–2608 (2000)
73. Protas, E.J., et al.: Supported treadmill ambulation training after spinal cord injury: a pilot study. Arch. Phys. Med. Rehabil. **82**, 825–831 (2001). doi:10.1053/apmr.2001.23198. S0003-9993(01)33942-4 [pii]
74. Colombo, G., Joerg, M., Schreier, R. and Dietz, V.: Treadmill training of paraplegic patients using a robotic orthosis. J. Rehabil. Res. Dev. **37**(6), 693–700 (2000)
75. Field-Fote, E.C.: Combined use of body weight support, functional electric stimulation, and treadmill training to improve walking ability in individuals with chronic incomplete spinal cord injury. Arch. Phys. Med. Rehabil. **82**, 818–824 (2001)
76. Ming, G., Henley, J., Tessier-Lavigne, M., Song, H. and Poo, M.: Electrical activity modulates growth cone guidance by diffusible factors. Neuron **29**, 441–452 (2001). S0896-6273(01)00217-3 [pii]
77. Lindholm, D.: Role of neurotrophins in preventing glutamate induced neuronal cell death. J. Neurol. **242**, S16–18 (1994)
78. Morrison, M.E. and Mason, C.A. Granule neuron regulation of Purkinje cell development: striking a balance between neurotrophin and glutamate signaling. The Journal of neuroscience: the official journal of the Society for Neuroscience **18**, 3563–3573 (1998)
79. Al-Majed, A.A., Brushart, T.M. and Gordon, T: Electrical stimulation accelerates and increases expression of BDNF and trkB mRNA in regenerating rat femoral motoneurons. Eur. J. Neurosci. **12,** 4381–4390 (2000). ejn1341 [pii]
80. Muir, G.D. and Steeves, J.D.: Sensorimotor stimulation to improve locomotor recovery after spinal cord injury. Trends Neurosci. **20**, 72–77 (1997). S0166-2236(96)10068-0 [pii]
81. McCaig, C.D., Rajnicek, A.M., Song, B. and Zhao, M.: Has electrical growth cone guidance found its potential? Trends Neurosci. **25**, 354–359 (2002). S0166-2236(02)02174-4 [pii]
82. Shi, R. and Borgens, R.B.: Three-dimensional gradients of voltage during development of the nervous system as invisible coordinates for the establishment of embryonic pattern. Dev. Dyn. **202**, 101–114 (1995). doi:10.1002/aja.1002020202.
83. Cogan, S. F.: Neural Stimulation and Recording Electrodes. Annu. Rev. Biomed. Eng. **10**, 275–309 (2008). doi:10.1146/annurev.bioeng.10.061807.160518
84. Venugopalan, S.: Kinetics of Hydrogen-Evolution Reaction on Lead and Lead-Alloy Electrodes in Sulfuric-Acid Electrolyte with Phosphoric-Acid and Antimony Additives. J. Power Sources **48**, 371–384 (1994)
85. Merrill, D. R., Bikson, M. and Jefferys, J. G.: Electrical stimulation of excitable tissue: design of efficacious and safe protocols. J. Neurosci. Methods **141**, 171–198 (2005). doi:10.1016/j.jneumeth.2004.10.020, S0165-0270(04)00382-6 [pii]

86. Bard, A.J., and Faulkner, L.R.: Electrochemical Methods: Fundamentals and Applications. 2nd. edn. John Wiley and Sons, Inc., New York, NY (1980)
87. Rand, D.A.J., and Woods, R.: The nature of adsorbed oxygen on rhodium, palladium and gold electrodes. Journal of Electroanalytical Chemistry and Interfacial Electrochemistry **31**, 29–38 (1971). http://dx.doi.org/10.1016/S0022-0728(71)80039-6
88. Conway, B.E., and Mozota, J.: Surface and bulk processes at oxidized iridium electrodes—II. Conductivity-switched behaviour of thick oxide films. Electrochim. Acta **28**, 9–16 (1983). http://dx.doi.org/10.1016/0013-4686(83)85080-4
89. Gileadi, E., Kirowa-Eisner, E. and Penciner, J. Interfacial Electrochemistry - An Experimental Approach. Addison-Wesley Publishing Company Inc., Reading, MA (1975)
90. Grupioni, A.A.F., Arashiro, E. and Lassali, T.A.F.: Voltammetric characterization of an iridium oxide-based system: the pseudocapacitive nature of the Ir0.3Mn0.7O2 electrode. Electrochim. Acta. **48**, 407–418 (2002). http://dx.doi.org/10.1016/S0013-4686(02)00686-2
91. Rose, T.L. and Robblee, L.S.: Electrical stimulation with Pt electrodes. VIII. Electrochemically safe charge injection limits with 0.2 ms pulses (neuronal application). Biomedical Engineering, IEEE Transactions on **37**, 1118–1120 (1990). doi:10.1109/10.61038
92. McCreery, D., Pikov, V. and Troyk, P.R.: Neuronal loss due to prolonged controlled-current stimulation with chronically implanted microelectrodes in the cat cerebral cortex. Journal of Neural Engineering **7**, 036005 (2010). doi:10.1088/1741-2560/7/3/036005
93. Cogan, S.F., Guzelian, A.A., Agnew, W.F., Yuen, T.G. and McCreery, D.B.: Over-pulsing degrades activated iridium oxide films used for intracortical neural stimulation. Journal of Neuroscience Methods **137**, 141–150 (2004). doi:10.1016/j.jneumeth.2004.02.019
94. Brummer, S.B. and Turner, M.J.: Electrochemical Considerations for Safe Electrical Stimulation of the Nervous System with Platinum Electrodes. Biomedical Engineering, IEEE Transactions on **BME-24**, 59–63 (1977). doi:10.1109/tbme.1977.326218
95. Short, G.D. and Bishop, E.: Concentration Overpotentials on Antimony Electrodes in Differential Electrolytic Potentiometry. Analytical Chemistry **37**, 962–967 (1965). doi:10.1021/ac60227a003
96. Wang, Q., Millard, D.C., Zheng, H.J. and Stanley, G.B.: Voltage-sensitive dye imaging reveals improved topographic activation of cortex in response to manipulation of thalamic microstimulation parameters. Journal of Neural Engineering **9**, 026008 (2012). doi:10.1088/1741-2560/9/2/026008
97. McIntyre, C.C. and Grill, W.M.: Selective microstimulation of central nervous system neurons. Ann Biomed Eng **28**, 219–233 (2000)
98. Elwassif, M.M., Datta, A., Rahman, A. and Bikson, M.: Temperature control at DBS electrodes using a heat sink: experimentally validated FEM model of DBS lead architecture. J. Neural Eng. **9**, 046009 (2012)
99. Mortimer, J.T., Shealy, C.N. and Wheeler, C.: Experimental nondestructive electrical stimulation of the brain and spinal cord. J. Neurosurg. **32**, 553–559 (1970). doi:10.3171/jns.1970.32.5.0553
100. Pudenz, R.H., Bullara, L.A., Jacques, S. and Hambrecht, F.T.: Electrical stimulation of the brain. III. The neural damage model. Surg. Neurol. **4**, 389–400 (1975)
101. Pudenz, R.H., Bullara, L.A., Dru, D. and Talalla, A.: Electrical stimulation of the brain. II. Effects on the blood-brain barrier. Surg. Neurol. **4**, 265–270 (1975)
102. Yuen, T.G., Agnew, W.F., Bullara, L.A., Jacques, S. and McCreery, D.B.: Histological evaluation of neural damage from electrical stimulation: considerations for the selection of parameters for clinical application. Neurosurgery **9**, 292–299 (1981)
103. Shannon, R.V.: A model of safe levels for electrical stimulation. IEEE Trans. Biomed. Eng. **39**, 424–426 (1992). doi:10.1109/10.126616
104. Mccreery, D.B., Agnew, W.F., Yuen, T.G. H. and Bullara, L.: Charge-Density and Charge Per Phase as Cofactors in Neural Injury Induced by Electrical-Stimulation. IEEE Trans. Biomed. Eng. **37**, 996–1001 (1990). doi:10.1109/10.102812
105. Agnew, W.F., McCreery, D.B., Yuen, T.G. and Bullara, L.A.: Histologic and physiologic evaluation of electrically stimulated peripheral nerve: considerations for the selection of parameters. Ann. Biomed. Eng. **17**, 39–60 (1989)

106. Bhargava, A.: Long-term effects of quasi-trapezoidal pulses on the structure and function of sacral anterior roots. Dissertation, Case Western Reserve University (1993)
107. Townsend, G., Peloquin, P., Kloosterman, F., Hetke, J.F. and Leung, L.S.: Recording and marking with silicon multichannel electrodes. Brain Res Brain Res Protoc **9**, 122–129 (2002). S1385299X02001393 [pii]
108. Schaldach, M., Hubmann, M., Weikl, A. and Hardt, R.: Sputter-Deposited TiN Electrode Coatings for Superior Sensing and Pacing Performance. PACE **13**, 1891–1895 (1990)
109. Negi, S., Bhandari, R. and Solzbacher, F.: A novel technique for increasing charge injection capacity of neural electrodes for efficacious and safe neural stimulation. In: Engineering in Medicine and Biology Society (EMBC), 2012 Annual International Conference of the IEEE, 5142–5145, San Diego 28 August–1 September (2012)
110. Tykocinski, M., Duan, Y., Tabor, B. and Cowan, R.S.: Chronic electrical stimulation of the auditory nerve using high surface area (HiQ) platinum electrodes. Hearing research **159**, 53–68 (2001)
111. Del Bufalo, A. G., Schlaepfer, J., Fromer, M. and Kappenberger, L.: Acute and long-term ventricular stimulation thresholds with a new, iridium oxide-coated electrode. Pacing Clin Electrophysiol **16**, 1240–1244 (1993)
112. Gradaus, R. et al.: Fractally coated defibrillation electrodes: is an improvement in defibrillation threshold possible? Europace **2**, 154–159 (2000). doi:10.1053/eupc.1999.0084
113. Frohlig, G. et al.: A fractally coated, 1.3 mm2 high impedance pacing electrode. Pacing Clin Electrophysiol **21**, 1239–1246 (1998)
114. Lau, C. et al.: Intraoperative study of polarization and evoked response signals in different endocardial electrode designs. Pacing Clin Electrophysiol **24**, 1055–1060 (2001)
115. Mond, H. et al.: The porous titanium steroid eluting electrode: a double blind study assessing the stimulation threshold effects of steroid. Pacing Clin Electrophysiol **11**, 214–219 (1988)
116. Norlin, A., Pan, J. and Leygraf, C.: Investigation of Electrochemical Behavior of Stimulation/Sensing Materials for Pacemaker Electrode Applications: I. Pt, Ti, and TiN Coated Electrodes. J. Electrochem. Soc. **152**, J7–J15 (2005). doi:10.1149/1.1842092
117. De Levie, R. The Influence of Surface Roughness of Solid Electrodes on Electrochemical Measurements. Electrochim. Acta **10**, 113–130 (1965)
118. Song, H., Jung, Y., Lee, K. and Dao, L.: Electrochemical impedance spectroscopy of porous electrodes: the effect of pore size distribution. Electrochim. Acta **44**, 3513–3519 (1999)
119. Santini, M., De Seta, F.: Do Steroid-Eluting Electrodes Really Have Better Performance Than Other State-of-the-Art Designs? The Italian Multicenter Study Group on Low Output Stimulation. Pacing. Clin. Electrophysiol. **16**, 722–728 (1993).
120. Johnson, M. D., Otto, K. J. and Kipke, D. R. Repeated Voltage Biasing Improves Unit Recordings by Reducing Resistive Tissue Impedances. IEEE Trans. Neural Syst. Rehabil. Eng. **13**, 160–165 (2005). doi:10.1109/TNSRE.2005.847373
121. Prasad, A. and Sanchez, J.C.: Quantifying long-term microelectrode array functionality using chronic in vivo impedance testing. J. Neural Eng. **9,** 026028 (2012). doi:10.1088/1741-2560/9/2/026028
122. Williams, J.C., Hippensteel, J.A., Dilgen, J., Shain, W. and Kipke, D.R.: Complex impedance spectroscopy for monitoring tissue responses to inserted neural implants. J. Neural Eng. **4**, 410–423 (2007)
123. Lempka, S.F., Miocinovic, S., Johnson, M.D., Vitek, J. and McIntyre, C.C.: In vivo impedance spectroscopy of deep brain stimulation electrodes. Journal of Neural Engineering **6**, 046001 (2009). doi:10.1088/1741-2560/6/4/046001
124. Gaylor, J.M. et al.: Cochlear implantation in adults: a systematic review and meta-analysis. JAMA Otolaryngol Head Neck Surg **139**, 265–272 (2013). doi:10.1001/jamaoto.2013.1744
125. Gustafsson, B. and Jankowska, E.: Direct and indirect activation of nerve cells by electrical pulses applied extracellularly. The Journal of Physiology **258**, 33–61 (1976)
126. Wei, X.F., and Grill, W.M.: Analysis of high-perimeter planar electrodes for efficient neural stimulation. Front. Neuroengineering **2**, 15 (2009). doi:10.3389/neuro.16.015.2009
127. Han, J. et al.: Elimination of nanovoids induced during electroforming of metallic nanostamps with high-aspect-ratio nanostructures by the pulse reverse current electroforming

process. Journal of Micromechanics and Microengineering **22**, 065004 (2012). doi:10.1088/0960-1317/22/6/065004
128. Babb, T.L., and Kupfer, W.: Phagocytic and metabolic reactions to chronically implanted metal brain electrodes. Exp. Neurol. **86**, 171–182 (1984). http://dx.doi.org/10.1016/0014-4886(84)90179-1
129. Miller, L., and Zhou, X.: Poly(N-methylpyrrolylium) poly(styrenesulfonate) - a conductive, electrically switchable cation exchanger that cathodically binds and anodically releases dopamine. Macromolecules **20**, 1594–1597 (1987)
130. Dymond, A.M., Kaechele, L.E., Jurist, J.M. and Crandall, P.H.: Brain tissue reaction to some chronically implanted metals. J. Neurosurg. **33**, 574–580 (1970). doi:10.3171/jns.1970.33.5.0574
131. Fischer, G., Sayre, G. and Bickford, R.: Histological Changes in the Cat's Brain after Introduction of Metallic and Plastic-Coated Wire. In: Sheer, D.E. (ed.) Electrical stimulation of the brain, pp. 55–59. University of Texas Press, Austin, TX (1961)
132. Sawyer, P., and Srinivasan, S.:Synthetics and Implants. In: Ray, C.D. (ed.), Medical engineering, pp. 1099–1100. Year Book Medical Publishers, Chicago, IL (1974)
133. Kozai, T.D.Y., et al.: Ultrasmall implantable composite microelectrodes with bioactive surfaces for chronic neural interfaces. Nat. Mater. **11**, 1065–1073 (2012). doi:10.1038/nmat3468
134. Kozai, T.D., et al.: Reduction of neurovascular damage resulting from microelectrode insertion into the cerebral cortex using in vivo two-photon mapping. J. Neural Eng. **7**, 046011 (2010) doi:10.1088/1741-2560/7/4/046011. S1741-2560(10)46092-9 [pii]
135. Beebe, X. and Rose, T.L.: Charge injection limits of activated iridium oxide electrodes with 0.2 ms pulses in bicarbonate buffered saline. IEEE Trans. Biomed. Eng. **35**, 494–495 (1988). doi:10.1109/10.2122
136. Blau, A., et al.: Characterization and optimization of microelectrode arrays for in vivo nerve signal recording and stimulation. Biosens. Bioelectron. **12**, 883–892 (1977). http://dx.doi.org/10.1016/S0956-5663(97)00017-1
137. Johnson, M.D., Langhals, N.B. and Kipke, D.R.: Neural interface dynamics following insertion of hydrous iridium oxide microelectrode arrays. Conf. Proc. IEEE Eng. Med. Biol. Soc. **1**, 3178–3181 (2006). doi:10.1109/IEMBS.2006.260521
138. Meyer, R.D., Cogan, S.F., Nguyen, T.H., and Rauh, R.D.: Electrodeposited iridium oxide for neural stimulation and recording electrodes. IEEE transactions on neural systems and rehabilitation engineering: a publication of the IEEE Engineering in Medicine and Biology Society **9**, 2–11 (2001). doi:10.1109/7333.918271
139. Mcintyre, J.D.E., Peck, W.F., and Nakahara, S.: Oxidation-State Changes and Structure of Electrochromic Iridium Oxide-Films. J. Electrochem. Soc. **127**, 1264–1268 (1980). doi:10.1149/1.2129868
140. Cogan, S.F. et al.: Sputtered Iridium Oxide Films for Neural Stimulation Electrodes. Journal of Biomedical Materials Research Part B-Applied Biomaterials **89B**, 353–361(2009). doi:10.1002/Jbm.B.31223
141. Klein, J.D., Clauson, S.L. and Cogan, S.F.: Reactive Iro2 Sputtering in Reducing Oxidizing Atmospheres. J. Mater. Res. **10**, 328–333 (1995). doi:10.1557/Jmr.1995.0328
142. Negi, S., Bhandari, R., Rieth, L., Van Wagenen, R. and Solzbacher, F.: Neural electrode degradation from continuous electrical stimulation: Comparison of sputtered and activated iridium oxide. J. Neurosci. Methods **186**, 8–17 (2010). doi:10.1016/j.jneumeth.2009.10.016
143. Klein, J.D., Clauson, S.L. and Cogan, S.F.: Morphology and charge capacity of sputtered iridium oxide films. Journal of Vacuum Science and Technology A: Vacuum, Surfaces, and Films **7**, 3043–3047 (1989)
144. Desai, S.A., Rolston, J.D., Guo, L. and Potter, S.M.: Improving impedance of implantable microwire multi-electrode arrays by ultrasonic electroplating of durable platinum black. Front Neuroeng **3**, 5 (2010). doi:10.3389/fneng.2010.00005
145. Cui, X. and Martin, D.C.: Fuzzy gold electrodes for lowering impedance and improving adhesion with electrodeposited conducting polymer films. Sensors and Actuators A: Physical **103**, 384–394 (2003). http://dx.doi.org/10.1016/S0924-4247(02)00427-2

146. Kim, J.H., Kang, G., Nam, Y. and Choi, Y.K.: Surface-modified microelectrode array with flake nanostructure for neural recording and stimulation. Nanotechnology **21**(8), 85303 (2010). doi:10.1088/0957-4484/21/8/085303
147. Park, S., Song, Y.J., Boo, H. and Chung, T.D.: Nanoporous Pt Microelectrode for Neural Stimulation and Recording: In Vitro Characterization. J. Phys. Chem. C **114**, 8721–8726 (2010). doi:10.1021/Jp911256h
148. Cogan, S.F., Troyk, P.R., Ehrlich, J., and Plante, T.D.: In vitro comparison of the charge-injection limits of activated iridium oxide (AIROF) and platinum-iridium microelectrodes. Biomedical Engineering, IEEE Transactions on **52**, 1612–1614 (2005). doi:10.1109/tbme.2005.851503
149. Johnson, M.D., Kao, O.E. and Kipke, D.R.: Spatiotemporal pH dynamics following insertion of neural microelectrode arrays. J Neurosci Methods **160**, 276–287 (2007). doi:10.1016/j.jneumeth.2006.09.023
150. Johnson, M.D., Franklin, R.K., Gibson, M.D., Brown, R.B. and Kipke, D.R.: Implantable microelectrode arrays for simultaneous electrophysiological and neurochemical recordings. J. Neurosci. Methods **174**, 62–70 (2008). doi:10.1016/j.jneumeth.2008.06.036, S0165-0270(08)00391-9 [pii]
151. McCreery, D.B., Agnew, W.F., Yuen, T.G. and Bullara, L.A.: Comparison of neural damage induced by electrical stimulation with faradaic and capacitor electrodes. Ann. Biomed. Eng. **16**, 463–481 (1988)
152. Rose, T.L., Kelliher, E.M., and Robblee, L.S.: Assessment of capacitor electrodes for intracortical neural stimulation. J. Neurosci. Methods **12**, 181–193 (1985)
153. Guyton, D.L., and Hambrecht, F.T.: Capacitor Electrode Stimulates Nerve or Muscle without Oxidation-Reduction Reactions. Science **181** (1973). 74–76, doi:10.1126/science.181.4094.74
154. Chouard, C.H., and Pialoux, P.: Biocompatibility of Cochlear Implants. Bull. Acad. Natl. Med. **179**(3), 549–555 (1995)
155. Weiland, J.D., Anderson, D.J. and Humayun, M.S.: In vitro electrical properties for iridium oxide versus titanium nitride stimulating electrodes. Biomedical Engineering, IEEE Transactions **49**, 1574–1579 (2002). doi:10.1109/tbme.2002.805487
156. Schmidt, E.M., Hambrecht, F.T. and McIntosh, J.S.: Intracortical capacitor electrodes: preliminary evaluation. J. Neurosci. Methods **5**, 33–39 (1982). doi:10.1016/0165-0270(82)90048-6
157. Fairbrother, F.: The Chemistry of Niobium and Tantalum, pp. 1–28. Elsevier Publishing Company, Amsterdam (1967)
158. Nashed, R., Hassan, W.M., Ismail, Y. and Allam, N.K.: Unravelling the interplay of crystal structure and electronic band structure of tantalum oxide (Ta2O5). Phys. Chem. Chem. Phys. **15**, 1352–1357 (2013). doi:10.1039/c2cp43492j
159. Posey, F.A. and Morozumi, T.: Theory of Potentiostatic and Galvanostatic Charging of the Double Layer in Porous Electrodes. J. Electrochem. Soc. **113**, 176–184 (1966) doi:10.1149/1.2423897
160. Goldberg, I.B. and Bard, A.J.: Resistive effects in thin electrochemical cells: Digital simulations of current and potential steps in thin layer electrochemical cells. Journal of Electroanalytical Chemistry and Interfacial Electrochemistry **38**, 313–322 (1972). http://dx.doi.org/10.1016/S0022-0728(72)80341-3
161. Schoen, I., and Fromherz, P.: Extracellular stimulation of mammalian neurons through repetitive activation of Na+ channels by weak capacitive currents on a silicon chip. J. Neurophysiol. **100**, 346–357 (2008). doi:10.1152/jn.90287.2008, 90287.2008 [pii]
162. Long, T.C., Saleh, N., Tilton, R.D., Lowry, G.V. and Veronesi, B. Titanium dioxide (P25) produces reactive oxygen species in immortalized brain microglia (BV2): implications for nanoparticle neurotoxicity. Environ. Sci. Technol. **40**, 4346–4352 (2006)
163. Fromherz, P. and Stett, A. Silicon-Neuron Junction: Capacitive Stimulation of an Individual Neuron on a Silicon Chip. Phys. Rev. Lett. **75**, 1670–1673 (1995)
164. Fromherz, P., Offenhausser, A., Vetter, T. and Weis, J.: A neuron-silicon junction: a Retzius cell of the leech on an insulated-gate field-effect transistor. Science **252**, 1290–1293 (1991)

165. Fromherz, P. Electrical interfacing of nerve cells and semiconductor chips. Chemphyschem: a European journal of chemical physics and physical chemistry **3**, 276–284 (2002). doi:10.1002/1439-7641(20020315)3:3<276::AID-CPHC276>3.0.CO;2-A
166. Hammerle, H., et al.: Biostability of micro-photodiode arrays for subretinal implantation. Biomaterials **23**, 797–804 (2002)
167. Janders, M., Egert, U., Stelzle, M. and Nisch, W.: Novel thin film titanium nitride micro-electrodes with excellent charge transfer capability for cell stimulation and sensing applications. In Engineering in Medicine and Biology Society, 1996. Bridging Disciplines for Biomedicine. Proceedings of the 18th Annual International Conference of the IEEE **241**, 245–247 (1996)
168. Kress, H., et al.: Cell stimulation with optically manipulated microsources. Nat. Methods **6**, 905–909 (2009). doi:10.1038/nmeth.1400
169. Hai, A., Shappir, J. and Spira, M.E.: Long-term, multisite, parallel, in-cell recording and stimulation by an array of extracellular microelectrodes. J. Neurophysiol. **104**, 559–568 (2010). doi:10.1152/jn.00265.2010
170. Hai, A. et al.: Changing gears from chemical adhesion of cells to flat substrata toward engulfment of micro-protrusions by active mechanisms. J. Neural Eng. **6**, 066009 (2009). doi:10.1088/1741-2560/6/6/066009
171. Braeken, D., et al.: Local electrical stimulation of single adherent cells using three-dimensional electrode arrays with small interelectrode distances. Proceedings of the Annual International Conference of the IEEE Engineering in Medicine and Biology Society. IEEE Engineering in Medicine and Biology Society. Conference **2009**, 2756–2759 (2009). doi:10.1109/IEMBS.2009.5333871
172. Hai, A. and Spira, M.E.: On-chip electroporation, membrane repair dynamics and transient in-cell recordings by arrays of gold mushroom-shaped microelectrodes. Lab. Chip **12**, 2865–2873 (2012). doi:10.1039/c2lc40091j
173. Xie, C., Lin, Z., Hanson, L., Cui, Y. and Cui, B.: Intracellular recording of action potentials by nanopillar electroporation. Nat. Nanotechnol. **7**, 185–190 (2012). doi:10.1038/nnano.2012.8
174. Hai, A., Shappir, J. and Spira, M.E.: In-cell recordings by extracellular microelectrodes. Nature methods **7** (2010). 200–202, doi:10.1038/nmeth.1420
175. Huys, R., et al.: Single-cell recording and stimulation with a 16k micro-nail electrode array integrated on a 0.18 mum CMOS chip. Lab. Chip **12**, 1274–1280 (2012). doi:10.1039/c2lc21037a
176. Robinson, J.T., et al.: Vertical nanowire electrode arrays as a scalable platform for intracellular interfacing to neuronal circuits. Nat. Nanotechnol. **7**, 180–184 (2012). doi:10.1038/nnano.2011.249, nnano.2011.249 [pii]
177. Ciofani, G., et al.: Enhancement of neurite outgrowth in neuronal-like cells following boron nitride nanotube-mediated stimulation. ACS Nano **4**, 6267–6277 (2010). doi:10.1021/nn101985a
178. Bredas, J. and Street, G.: Polarons, Bipolarons, and Solitons in Conducting Polymers. Acc. Chem. Res. **18**, 309–315 (1985)
179. Svirskis, D., Travas-Sejdic, J., Rodgers, A. and Garg, S. Electrochemically controlled drug delivery based on intrinsically conducting polymers. J. Control. Release **146**, 6–15 (2010). doi:10.1016/j.jconrel.2010.03.023
180. Green, R. A., Lovell, N. H., Wallace, G. G. and Poole-Warren, L. A. Conducting polymers for neural interfaces: Challenges in developing an effective long-term implant. Biomaterials **29**, 3393–3399 (2008). doi:10.1016/j.biomaterials.2008.04.047
181. Cui, X., Hetke, J.F., Wiler, J.A., Anderson, D.J., and Martin, D.C.: Electrochemical deposition and characterization of conducting polymer polypyrrole/PSS on multichannel neural probes. Sensors and Actuators A: Physical **93**, 8–18 (2001)
182. Vernitskaya, T. and Efimov, O.: Polypyrrole: a conducting polymer; its synthesis, properties and applications. Russ. Chem. Rev. **66**, 443–457 (1997)
183. Cui, X., and Martin, D.C.: Electrochemical deposition and characterization of poly (3, 4-ethylenedioxythiophene) on neural microelectrode arrays. Sensors and Actuators B: Chemical **89**, 92–102 (2003)

184. Cui, X.T., and Zhou, D.D.: Poly (3,4-Ethylenedioxythiophene) for Chronic Neural Stimulation. IEEE Trans. Neural Syst. Rehabil. Eng. **15**, 502–508 (2007). doi:10.1109/TNSRE.2007.909811
185. Wilks, S., Richardson-Burns, S. M., Hendricks, J. L., Martin, D. C. and Otto, K. J.: Poly(3,4-ethylene dioxythiophene) (PEDOT) as a micro-neural interface material for electrostimulation. Front. Neuroengineering **2**, 7 (2009). doi:10.3389/neuro.16.007.2009
186. Xiao, Y. et al.: Electrochemical polymerization of poly (hydroxymethylated-3, 4-ethylenedioxythiophene) (PEDOT-MeOH) on multichannel neural probes. Sensors and Actuators B: Chemical **99**, 437–443 (2004)
187. Xiao, Y., Cui, X. and Martin, D.C.: Electrochemical polymerization and properties of PEDOT/S-EDOT on neural microelectrode arrays. J. Electroanal. Chem. **573**, 43–48 (2004)
188. Kotwal, A., and Schmidt, C.E.: Electrical stimulation alters protein adsorption and nerve cell interactions with electrically conducting biomaterials. Biomaterials **22**, 1055–1064 (2001)
189. Schmidt, C.E., Shastri, V.R., Vacanti, J.P. and Langer, R.: Stimulation of neurite outgrowth using an electrically conducting polymer. Proc. Natl. Acad. Sci. U. S. A. **94**, 8948–8953 (1997)
190. Thaning, E.M., Asplund, M.L.M., Nyberg, T.A., Inganäs, O.W., and von Holst, H.: Stability of poly(3,4-ethylene dioxythiophene) materials intended for implants. Journal of Biomedical Materials Research Part B: Applied Biomaterials **93B**, 407–415 (2010). doi:10.1002/jbm.b.31597
191. Soforo, E., et al.: Induction of systemic lupus erythematosus with tumor necrosis factor blockers. J Rheumatol **37**, 204–205 (2010). doi:10.3899/jrheum.081312, 37/1/204 [pii]
192. Yamato, H., Ohwa, M. and Wernet, W.: Stability of polypyrrole and poly(3,4-ethylenedioxythiophene) for biosensor application. J. Electroanal. Chem. **397**, 163–170 (1995). doi:10.1016/0022-0728(95)04156-8
193. Green, R., Baek, S., Poole-Warren, L. and Martens, P.: Conducting polymer-hydrogels for medical electrode applications. Sci Technol Adv Mater. **11**, 014107 (2010)
194. Hardy, J.G., Lee, J.Y., and Schmidt, C.E.: Biomimetic conducting polymer-based tissue scaffolds. Curr. Opin. Biotechnol. **24**,847–854 (2013). doi:10.1016/j.copbio.2013.03.011
195. Li, C., Sun, C., Chen, W., and Pan, L.: Electrochemical thin film deposition of polypyrrole on different substrates. Surf. Coat. Technol. **198**, 474–477 (2005)
196. Fonner, J. M. et al. Biocompatibility implications of polypyrrole synthesis techniques. Biomed Mater **3**, 034124 (2008). doi:10.1088/1748-6041/3/3/034124
197. Baker, C., Qiu, Y., and Reynolds, J.: Electrochemically Induced Charge and Mass Transport In Polypyrrole/Poly(styrenesulfonate) Molecular Composites. J. Phys. Chem. **95**, 4446–4452 (1991)
198. Ludwig, K.A., et al.: Poly (3, 4-ethylenedioxythiophene) (PEDOT) polymer coatings facilitate smaller neural recording electrodes. J. Neural Eng. **8**, 014001 (2011)
199. Ludwig, K.A., Uram, J.D., Yang, J., Martin, D.C., and Kipke, D.R.: Chronic neural recordings using silicon microelectrode arrays electrochemically deposited with a poly(3,4-ethylenedioxythiophene) (PEDOT) film. J. Neural Eng. **3**, 59–70 (2006). doi:10.1088/1741-2560/3/1/007
200. Yang, J., and Martin, D.C.: Microporous conducting polymers on neural microelectrode arrays. Sensors and Actuators A: Physical **113**, 204–211 (2004). doi:10.1016/j.sna.2004.02.029
201. Yang, J. and Martin, D. C. Microporous conducting polymers on neural microelectrode arrays. Sensors and Actuators B: Chemical **101**, 133–142 (2004). doi:10.1016/j.snb.2004.02.056
202. Venkatraman, S., et al.: In Vitro and In Vivo Evaluation of PEDOT Microelectrodes for Neural Stimulation and Recording. IEEE Trans. Neural Syst. Rehabil. Eng. **19**, 307–316 (2011). doi:10.1109/TNSRE.2011.2109399
203. Silk, T., Hong, Q., Tamm, J., and Compton, R.: AFM studies of polypyrrole film surface morphology I. The influence of film thickness and dopant nature. Synth. Met. **93**, 59–64 (1998)
204. Kim, J., Jung, J., Lee, D. and Joo, J.: Enhancement of electrical conductivity of poly(3,4-ethylenedioxythiophene)/poly(4-styrenesulfonate) by a change of solvents. Synthetic Metals **126**, 311–316 (2002)

205. Nyberg, T., Shimada, A. and Torimitsu, K. Ion conducting polymer microelectrodes for interfacing with neural networks. J. Neurosci. Methods **160**, 16–25 (2007). doi:10.1016/j.jneumeth.2006.08.008
206. Cogan, S., et al.: Polyethylenedioxythiophene (PEDOT) coatings for neural stimulation and recording electrodes. Mater. Res. Soc. Meet., Abstr. QQ2. 7 (2007)
207. Cogan, S.F., Troyk, P.R., Ehrlich, J., Plante, T.D., and Detlefsen, D.E.: Potential-biased, asymmetric waveforms for charge-injection with activated iridium oxide (AIROF) neural stimulation electrodes. IEEE Transactions on Bio-medical Engineering **53**, 327–332 (2006). doi:10.1109/TBME.2005.862572
208. Kelliher, E.M., and Rose, T.L.: Evaluation of Charge Injection Properties of Thin Film Redox Materials for use as Neural Stimulation Electrodes. MRS Online Proceedings Library **110**, 23–27 (1987). doi:10.1557/PROC-110-23
209. Boretius, T., Schuettler, M. and Stieglitz, T.: On the Stability of Poly-Ethylenedioxythiopene as Coating Material for Active Neural Implants. Artif. Organs **35**, 245–248 (2011)
210. Xue, F., Su, Y., and Varahramyan, K.: Modified PEDOT-PSS Conducting Polymer as S/D Electrodes for Device Performance Enhancement of P3HT TFTs. IEEE Trans. Electron Devices **52**, 1982–1987 (2005)
211. Drillet, J., Dittmeyer, R., and Jüttner, K.: Activity and long-term stability of PEDOT as Pt catalyst support for the DMFC anode. J. Appl. Electrochem. **37**, 1219–1226 (2007)
212. Green, R.A., et al.: Substrate dependent stability of conducting polymer coatings on medical electrodes. Biomaterials **33**, 5875–5886 (2012). doi:10.1016/j.biomaterials.2012.05.017
213. Isaksson, J., et al.: Electronic control of Ca2+ signalling in neuronal cells using an organic electronic ion pump. Nat. Mater. **6**, 673–679 (2007). doi:10.1038/nmat1963
214. Cui, X., et al.: Surface modification of neural recording electrodes with conducting polymer/biomolecule blends. J. Biomed. Mater. Res. **56**, 261–272 (2001)
215. Cui, X., Wiler, J., Dzaman, M., Altschuler, R.A., and Martin, D.C.: In vivo studies of polypyrrole/peptide coated neural probes. Biomaterials **24**, 777–787 (2003)
216. Green, R.A., Lovell, N.H., and Poole-Warren, L.A.: Impact of co-incorporating laminin peptide dopants and neurotrophic growth factors on conducting polymer properties. Acta Biomater. **6**, 63–71 (2010). doi:10.1016/j.actbio.2009.06.030
217. Stauffer, W.R., and Cui, X.T.: Polypyrrole doped with 2 peptide sequences from laminin. Biomaterials **27**, 2405–2413 (2006)
218. Collier, J.H., Camp, J.P., Hudson, T.W., and Schmidt, C.E.: Synthesis and characterization of polypyrrole-hyaluronic acid composite biomaterials for tissue engineering applications. J. Biomed. Mater. Res. **50**, 574–584 (2000)
219. Asplund, M., von Holst, H., and Inganas, O.: Composite biomolecule/PEDOT materials for neural electrodes. Biointerphases **3**, 83–93 (2008). doi:10.1116/1.2998407
220. Ru, X., et al.: Synthesis of polypyrrole nanowire network with high adenosine triphosphate release efficiency. Electrochim. Acta **56**, 9887–9892 (2011)
221. Kim, D.H., Richardson-Burns, S.M., Hendricks, J.L., Sequera, C., and Martin, D.C.: Effect of Immobilized Nerve Growth Factor on Conductive Polymers: Electrical Properties and Cellular Response. Adv. Funct. Mater. **17**, 79–86 (2007). doi:10.1002/adfm.200500594
222. Wadhwa, R., Lagenaur, C.F., and Cui, X.T.: Electrochemically controlled release of dexamethasone from conducting polymer polypyrrole coated electrode. J. Control. Release **110**, 531–541 (2006)
223. Richardson, R.T., et al.: The effect of polypyrrole with incorporated neurotrophin-3 on the promotion of neurite outgrowth from auditory neurons. Biomaterials **28**, 513–523 (2007)
224. Richardson, R.T., et al. Polypyrrole-coated electrodes for the delivery of charge and neurotrophins to cochlear neurons. Biomaterials **30**, 2614–2624 (2009). doi:10.1016/j.biomaterials.2009.01.015
225. Thompson, B.C., et al.: Optimising the incorporation and release of a neurotrophic factor using conducting polypyrrole. J. Control. Release **116**, 285–294 (2006). doi:10.1016/j.jconrel.2006.09.004

226. Thompson, B.C., Moulton, S.E., Richardson, R.T., and Wallace, G.G.: Effect of the dopant anion in polypyrrole on nerve growth and release of a neurotrophic protein. Biomaterials **32**, 3822–3831 (2011). doi:10.1016/j.biomaterials.2011.01.053
227. Thompson, B.C., et al.: Conducting polymers, dual neurotrophins and pulsed electrical stimulation - dramatic effects on neurite outgrowth. J. Control. Release **141**, 161–167 (2010). doi:10.1016/j.jconrel.2009.09.016
228. Bidez, P.R., et al.: Polyaniline, an electroactive polymer, supports adhesion and proliferation of cardiac myoblasts. Journal of Biomaterials Science, Polymer Edition **17**, 199–212 (2006)
229. Di, L., et al.: Protein adsorption and peroxidation of rat retinas under stimulation of a neural probe coated with polyaniline. Acta Biomater. **7**, 3738–3745 (2011). doi:10.1016/j.actbio.2011.06.009
230. Lee, H., Dellatore, S.M., Miller, W.M., and Messersmith, P.B.: Mussel-Inspired Surface Chemistry for Multifunctional Coatings. Science **318**, 426–430 (2007). doi:10.1126/science.1147241
231. Kang, K., Choi, I. and Nam, Y.: A biofunctionalization scheme for neural interfaces using polydopamine polymer. Biomaterials **32**, 6374–6380 (2011)
232. Abidian, M.R., Corey, J.M., Kipke, D.R., and Martin, D.C.: Conducting-Polymer Nanotubes Improve Electrical Properties, Mechanical Adhesion, Neural Attachment, and Neurite Outgrowth of Neural Electrodes. Small **6**, 421–429 (2010)
233. Abidian, M.R., Ludwig, K.A., Marzullo, T.C., Martin, D.C., and Kipke, D.R.: Interfacing Conducting Polymer Nanotubes with the Central Nervous System: Chronic Neural Recording using Poly(3,4-ethylenedioxythiophene) Nanotubes. Adv. Mater. **21**, 3764–3770 (2009). doi:10.1002/adma.200900887
234. Abidian, M.R., and Martin, D.C.: Experimental and theoretical characterization of implantable neural microelectrodes modified with conducting polymer nanotubes. Biomaterials **29**, 1273–1283 (2008)
235. González, M.B., and Saidman, S.B.: Electrosynthesis of hollow polypyrrole microtubes with a rectangular cross-section. Electrochem. Commun. **13**, 513–516 (2011). http://dx.doi.org/10.1016/j.elecom.2011.02.037
236. Lee, J.Y., Lee, J.W., and Schmidt, C.E.: Neuroactive conducting scaffolds: nerve growth factor conjugation on active ester-functionalized polypyrrole. J. R. Soc. Interface **6**, 801–810 (2009). doi:10.1098/rsif.2008.0403
237. Xie, J., et al.: Conductive Core-Sheath Nanofibers and Their Potential Application in Neural Tissue Engineering. Adv. Funct. Mater. **19**, 2312–2318 (2009). doi:10.1002/adfm.200801904
238. Bolin, M.H., et al.: Nano-fiber scaffold electrodes based on PEDOT for cell stimulation. Sensors and Actuators B: Chemical **142**, 451–456 (2009). http://dx.doi.org/10.1016/j.snb.2009.04.062
239. Ghasemi-Mobarakeh, L., Prabhakaran, M. P., Morshed, M., Nasr-Esfahani, M. H. and Ramakrishna, S. Electrical stimulation of nerve cells using conductive nanofibrous scaffolds for nerve tissue engineering. Tissue Eng. Part A **15**, 3605–3619 (2009). doi:10.1089/ten.TEA.2008.0689
240. Huang, J., et al.: Electrical regulation of Schwann cells using conductive polypyrrole/chitosan polymers. J. Biomed. Mater. Res. A **93**, 164–174 (2010). doi:10.1002/jbm.a.32511
241. Jeong, S.I., et al.: Development of electroactive and elastic nanofibers that contain polyaniline and poly(L-lactide-co-epsilon-caprolactone) for the control of cell adhesion. Macromol. Biosci. **8**, 627–637 (2008). doi:10.1002/mabi.200800005
242. Li, M., Guo, Y., Wei, Y., MacDiarmid, A.G., and Lelkes, P.I.: Electrospinning polyaniline-contained gelatin nanofibers for tissue engineering applications. Biomaterials **27**, 2705–2715 (2006). doi:10.1016/j.biomaterials.2005.11.037
243. Rowlands, A.S., and Cooper-White, J.J.: Directing phenotype of vascular smooth muscle cells using electrically stimulated conducting polymer. Biomaterials **29**, 4510–4520 (2008). doi:10.1016/j.biomaterials.2008.07.052
244. Shi, G., Rouabhia, M., Wang, Z., Dao, L.H., and Zhang, Z.: A novel electrically conductive and biodegradable composite made of polypyrrole nanoparticles and polylactide. Biomaterials **25**, 2477–2488 (2004)

245. Shi, G., Zhang, Z., and Rouabhia, M.: The regulation of cell functions electrically using biodegradable polypyrrole-polylactide conductors. Biomaterials **29**, 3792–3798 (2008). doi:10.1016/j.biomaterials.2008.06.010
246. Zhang, Z., et al.: Electrically conductive biodegradable polymer composite for nerve regeneration: electricity-stimulated neurite outgrowth and axon regeneration. Artif. Organs **31**, 13–22 (2007). doi:10.1111/j.1525-1594.2007.00335.x
247. Ghasemi-Mobarakeh, L., et al.: Application of conductive polymers, scaffolds and electrical stimulation for nerve tissue engineering. J. Tissue Eng. Regen. Med. **5**, e17–35 (2011). doi:10.1002/term.383
248. Heo, C., et al.: The control of neural cell-to-cell interactions through non-contact electrical field stimulation using graphene electrodes. Biomaterials **32**, 19–27 (2011). doi:10.1016/j.biomaterials.2010.08.095
249. Treacy, M.M.J., Ebbesen, T.W., and Gibson, J.M.: Exceptionally high Young's modulus observed for individual carbon nanotubes. Nature **381**, 678–680 (1996). doi:10.1038/381678a0
250. Yu, M. F. et al. Strength and breaking mechanism of multiwalled carbon nanotubes under tensile load. Science **287**, 637–640 (2000).
251. Charlier, J.-C., Blase, X., and Roche, S.: Electronic and transport properties of nanotubes. Reviews of Modern Physics **79**, 677–732 (2007)
252. Gabay, T., et al.: Electro-chemical and biological properties of carbon nanotube based multi-electrode arrays. Nanotechnology **18**, 035201 (2007). doi:10.1088/0957-4484/18/3/035201, S0957-4484(07)32481-1 [pii]
253. Wang, K., Fishman, H.A., Dai, H., and Harris, J.S.: Neural stimulation with a carbon nanotube microelectrode array. Nano Lett. **6**, 2043–2048 (2006). doi:10.1021/nl061241t
254. Cannizzaro, C. et al.: Tissue Engineering. In: Hauser, H., and Fussenegger, M.M. (eds.) Methods Molecular Medicine, pp. 291–307. Humana Press, Totowa, NJ (2007)
255. Lee, C.Y., Choi, W., Han, J.H., and Strano, M.S.: Coherence resonance in a single-walled carbon nanotube ion channel. Science **329**, 1320–1324 (2010). doi:10.1126/science.1193383
256. Fang, W.-C., et al.: Carbon Nanotubes Grown Directly on Ti Electrodes and Enhancement of Their Electrochemical Properties by Nitric Acid Treatment. Electrochem. Solid-State Lett. **9**, A5–A8 (2006). doi:10.1149/1.2128123
257. Li, J., Cassell, A., Delzeit, L., Han, J., and Meyyappan, M.: Novel Three-Dimensional Electrodes: Electrochemical Properties of Carbon Nanotube Ensembles. The Journal of Physical Chemistry B **106**, 9299–9305 (2002). doi:10.1021/jp021201n
258. Chmiola, J., et al.: Anomalous Increase in Carbon Capacitance at Pore Sizes Less Than 1 Nanometer. Science **313**, 1760–1763 (2006). doi:10.1126/science.1132195
259. Barisci, J. N., Wallace, G. G., and Baughman, R.H.: Electrochemical studies of single-wall carbon nanotubes in aqueous solutions. J. Electroanal. Chem. **488**, 92–98 (2000). http://dx.doi.org/10.1016/S0022-0728(00)00179-0
260. Lu, X., and Chen, Z.: Curved pi-conjugation, aromaticity, and the related chemistry of small fullerenes (< C60) and single-walled carbon nanotubes. Chem. Rev. **105**, 3643–3696 (2005). doi:10.1021/cr030093d
261. Hong, S., and Myung, S.: Nanotube electronics: a flexible approach to mobility. Nat. Nanotechnol. **2**, 207–208 (2007). doi:10.1038/nnano.2007.89, nnano.2007.89 [pii]
262. Jiang, L.Q., and Gao, L.: Fabrication and characterization of carbon nanotube-titanium nitride composites with enhanced electrical and electrochemical properties. J. Am. Ceram. Soc. **89**, 156–161 (2006). doi:10.1111/j.1551-2916.2005.00687.x
263. Liopo, A.V., Stewart, M.P., Hudson, J., Tour, J.M., and Pappas, T.C.: Biocompatibility of native and functionalized single-walled carbon nanotubes for neuronal interface. J. Nanosci. Nanotechnol. **6**, 1365–1374 (2006). doi:10.1166/Jnn.2006.155
264. Jan, E., et al.: Layered carbon nanotube-polyelectrolyte electrodes outperform traditional neural interface materials. Nano Lett **9**, 4012–4018 (2009). doi:10.1021/nl902187z
265. Mazzatenta, A., et al.: Interfacing Neurons with Carbon Nanotubes: Electrical Signal Transfer and Synaptic Stimulation in Cultured Brain Circuits. J. Neurosci. **27**, 6931–6936 (2007). doi:10.1523/JNEUROSCI.1051-07.2007

266. de Asis, E.D., et al.: High efficient electrical stimulation of hippocampal slices with vertically aligned carbon nanofiber microbrush array. Biomed Microdevices **11**, 801–808 (2009). doi:10.1007/s10544-009-9295-7
267. Mattson, M.P., Haddon, R.C., and Rao, A.M.: Molecular functionalization of carbon nanotubes and use as substrates for neuronal growth. J. Mol. Neurosci. **14**, 175–182 (2000). doi:10.1385/JMN:14:3:175
268. Hu, H., et al.: Polyethyleneimine functionalized single-walled carbon nanotubes as a substrate for neuronal growth. J. Phys. Chem. B **109**, 4285–4289 (2005). doi:10.1021/jp0441137
269. Voge, C. M. and Stegemann, J. P. Carbon nanotubes in neural interfacing applications. J. Neural Eng. **8**, 011001 (2011). doi:10.1088/1741-2560/8/1/011001, S1741-2560(11)68242-0 [pii]
270. Guitchounts, G., Markowitz, J.E., Liberti, W.A., and Gardner, T.J.: A carbon-fiber electrode array for long-term neural recording. J. Neural Eng. **10**, 046016 (2013)
271. Decher, G., Hong, J.D., and Schmitt, J.: Buildup of Ultrathin Multilayer Films by a Self-Assembly Process. 3. Consecutively Alternating Adsorption of Anionic and Cationic Polyelectrolytes on Charged Surfaces. Thin Solid Films **210**, 831–835 (1992). doi:10.1016/0040-6090(92)90417-A
272. Iler, R.K.: Multilayers of Colloidal Particles. J. Colloid Interface Sci. **21**, 569–594 (1966). doi:10.1016/0095-8522(66)90018-3
273. Tang, Z., Wang, Y., Podsiadlo, P., and Kotov, N.A.: Biomedical applications of layer-by-layer assembly: from biomimetics to tissue engineering. Adv. Mater. **18**, 3203–24 (2007)
274. Cogan, S.F.: Neural Stimulation and Recording Electrodes. Annu. Rev. Biomed. Eng. **10**, 275–309 (2008)
275. Shim, B.S., et al.: Multiparameter Structural Optimization of Single-Walled Carbon Nanotube Composites: Toward Record Strength, Stiffness, and Toughness. ACS Nano **3**, 1711–1722 (2009)
276. Kozai, T.D., and Kipke, D.R.: Insertion shuttle with carboxyl terminated self-assembled monolayer coatings for implanting flexible polymer neural probes in the brain. J. Neurosci. Methods **184**, 199–205 (2009). doi:10.1016/j.jneumeth.2009.08.002, S0165-0270(09)00434-8 [pii]
277. Subbaroyan, J., Martin, D. C. and Kipke, D. R. A finite-element model of the mechanical effects of implantable microelectrodes in the cerebral cortex. J. Neural Eng. **2**, 103–113 (2005). doi:10.1088/1741-2560/2/4/006, S1741-2560(05)99280-X [pii]
278. Bai, Y.X., Ho, S.S., and Kotov, N.A.: Direct-write maskless lithography of LBL nanocomposite films and its prospects for MEMS technologies. Nanoscale **4**, 4393–4398 (2012). doi:10.1039/C2nr30197k
279. Jan, E., and Kotov, N.A.: Successful Differentiation of Mouse Neural Stem Cells on Layer-by-Layer Assembled Single-Walled Carbon Nanotube Composite. Nano Lett. **7**, 1123–1128 (2007)
280. Gheith, M.K., et al.: Stimulation of Neural Cells by Lateral Currents in Conductive Layer-by-Layer Films of Single-Walled Carbon Nanotubes. Adv. Mater. **18**, 2975–2979 (2006). doi:10.1002/adma.200600878
281. Jan, E., et al.: Layered carbon nanotube-polyelectrolyte electrodes outperform traditional neural interface materials. Nano Lett **9**, 4012–4018 (2009). doi:10.1021/nl902187z
282. Zhang, H., et al.: Tissue-Compliant Neural Implants from Microfabricated Carbon Nanotube Multilayer Composite. ACS Nano **7**, 7619–7629 (2013). doi:10.1021/nn402074y
283. Zhang, H., Shih, J., Zhu, J. and Kotov, N.A.: Layered Nanocomposites from Gold Nanoparticles for Neural Prosthetic Devices. Nano Lett. **12**, 3391–3398 (2012). doi:10.1021/Nl3015632
284. Kam, N.W.S., Jan, E., and Kotov, N.A.: Electrical Stimulation of Neural Stem Cells Mediated by Humanized Carbon Nanotube Composite Made with Extracellular Matrix Protein. Nano Lett. **9**, 273–278 (2009)
285. Jan, E., Pereira, F.N., Turner, D.L., and Kotov, N.A.: In situ gene transfection and neuronal programming on electroconductive nanocomposite to reduce inflammatory response. J. Mater. Chem. **21**, 1109–1114 (2011). doi:10.1039/C0jm01895c

286. Chen, G. Z., et al.: Carbon Nanotube and Polypyrrole Composites: Coating and Doping. Adv. Mater. **12**, 522–526 (2000).
287. Lee, Y., Lee, K., Kim, D., Lee, D., and Kim, J.: Polypyrrole-carbon nanotube composite films synthesized through gas-phase polymerization. Synth. Met. **160**, 814–818 (2010)
288. Lu, Y. et al. Electrodeposited polypyrrole/carbon nanotubes composite films electrodes for neural interfaces. Biomaterials **31**, 5169–5181 (2010). doi:10.1016/j.biomaterials.2010.03.022
289. Bhandari, S., Deepa, M., Srivastava, A.K., Joshi, A.G., and Kant, R.: Poly(3,4-ethylenedioxythiophene)-multiwalled carbon nanotube composite films: structure-directed amplified electrochromic response and improved redox activity. J. Phys. Chem. B **113**, 9416–9428 (2009). doi:10.1021/jp9012976
290. Gerwig, R., et al.: PEDOT-CNT Composite Microelectrodes for Recording and Electrostimulation Applications: Fabrication, Morphology, and Electrical Properties. Front. Neuroengineering **5,** 8 (2012). doi:10.3389/fneng.2012.00008
291. Luo, X., Weaver, C.L., Zhou, D.D., Greenberg, R., and Cui, X.T.: Highly stable carbon nanotube doped poly(3,4-ethylenedioxythiophene) for chronic neural stimulation. Biomaterials **32**, 5551–5557 (2011). doi:10.1016/j.biomaterials.2011.04.051
292. Zhou, H., Cheng, X., Rao, L., Li, T., and Duan, Y.Y.: Poly(3,4-ethylenedioxythiophene)/multiwall carbon nanotube composite coatings for improving the stability of microelectrodes in neural prostheses applications. Acta Biomater. **9**, 6439–6449 (2013). doi:10.1016/j.actbio.2013.01.042
293. Zou, J., Tran, B., Huo, Q., and Zhai, L.: Transparent carbon nanotube/poly (3,4-ethylenedioxythiophene) composite electrical conductors. Soft Materials **7**, 355–365 (2009)
294. Luo, X., Matranga, C., Tan, S., Alba, N., and Cui, X.T.: Carbon nanotube nanoreservior for controlled release of anti-inflammatory dexamethasone. Biomaterials **32**, 6316–6323 (2011). doi:10.1016/j.biomaterials.2011.05.020
295. Starovoytov, A., Choi, J., and Seung, H.S.: Light-directed electrical stimulation of neurons cultured on silicon wafers. J. Neurophysiol. **93**, 1090–1098 (2005). doi:10.1152/jn.00836.200400836.2004
296. Kimura, M., et al.: Evaluation of thin-film photodevices and application to an artificial retina. Journal of the Society for Information Display **16**, 661–667 (2008). doi:10.1889/1.2938867
297. Bucher, V., Brunner, B., Leibrock, C., Schubert, M., and Nisch, W.: Electrical properties of a light-addressable microelectrode chip with high electrode density for extracellular stimulation and recording of excitable cells. Biosens. Bioelectron. **16**, 205–210 (2001). S095656630100135X [pii]
298. Zochowski, M., et al.: Imaging membrane potential with voltage-sensitive dyes. The Biological bulletin **198**, 1–21 (2000)
299. Gross, G.W., Rhoades, B.K., Reust, D.L., and Schwalm, F.U.: Stimulation of monolayer networks in culture through thin-film indium-tin oxide recording electrodes. J. Neurosci. Methods **50**, 131–143 (1993)
300. Fraser, D.B., and Cook, H.D.: Highly Conductive, Transparent Films of Sputtered In2-Xsnxo3-Y. J. Electrochem. Soc. **119**, 1368–1374 (1972). doi:10.1149/1.2403999
301. Pappas, T.C., et al.: Nanoscale engineering of a cellular interface with semiconductor nanoparticle films for photoelectric stimulation of neurons. Nano Lett. **7**, 513–519 (2007). doi:10.1021/Nl062513v
302. Zhao, Y., Larimer, P., Pressler, R.T., Strowbridge, B.W., and Burda, C.: Wireless Activation of Neurons in Brain Slices Using Nanostructured Semiconductor Photoelectrodes. Angewandte Chemie International Edition **48**, 2407–2410 (2009). doi:10.1002/anie.200806093
303. Yue, Z., Moulton, S.E., Cook, M., O'Leary, S., and Wallace, G.G.: Controlled delivery for neuro-bionic devices. Adv Drug Deliv Rev **65**, 559–569 (2013). doi:10.1016/j.addr.2012.06.002
304. Otero, T.F., Martinez, J.G., and Arias-Pardilla, J.: Biomimetic electrochemistry from conducting polymers. A review Artificial muscles, smart membranes, smart drug delivery and computer/neuron interfaces. Electrochim. Acta. **84**, 112–128 (2012)

305. Abidian, M.R., Kim, D.H., and Martin, D.C.: Conducting-Polymer Nanotubes for Controlled Drug Release. Adv. Mater. **18**, 405–409 (2006). doi:10.1002/adma.200501726
306. Hepel, M., and Mahdavi, F.: Application of the Electrochemical Quartz Crystal Microbalance for Electrochemically Controlled Binding and Release of Chlorpromazine from Conductive Polymer Matrix. Microchem. J. **56**, 54–64 (1997)
307. Gade, V., et al.: Synthesis and Characterization of Ppy-PVS, Ppy-pTS, and Ppy-DBS Composite Films. International Journal of Polymeric Materials and Polymeric Biomaterials **56**, 107–114 (2007)
308. Svirskis, D. and Garg, S.: Polypyrrole Film as a Drug Delivery System for the Controlled Release of Risperidone. In: Hendy, S.C., and Brown I.W.M. (eds.) Proceedings of the International Conference on Advanced Materials and Nanotechnology, Dunedin, New Zeland, February 2009. AIP Conference Proceedings, vol. 1151, pp. 36–39. Amer. Inst. Physics, New York (2009)
309. Sharma, M., Waterhouse, G.I., Loader, S.W., Garg, S., and Svirskis, D.: High surface area polypyrrole scaffolds for tunable drug delivery. Int. J. Pharm. **443**, 163–168 (2013). doi:10.1016/j.ijpharm.2013.01.006
310. Svirskis, D., Sharma, M., Yu, Y., and Garg, S.: Electrically switchable polypyrrole film for the tunable release of progesterone. Ther Deliv **4**, 307–313 (2013). doi:10.4155/tde.12.166
311. Zhou, Q., Miller, L., and Valentine, J.: Electrochemically controlled binding and release of protonated dimethyldopamine and other cations from poly(N-methyl-pyrrole)/polyanion composite redox polymers. J. Electroanal. Chem. **261**, 147–164 (1989)
312. Pyo, M., and Reynolds, J.: Electrochemically Stimulated Adenosine 5'-Triphosphate (ATP) Release through Redox Switching of Conducting Polypyrrole Films and Bilayers. Chem. Mater. **8**, 128–133 (1996)
313. Stauffer, W.R., Lau, P.M., Bi, G.Q., and Cui, X.T.: Rapid modulation of local neural activity by controlled drug release from polymer-coated recording microelectrodes. J. Neural Eng. **8**, 044001 (2011). doi:10.1088/1741-2560/8/4/044001
314. Kontturi, K., Pentti, P., and Sundholm, G.: Polypyrrole as a model membrane for drug delivery. J Electroanal Chem **453**, 231–238 (1998). http://dx.doi.org/10.1016/S0022-0728(98)00246-0
315. Bidan, G., Lopez, C., Mendes-Viegas, F., Vieil, E., and Gadelle, A.: Incorporation of sulphonated cyclodextrins into polypyrrole: an approach for the electro-controlled delivering of neutral drugs. Biosens. Bioelectron. **10**, 219–229 (1995). doi:10.1016/0956-5663(95)96808-C
316. Sirivisoot, S., Pareta, R.A., and Webster, T.J.: Electrically-Controlled Penicillin/Streptomycin Release from Nanostructured Polypyrrole Coated on Titanium for Orthopedic Implants. Solid State Phenomena **151**, 197–202 (2009)
317. Zinger, B., and Miller, L. Timed release of chemicals from polypyrrole films. J. Am. Chem. Soc. **106**, 6861–6863 (1984)
318. Cho, Y., Shi, R., Ivanisevic, A., and Ben Borgens, R.:A mesoporous silica nanosphere-based drug delivery system using an electrically conducting polymer. Nanotechnology **20,** 275102 (2009). doi:10.1088/0957-4484/20/27/275102
319. Jiang, S., et al.: Enhanced drug loading capacity of polypyrrole nanowire network for controlled drug release. Synth. Met. **163**, 19–23 (2013)
320. Luo, X., and Cui, X.T.: Electrochemically controlled release based on nanoporous conducting polymers. Electrochem. Commun. **11**, 402–404, (2009). doi:10.1016/j.elecom.2008.11.052
321. Luo, X., and Cui, X.T.: Sponge-like nanostructured conducting polymers for electrically controlled drug release. Electrochem. Commun. **11**, 1956–1959 (2009). doi:10.1016/j.elecom.2009.08.027
322. Luo, X., Matranga, C., Tan, S., Alba, N., and Cui, X.T.: Carbon nanotube nanoreservior for controlled release of anti-inflammatory dexamethasone. Biomaterials **32**, 6316–6323 (2011). doi:10.1016/j.biomaterials.2011.05.020
323. Xiao, Y., Ye, X., He, L., and Che, J.: New carbon nanotube-conducting polymer composite electrodes for drug delivery applications. Polym. Int. **61**, 190–196 (2012)

324. Sirivisoot, S., Pareta, R., and Webster, T.J.: Electrically controlled drug release from nanostructured polypyrrole coated on titanium. Nanotechnology **22**, 085101 (2011). doi:10.1088/0957-4484/22/8/085101
325. Leprince, L., Dogimont, A., Magnin, D. and Demoustier-Champagne, S. Dexamethasone electrically controlled release from polypyrrole-coated nanostructured electrodes. J. Mater. Sci. Mater. Med. **21**, 925–930 (2010). doi:10.1007/s10856-010-4008-6
326. Simon, D.T., et al.: Organic electronics for precise delivery of neurotransmitters to modulate mammalian sensory function. Nat. Mater. **8**, 742–746 (2009). doi:10.1038/nmat2494
327. Kopell, B.H., Machado, A., and Butson, C. Stimulation technology in functional neurosurgery. In: Lozano, A.M., Gildenberg, P.L., Tasker, R.R. (eds.) Textbook of Stereotactic and Functional Neurosurgery, pp. 1401–1425. Springer, Berlin (2009)
328. Yoo, J.-M., et al.: Excimer laser deinsulation of Parylene-C on iridium for use in an activated iridium oxide film-coated Utah electrode array. J. Neurosci. Methods **215**, 78–87 (2013). http://dx.doi.org/10.1016/j.jneumeth.2013.02.010
329. Eick, S., et al.: Iridium oxide microelectrode arrays for in vitro stimulation of individual rat neurons from dissociated cultures. Front Neuroeng **2**, 16 (2009). doi:10.3389/neuro.16.016.2009
330. Guimard, N.K., Gomez, N., and Schmidt, C.E.: Conducting polymers in biomedical engineering. Prog. Polym. Sci. **32**, 876–921 (2007). doi:10.1016/j.progpolymsci.2007.05.012
331. Wang, L., Wang, W., Di, L., Lu, Y., and Wang, J.: Protein Adsorption under electrical stimulation of neural probe coated with polyaniline. Colloids and Surfaces B: Biointerfaces **80**, 72–78 (2010)

Chapter 5
Micromachining Techniques for Realization of Three-Dimensional Microelectrode Arrays

Swaminathan Rajaraman

5.1 Introduction

Cellular function and response has been a significant subject of human fascination since time immemorial and a major field of study that has improved understanding of the mechanics of the human body. Specifically the functioning of electrogenic or electrically active cells is of particular interest as these cells control several important physiological functions such as visualization, locomotion, and activities of key organs such as the brain, heart, eyes, ears and the spinal cord. Advances in both engineering (including microelectronics, signal processing, microelectronic and biomedical packaging techniques, and micromachining technologies) and biology (including electrophysiology, neuroscience, cardiology, etc.) have contributed toward a better understanding of this field by introducing instrumentation and devices capable of interfacing with cells and tissue. This chapter summarizes the technological achievements in the development of one such instrument which has been fundamental toward electrical interfacing with biological constructs—three-dimensional microelectrode arrays (3-D MEAs), also called 3-D multielectrode arrays or 3-D micromachined probes. These electrode arrays are utilized in stimulating and recording applications both in vitro (outside the body) and in vivo (within the body) from neural tissue, neural cultures, neuromuscular tissue, cardiac tissue, cardiac cultures, 3-D cocultures of electrically active cells, stem cell cultures, and cultured networks of various electrically active cells (e.g., retinal cells).

If we consider neuroscience as an example area of interest (since most of the tools described in this chapter have a neuroscience focus), most physiological functions controlled by the brain involve a coordinated activity of networks of cells

S. Rajaraman, Ph.D. (✉)
Axion BioSystems Inc., ATDC Bioscience Center at the Georgia Institute of Technology, 1819 Peachtree Road, Suite 350, Atlanta, GA 30309, USA
e-mail: srajaraman@axionbio.com; srajaraman@gatech.edu

M. De Vittorio et al. (eds.), *Nanotechnology and Neuroscience: Nano-electronic, Photonic and Mechanical Neuronal Interfacing*, DOI 10.1007/978-1-4899-8038-0_5,

making it imperative that network activity be studied in addition to single cell activity. Studying such single cell activity has dominated the field for a long time with tools such as wire electrodes, tetrodes, glass micropipette electrodes, and patch clamps [1–4]. In fact to this day, these tools serve the very important function of studying isolated ion channels in a single cell (with high fidelity). These ion channels are responsible for controlling the various signaling pathways in a cell.

Microelectrode arrays (MEAs) can be considered both complimentary and competitive to the tools such as wire electrodes and patch clamps. Network-level activity and *extracellular* measurements can be performed with MEAs, while patch clamps and micropipette electrodes measure *intracellular* electrical signals. Specifically, MEAs are tiny electrodes arranged in a geometrically repeatable fashion either in two dimensions (2-D) or three dimensions (3-D). In some cases additional functionality such as multiplexing circuitry or microfluidic ports are integrated with these electrodes. Research has demonstrated that network-level activity and how "cells talk to one another" are the key to understanding not just the healthy state of various organs (such as the brain and the heart) but also to understanding the various disease states [5–7]. MEAs are tools that enable such an understanding.

Since the 1960s, MEAs have become an invaluable tool for scientific discovery and medical advancement. Because they can actively manipulate and monitor cellular activity at both the single cell and tissue level, these tools provide extraordinary insight into complex neural interactions [8]. Today, first-generation MEAs are used in applications as far ranging as drug screening, biosensing, cardiac pacing, and epilepsy research [9, 10]. For example, MEAs were instrumental in the landmark discovery of spontaneous waves in the developing retina [11]. They have also been used to investigate the role of extracellular stimulation in the suppression of epileptic activity [5, 12] and in the study of novel plasticity mechanisms in cultured neural networks [13–15]. Recently, in a wide variety of tissue and culture preparations, MEAs have shown great promise for drug screening [16, 17]; safety pharmacology [18–20]; biosensing [21, 22]; detection of chemical, biological, and environmental toxins [9, 23]; detection of biohazards and bioterrorism agents [24]; neural prosthetics [25, 26]; and acting as bioelectrodes for measurements of various biopotentials such as electromyographic signals, electrocardiographic signals, and nerve conduction studies [27–29].

Figure 5.1 summarizes the application areas for 2-D and 3-D microelectrode arrays. It can be inferred from the figure, the application space for 3-D MEAs spans a wide variety of areas in life sciences making these tools indispensable for advances in several fields.

This chapter summarizes the advances in the microfabrication technologies used to develop 3-D MEAs. These devices have been fabricated out of traditional substrates such as silicon and glass as well as nontraditional substrates such as parylene, SU-8, various metals, polyimides, etc. A variety of both traditional and nontraditional approaches are enumerated. The focus of the chapter is in the micromachining technological advancements and not the applications that have been developed with MEAs. Readers are referred to the various papers cited in the text for details of the applications that have been developed using these tools.

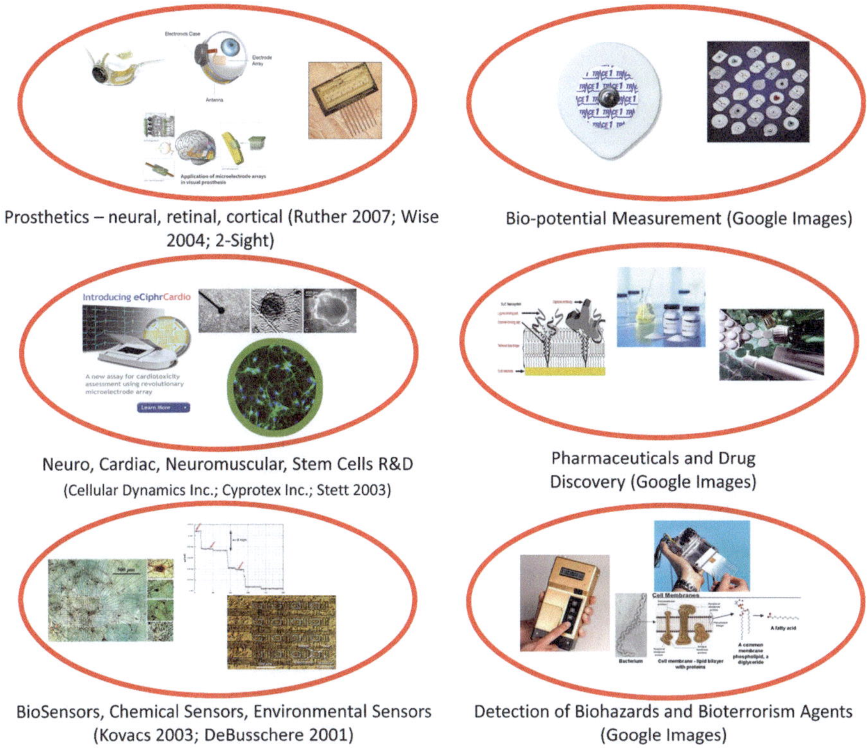

Fig. 5.1 Application space for 3-D MEAs (*top left*, [37], 2-sight) (*center left*, reproduced with permission from [19], © (2003) Springer) (*bottom left*, [23]; reproduced with permission from [24], © (2001) Elsevier)

5.2 Silicon Probes as 3-D MEAs

Prof. Kensall Wise's group at the University of Michigan (Ann Arbor, MI, USA) and Prof. Richard Normann's group at the University of Utah (Salt Lake City, UT, USA) have been at the forefront of 3-D MEAs for in vivo and in vitro applications utilizing silicon probes. Since the introduction of the first integrated circuits (ICs) in the early 1960s, silicon-based technologies have been applied toward biomedical applications. Some of this early work was performed in the development of reading aid for the blind [30], blood flow monitoring [31], and implantable pressure sensors [32].

Silicon-based technologies have been pursued by both the Michigan and Utah groups (and others) due to the several advantages that these technologies have to offer—(a) highly developed microfabrication processes that can produce probes with high yields; (b) batch fabrication on silicon wafers; (c) the ability to create a dense set of microelectrodes that permit acute and semi-chronic single unit and cellular network recordings and stimulation; (d) the ability to be complementary metal-oxide semiconductor (CMOS) compatible for on-chip electronics integration;

(e) the ability to integrate site selection, amplification, and multiplexing circuitry in close proximity to the probe arrays; and (f) the ability to wirelessly transfer information from hundreds of these channels [25].

5.2.1 The Michigan Probes

Prof. Kensall Wise started applying silicon-based technologies toward the creating of thin-film electrodes for single unit recordings in the nervous system when he was a graduate student at Stanford University under the direction of Prof. James Angell in 1966. This microprobe was designed specifically for extracellular biopotential measurement in the brain and consisted of a gold electrode formed on a silicon dioxide (SiO_2) surface, which was defined on a silicon carrier [33, 34]. The insulation in this process was also fabricated from a photolithographically defined layer of SiO_2. Silicon etching which was revolutionary at that time was being used by Bell Telephone Laboratories and was adapted in the original Michigan probe process to define silicon mesas on which the electrodes were defined. The technological achievements of such a process (given that it was performed in the 1960s) are very impressive. From the original silicon wafer (~50 μm), silicon etching was performed to a depth of 25–35 μm on which 2 μm recording sites were defined photolithographically. Figure 5.2 depicts a schematic of a neural probe and an optical image of the first silicon probe as a 3-D MEA and one of the first micromachined arrays that were created anywhere. These probes were utilized in recording extracellular action potentials for the auditory cortex of cats (Fig. 5.3).

This technology was the earliest prototype for a micromachined probe but was not something that could be repeated in a reproducible fashion. Over the next few decades, the development of etch stops, deep reactive-ion etching (DRIE), and anisotropic silicon etching amongst other techniques has created a wide variety of Michigan probes and dramatically advanced this technology. Additionally three-dimensional stacking or assembly technologies were developed to truly fabricate 3-D MEAs with integrated on-chip electronics and wireless data transmission.

Boron diffusion to define etch stops in silicon was one of the key technological breakthroughs that allowed for arbitrary definition of probe thicknesses to less than 15 μm [35, 36]. Conductors such as polysilicon, metal silicides, or metals ranging from gold to tantalum to platinum to iridium are used to define metal tracks and as recording sites in the various Michigan probe microfabrication processes. The choice of the recording site depends on the fabrication process and the charge delivery of the material used. Gold, platinum, and iridium (iridium oxide) are common choices for recording and stimulation sites on the Michigan probes [37].

A combination of silicon dioxide and silicon nitrides deposited utilizing a chemical vapor deposition (CVD) process defined the insulation in this process. The relative proportions of the nitride and oxide in the insulation layer are key to achieving a composite insulator whose thermal expansion coefficient approximately matches that of silicon. This is performed in order to minimize warpage of the structure. Figure 5.4 depicts a very basic fabrication process flow that details the key steps used in the fabrication of the Michigan probes.

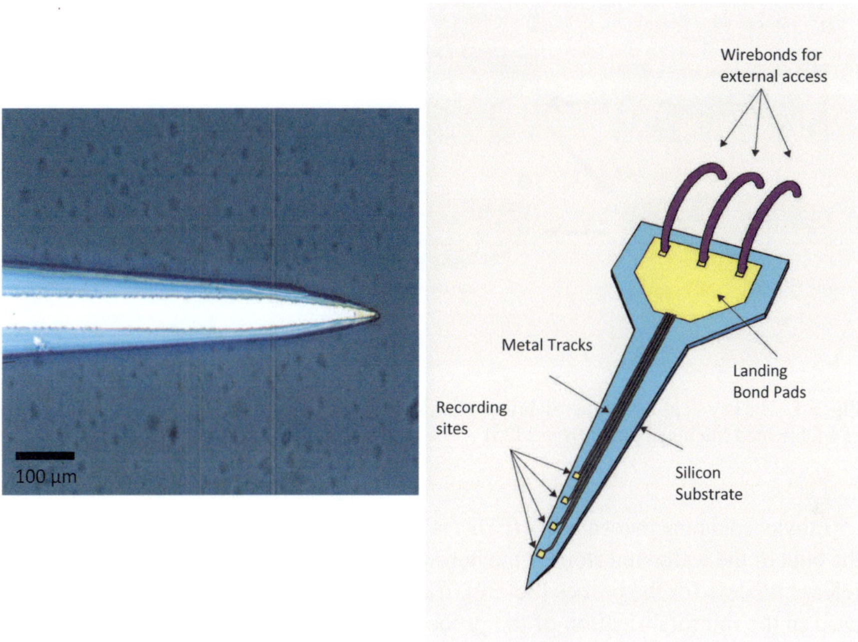

Fig. 5.2 Schematic of a neural probe and the earliest demonstration of a micromachined neural probe (*left*, [25]; *right*, [37])

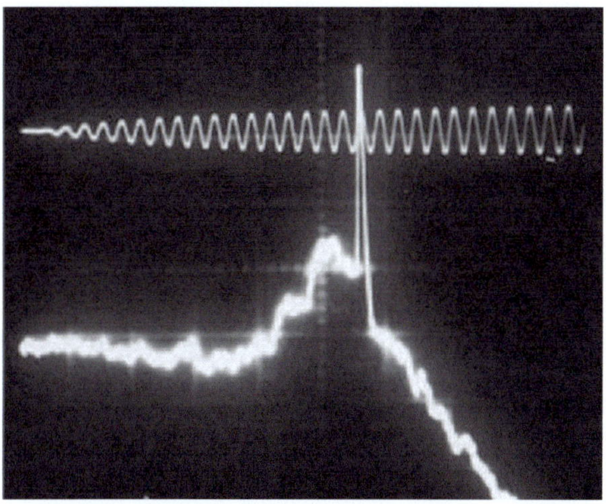

Fig. 5.3 One of the earliest demonstrations of the functioning of a neural probe. Signal recorded from the auditory cortex of a rat [25]

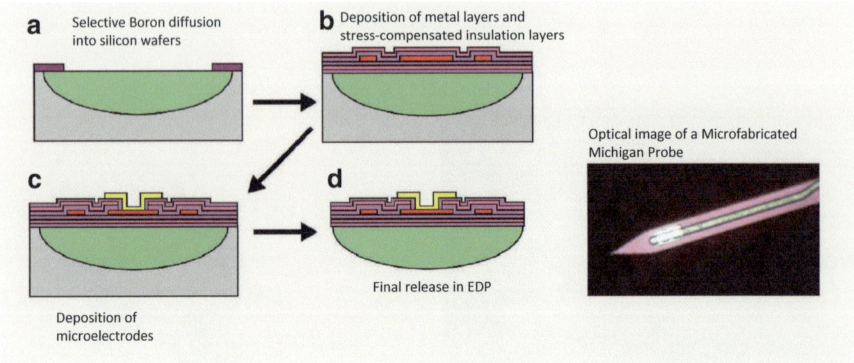

Fig. 5.4 The key steps in the microfabrication of Michigan probes (*left*) and an optical micrograph of a fabricated Michigan probe (*right*) [25]

Ethylenediamine pyrocatechol (EDP) is used as the selective etchant that dissolves the bulk of the wafer and stops at the boron-diffused silicon areas thus serving as the release process for the probes [36–39]. It does not attack any of the other materials used in the microfabrication of the probes. This fabrication technique is versatile to produce arbitrary 2-D probe shapes and site configurations with dimensions controlled to an accuracy of ±1 μm.

In order to fabricate flexible probes for applications such as a cochlear implant, the probe fabrication has been adopted with multiple boron diffusions in specific areas to realize probes that are ultrathin (nanometers) to less than 5 μm in thickness [39, 40].

Both integrated (a CMOS process integrated with the probe fabrication) and hybrid approaches have been taken by Michigan researchers to integrate on-chip circuitry with the probes themselves. A simplified process flow for this CMOS integrated fabrication process flow is depicted schematically in Fig. 5.5. This integrated approach overcomes the external leads issue, which is one of the most difficult problems with any 3-D MEA fabrication, but comes at the expense of a more complicated microfabrication process [37, 41]. Typically a standard CMOS process is first implemented on standard silicon wafers (or silicon-on-insulator (SOI) wafers) with recording/stimulation site metallization performed as one of the last steps in the CMOS fabrication. Finally the field dielectrics are trimmed away, and the silicon in the field area is recessed using a dry etch. The wafer is typically thinned to around 150 μm and then released in EDP. Optical micrographs of Michigan probe arrays with CMOS circuitry are shown in Fig. 5.6. Michigan probe arrays with CMOS circuits are depicted in Fig. 5.7.

In order to develop a truly three-dimensional electrode array with multiple 2-D probe shanks and thousands of recording sites, Michigan researchers have developed several techniques. Some commonalities between these techniques are the integration of a gold beam electroplating process at the end of probe fabrication [42]. These electroplated gold beams are soft enough to be bent with the aid of micromachined assembly devices (which are also constructed on silicon wafers).

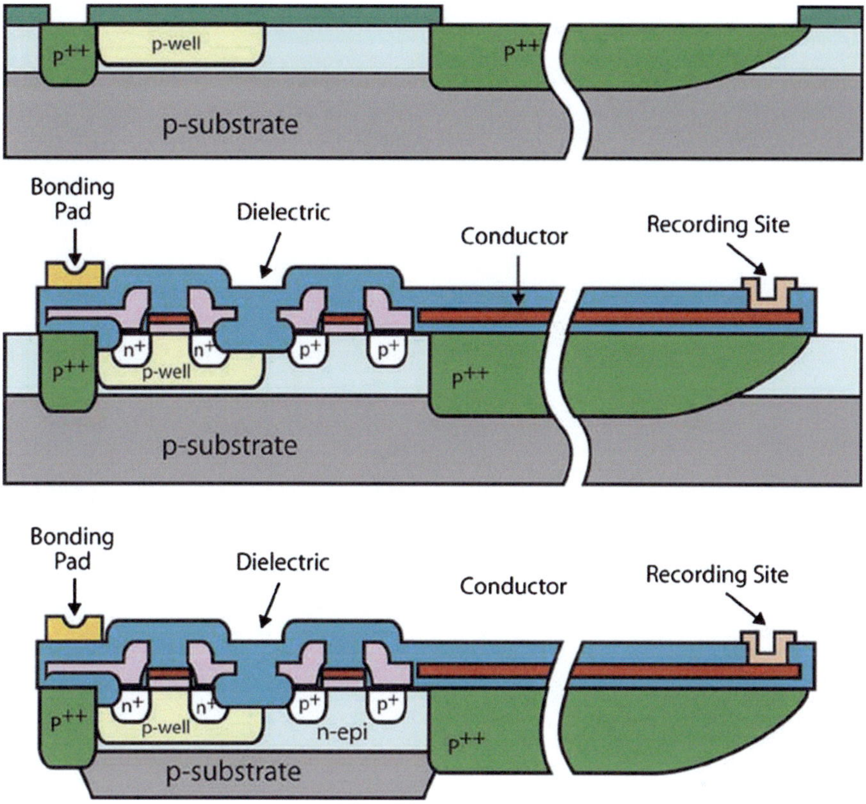

Fig. 5.5 The key steps in a CMOS integrated Michigan probes fabrication process [37]

Ultrasonic wire-free wedge bonding is performed to attach the gold pads of the probe arrays with landing pads of a platform created for assembling multiple probe arrays [43]. Simpler stacking methodologies and more involved dynamic motion-based techniques have been developed to automate this process. Examples of fully assembled 3-D Michigan probe arrays are shown in Fig. 5.8. More recently simple, rapid folding-based techniques have been demonstrated to fabricate 3-D MEAs from planar shanks [44].

The monitoring of electrical activity of the neural network is further enhanced by the ability to monitor and control the chemical "microenvironment" surrounding neural tissue. This first step in achieving such control is adding microfluidics to the Michigan probes. Papageorgiou et al. [45] demonstrate that by adding one mask to the standard Michigan probe fabrication process, microchannels can be created. The combination of a shallow (2 μm deep) unmasked boron etch stop followed by a DRIE step is utilized to create a grid of holes or slots through the etch stop region on top of the intended channel region. The microchannels are then formed utilizing either a dry or wet silicon etch to undercut this grid of slots. The channel access holes are subsequently sealed using a CVD layer of silicon dioxide and silicon nitride. Normal probe process to form interconnects and stimulation/recording sites

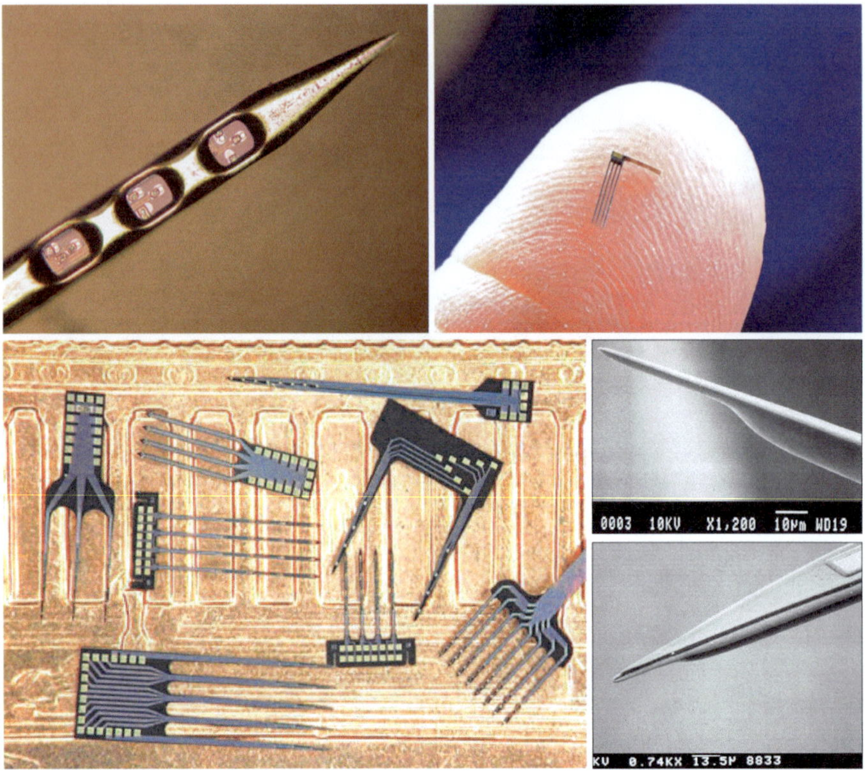

Fig. 5.6 Optical micrographs of fabricated Michigan probes without CMOS circuits integrated. SEM images of single unit probes. *Bottom right* and *top left*, [37], *bottom left*, [25], *top right*, [44]

occur afterward. Figure 5.9 depicts a schematic of Michigan probe arrays with integrated microfluidics and SEM images of probes with such channels.

Michigan probe arrays are well characterized for in vivo applications including retinal prosthesis, cochlear implants, and fundamental neural interfacing studies. The probe fabrication process has proved widely successful providing over 7,000 arrays (up until 2004) to researchers worldwide resulting in over 350 publications [25]. NeuroNexus has commercialized the Michigan probe array process and cites roughly 1,000 publications (from 2005 to 2013) where their probes have been utilized [46]. Figure 5.10 presents a few examples of the data (from the paper references in this chapter) collected from the Michigan probe arrays, which have been used exhaustively in cochlear implants and neural prosthetic applications.

5.2.2 The Utah Array

The first reported microfabrication processing details for the now well-established Utah Electrode Array (UEA) comes from Campbell et al. in 1991 [47]. It reports a

Fig. 5.7 Optical micrographs of CMOS integrated, fully assembled Michigan probes [37]

microfabrication process that has remained remarkably consistent over the years [26] that has the following major steps: (a) thermomigration of p+ silicon into n-type silicon wafers, (b) a combination of mechanical and chemical micromachining to achieve sharp microneedles or 3-D microelectrodes in the p+ silicon areas, (c) metal definition to create active recording sites and contact pads, and (d) encapsulation processes.

The thermomigration process is carried out by defining aluminum pads on the bottom side of a silicon wafer with a lithographic approach. A silicon–aluminum eutectic is driven through the thickness of the wafer with a created temperature gradient [47, 48]. This process creates p+ islands, which are isolated to the region of the aluminum definition and isolated from one another due to the n-type substrate. Aluminum pads are defined for final packaging of the arrays at the bottom of the wafer such that each pad overlaps one island. The nominal design for the Utah array is a 10×10 microneedle structure that is roughly 1–1.5 mm tall. In recent years, up to roughly 8 mm tall needles have been reported with similar processes [49].

In order to create the 3-D microneedles across the thickness of the silicon wafer, a dicing process is used to create hatched grooves in the silicon wafer as

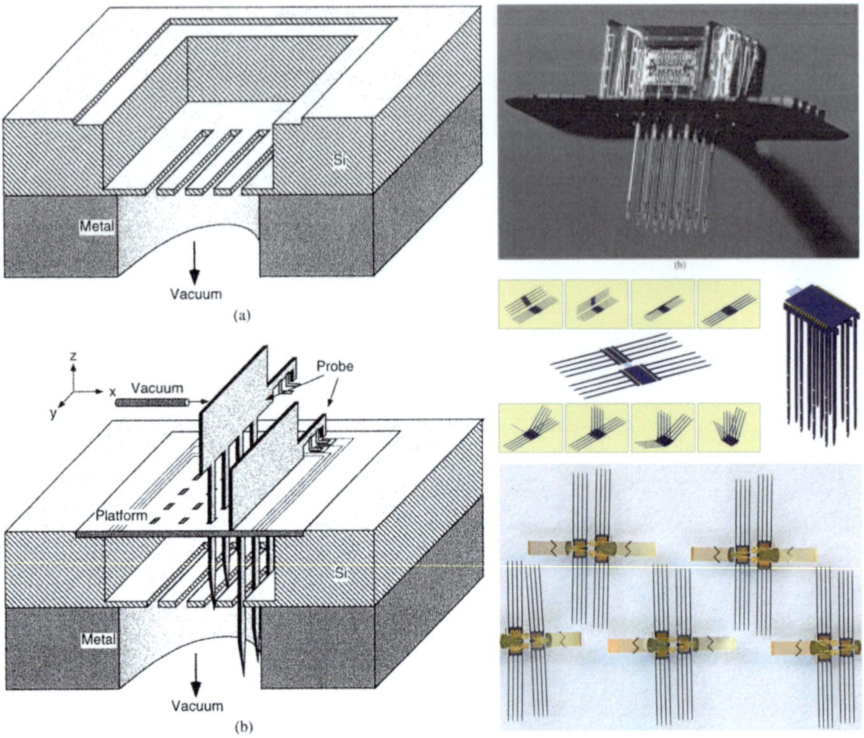

Fig. 5.8 Assembly processes involved in the fabrication of Michigan probes and the resulting assembled probes. *Left* and *top right* [42]; *center right* and *bottom right* [44]

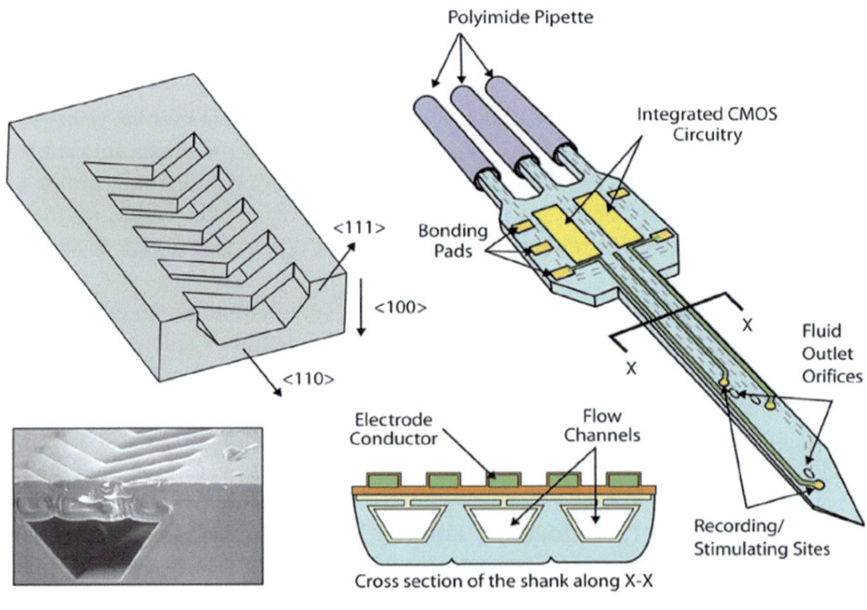

Fig. 5.9 Schematic and SEM images of Michigan probes with integrated microfluidic ports [45]

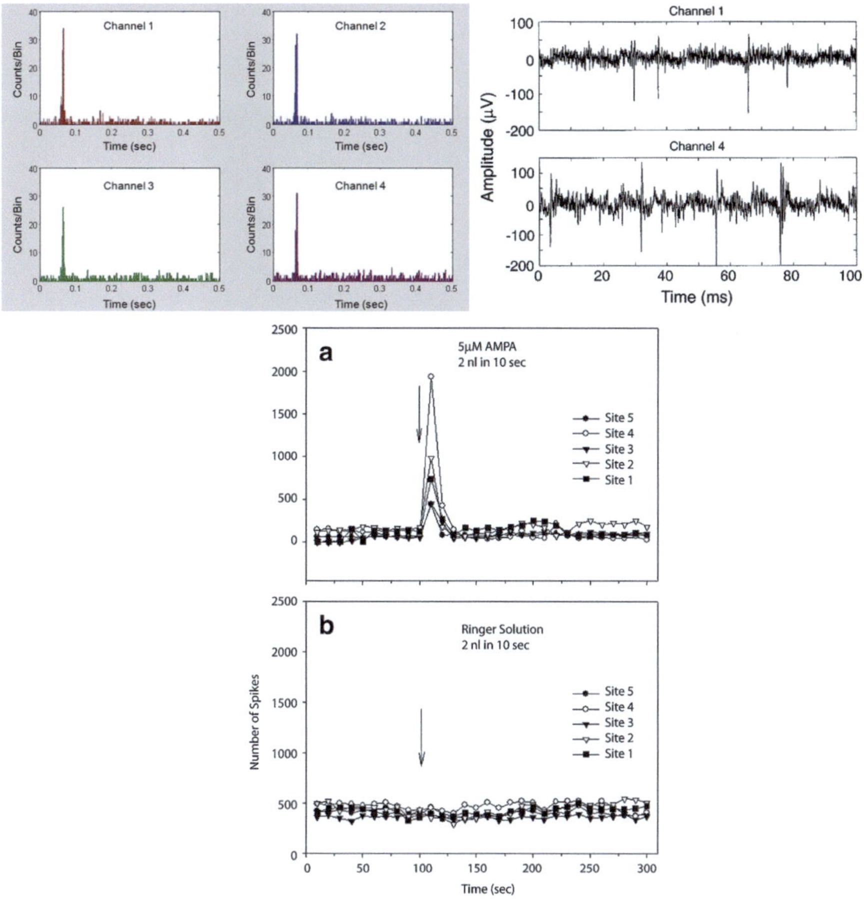

Fig. 5.10 Spike counts on four specific channels on a 3-D array implanted in the auditory cortex of a guinea pig (*top left*); neural discharges from five recording sites in the inferior colliculus of a guinea pig before and after the injection of a neurotransmitter agonist and control (*left*); single unit activity recorded from two different channels from a guinea pig auditory cortex (*top right*). *Top left* [44]; *left* [37]; *top right* [42]

shown in Fig. 5.11. Sometimes a layer of glass is deposited onto these grooves in silicon to effectively isolate the 3-D microelectrodes. Isotropic silicon etching is then carried out in a solution of 5 % hydrofluoric acid (HF) and 95 % nitric acid (HNO_3) [47]. Both static and dynamic modes of etching have been demonstrated to create various shapes and sizes of the silicon microelectrodes as shown in Fig. 5.12.

For the definition of recording sites on the individual 3-D microelectrodes, a couple of techniques are reported—rudimentary metal foil isolation followed by the deposition of gold and platinum in a sputter coater [47, 48] and photoresist masked metal sputtering (metals like gold, platinum, and iridium are reported as recording

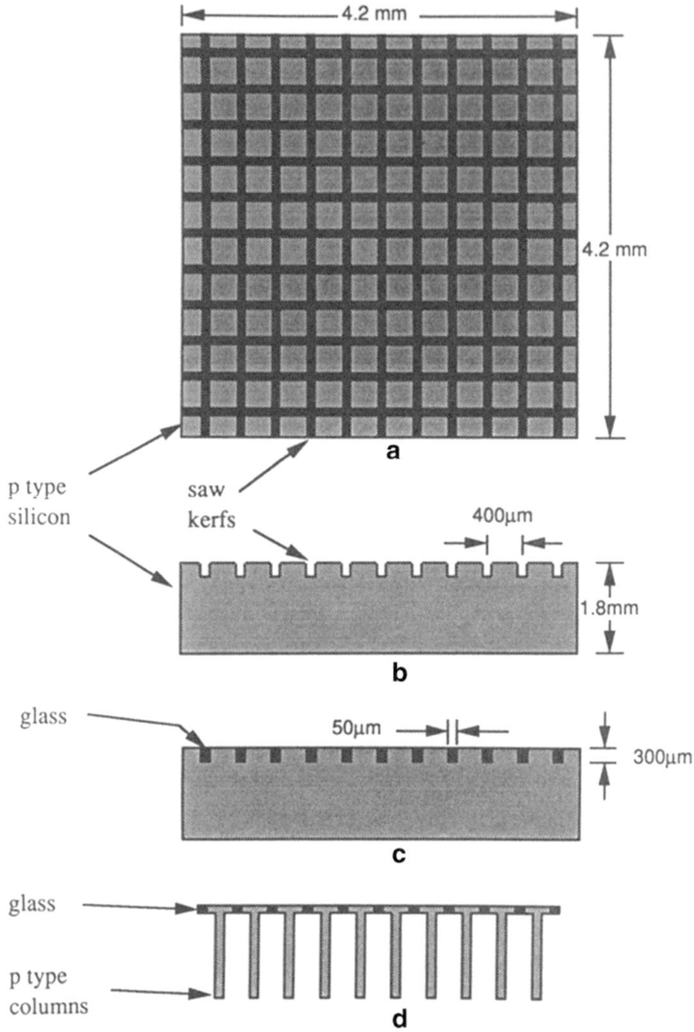

Fig. 5.11 Top and side views of a basic process flow for the Utah array microfabrication process. Reproduced with permission from [48], © (1992) Springer

sites for Utah arrays) followed by the deposition of parylene-C and etching of the parylene layer using a photoresist mask to de-insulate just the tips of the 3-D MEAs [51]. Figure 5.13 illustrates SEM images of the tips of Utah arrays after de-insulation with the techniques described above. Additionally Fig. 5.13 depicts SEM images of Utah arrays with slanted electrodes at various heights.

Encapsulation of the arrays is accomplished using silicon dioxide, silicon nitride, polyimide, and Parylene-C [49, 52]. For de-insulating the tips of the 3-D

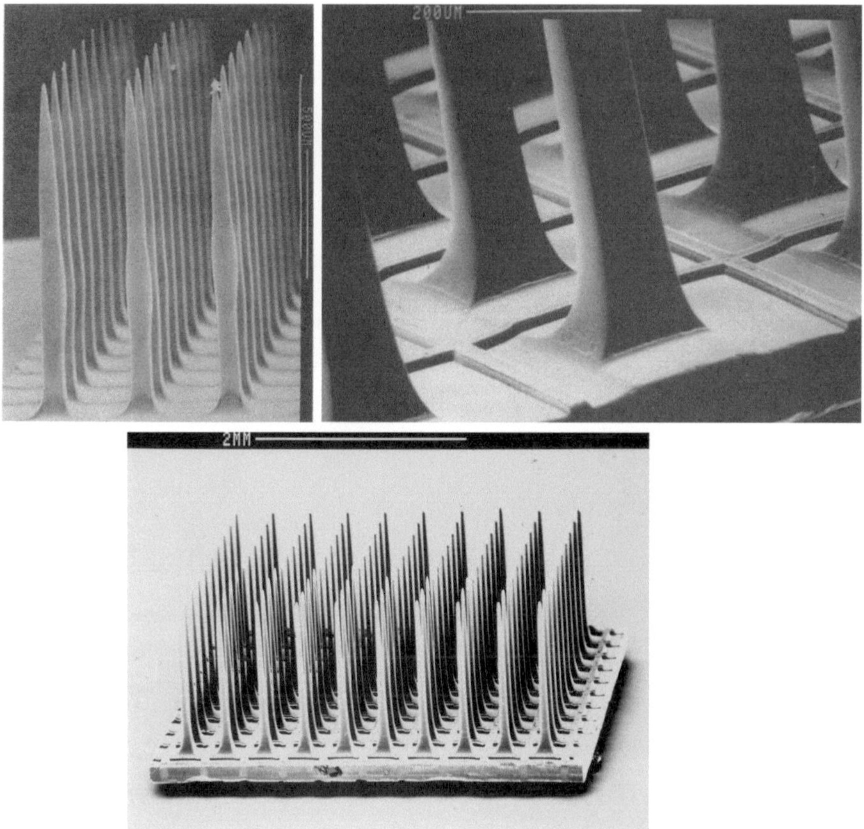

Fig. 5.12 Silicon microneedles of various shapes and sizes created using the Utah array microfabrication process. *Top (right* and *left)*, reproduced with permission from [48] © (1992); *bottom*, reproduced with permission from [50], © (1998) Elsevier

electrodes, rudimentary techniques such as aluminum foil attachment and etching of the polymer in a reactive-ion etch (RIE) to more sophisticated flooding of the array with photoresist with thickness controlled using spin speeds have been demonstrated.

These arrays have been successfully inserted in several studies for neural, retinal, and motion-based prosthetic applications. The Utah array was originally developed as means of restoring limited but useful sight to individuals with profound blindness but has since been applied to produce a sit-to-stand maneuver and other promising prosthetic applications. Figure 5.14 depicts some of these promising applications. Blackrock Microsystems [53] has been able to successfully commercialize this rather complicated process and also provides electronics to interface with these arrays.

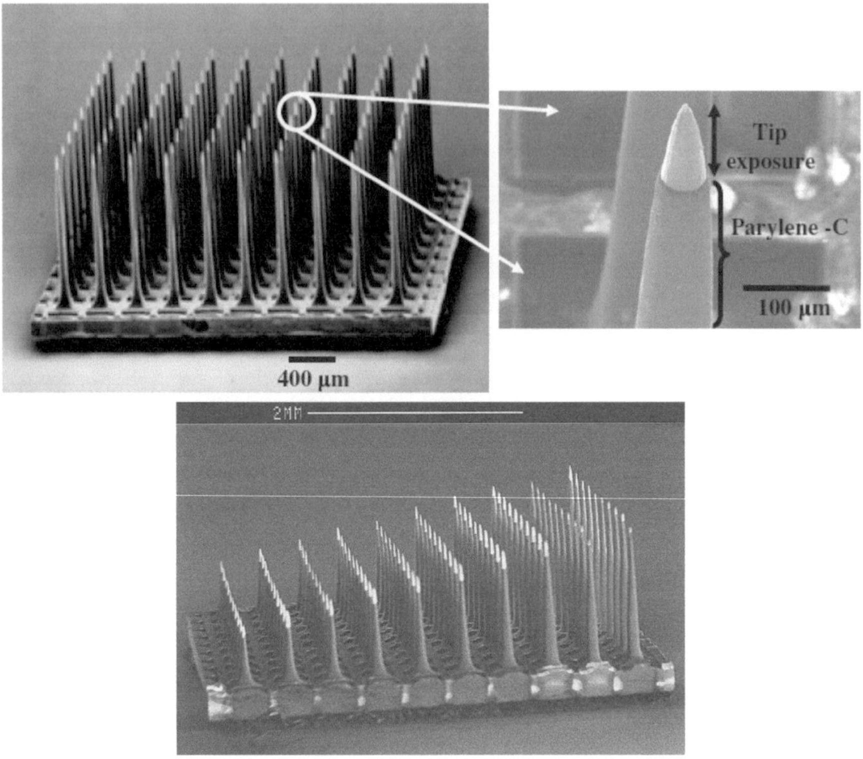

Fig. 5.13 SEM images of the completed Utah array 3-D MEAs. *Top* [51]; *bottom* [54]

5.2.3 EU NeuroProbes

European Union's consortium has been working toward a solution for silicon-based intracortical 3-D microelectrodes. This project is called the European NeuroProbes consortium (www.neuroprobes.org). It was a 4-year integrated project funded by the European Commission with technology development happening in the following institutes: Interuniversity Microelectronics Center (IMEC), Belgium; Department of Microsystems, IMTEK, University of Freiburg, Germany; Hahn-Schickard-Gesellschaft Institute of Micromachining and Information Technology (HSG-IMIT), Germany; Institute of Microtechnology (IMT), University of Neuchatel, Switzerland; and Malardalen University, Sweden [55].

The goal of the EU NeuroProbes effort was to overcome some of the limitations of the Utah arrays and Michigan probes by developing a 3-D implementation of the probe arrays conducive to chronic applications [55] by the development of a platform with emphasis on elegant assembly that would allow for electrode sites for recording and stimulation of cortical neurons, biosensors for monitoring glutamate and dopamine, and integrated microfluidic channels for sampling or drug delivery [56].

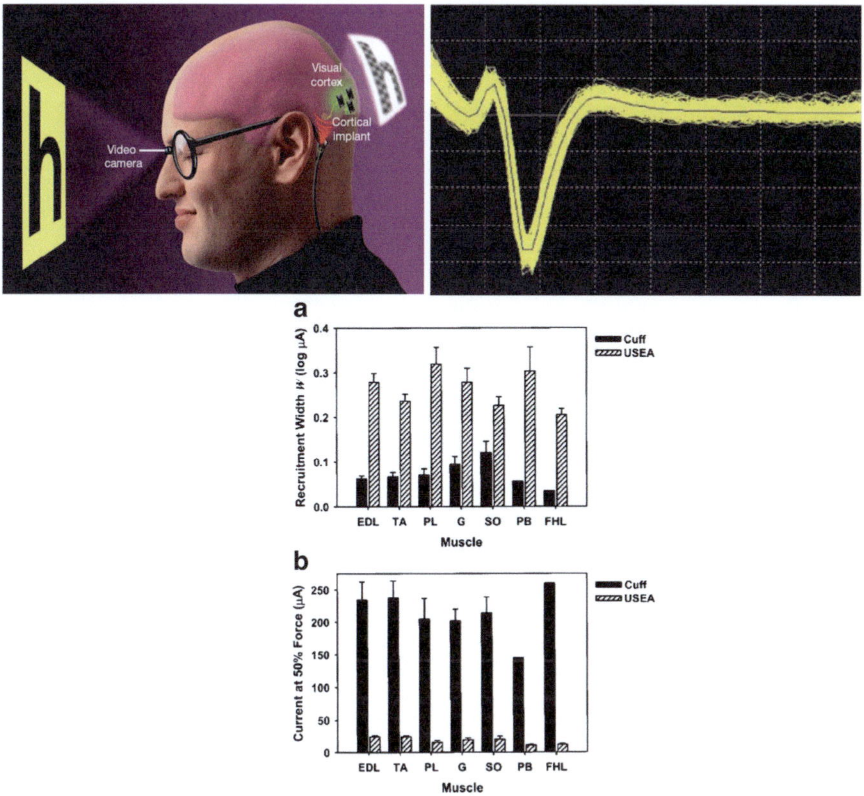

Fig. 5.14 Schematic of a cortex-based artificial vision systems with the Utah array electrodes implanted in the primary visual cortex (*top left* [26]); 285 superimposed recordings of action potentials from a cat's sciatic nerve recorded when the cat's ankle was flexed and extended (*top right* [26]); summary of the recruitment properties of the Utah array and conventional electrodes both measured on the cat sciatic nerve (*bottom* [54])

One of the first papers out of this effort comes from Kisban et al. [57]. This paper describes the fabrication of single-shaft probes with a pitch of 500 µm and lengths of 2, 4, and 8 mm, respectively. To facilitate insertion into the brain tissue, a sharp tip with an opening angle of 17° and a tapered shaft of 0.5°, 0.3°, and 0.1°, respectively, was designed. The thickness of the probes is targeted to be around 120 µm to provide sufficient mechanical stiffness for insertion into brain tissue. Each shaft comprises of eight circular electrodes with a 20 µm diameter and a tip electrode made of platinum. Figure 5.15 represents a basic process flow for these silicon probes. The process starts with the deposition of 500 nm of stress-compensated dielectrics on both sides of a 300 µm-thick silicon wafer. An additional 1.5 µm thick Plasma Enhanced Chemical Vapor Deposition (PECVD) SiO_2 is deposited on the backside of the wafer. The next step is to define metal tracks and the electrodes using a lift-off process. The metal track consists of Ti (30 nm), Au (200 nm), Pt (100 nm), and Ti (30 nm), respectively. The metallization is electrically isolated

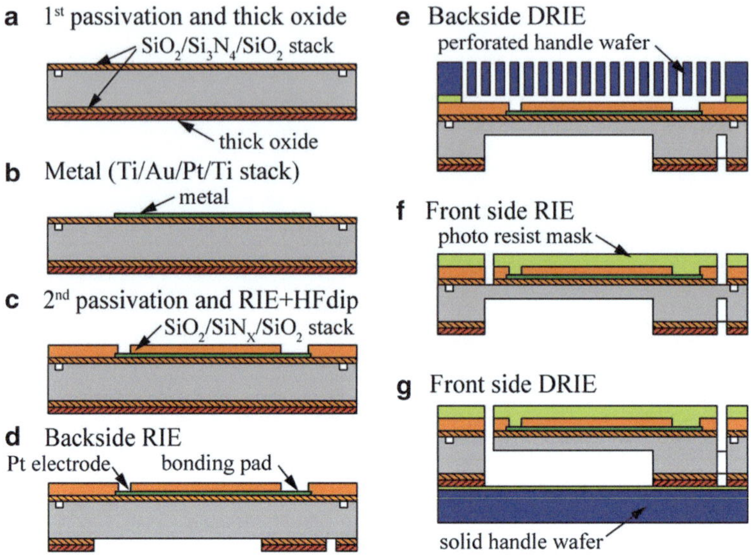

Fig. 5.15 Basic process flow for the EU NeuroProbes approach to the fabrication of silicon 3-D MEAs [57]

with a second stress-compensated PECVD layer, which is defined using an RIE etch and a HF dip. DRIE is now performed from the backside (after an RIE etch of the dielectric layer) to define the probe shaft to its final thickness of around 100 μm. A perforated handle wafer is attached to the front side of the wafer during this process. Front-side processing is performed to release the probe shaft as the final step. This includes removal of the dielectrics with an RIE step and silicon release using a DRIE step. As in the backside-etching step, a handle wafer is used for this step as well. Figure 5.16 depicts optical and SEM images of the silicon probe arrays. A custom ribbon cable is fabricated utilizing two layers of polyimide (PI) with an intermediate layer of metal. The silicon probe array and the PI cable are brought together using a gold bump bonding process (Fig. 5.16). Completed devices are depicted in Fig. 5.17. Herwik et al. [58] describe a variation of this process to include electroplated gold leads for packaging purposes.

Aarts et al. [60] report a *slim-base platform* technology for assembly of the EU NeuroProbes. This technology utilizes a DRIE-etched silicon groove onto which the silicon probes are assembled. Electroplated gold is used as the bonding material between the platform and the probe arrays, which are positioned onto the platform using a flip-chip bonder. Figure 5.18 depicts the schematic, SEM, and optical images of such a platform technology used in assembling the EU NeuroProbes.

Microfluidic ports are integrated with the probes as described by Spieth et al. [62]. Individual probe shafts have an area of 250 × 250 μm^2, and this includes a microfluidic channel with a cross section of 50 × 50 μm^2. The basic process flow for this integration is depicted in Fig. 5.19. First a two-step DRIE process defines the

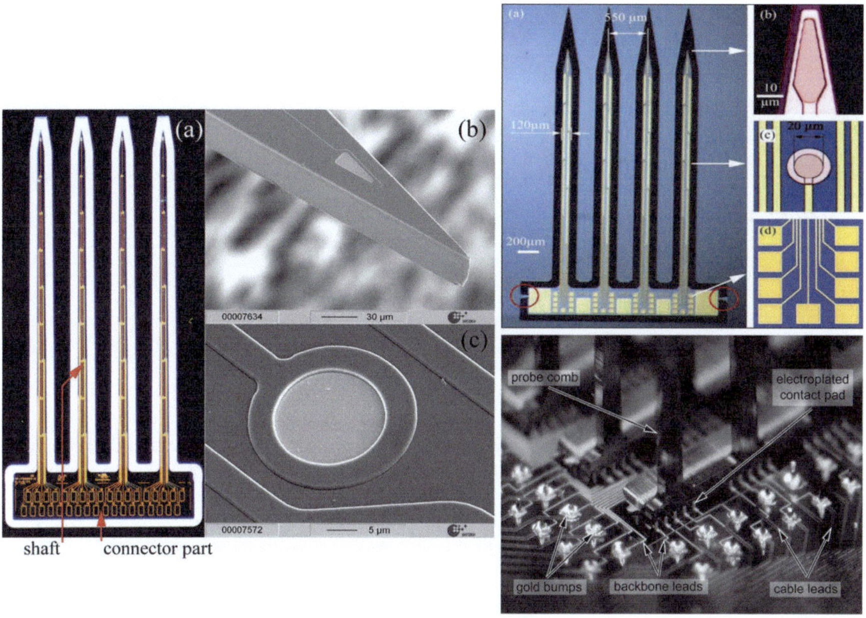

Fig. 5.16 Optical and SEM images of the fabricated probe arrays (*top left* and *right*). Gold bump bonding process for packaging (*bottom right*). *Left* [57]; *top right* [59]; *bottom right*, reproduced with permission from [58], © (2009) IOP Publishing. All Rights reserved

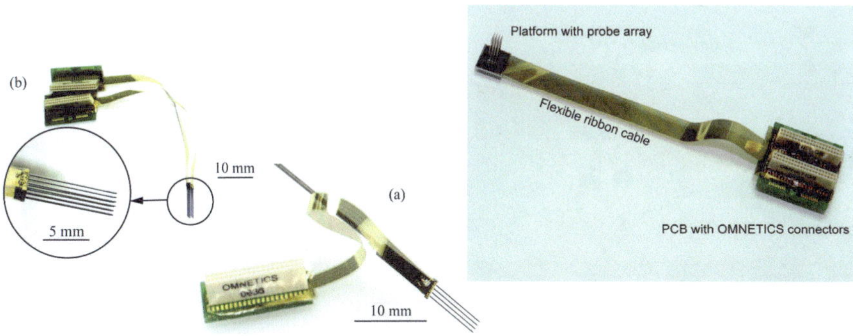

Fig. 5.17 Optical images of assembled probe arrays with flexible polyimide cables. *Left* [57]; *right* [59]

fluidic channel and the outer shape of the probes. The fluidic channel is shaped in the second step with the bonding of a ground silicon wafer. The entire wafer is then passivated on both sides and platinum electrodes are defined. The electrodes are passivated to define the insulation layer and finally thinned down from the bottom side to define a probe shape. Topside silicon etching is also performed to define the fluidic outlet ports. Figure 5.20 depicts SEM images of the constructed devices.

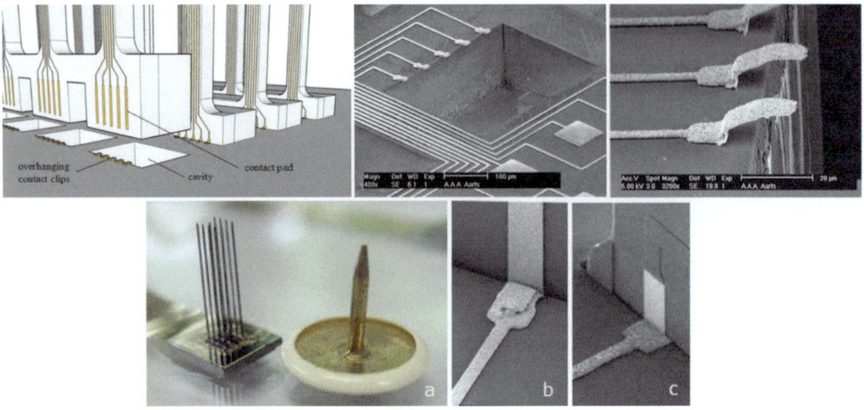

Fig. 5.18 Schematic, optical, and SEM images of assembled probe arrays with a high-throughput assembly technology. *Top left* and *top right* [60], *bottom* [61]

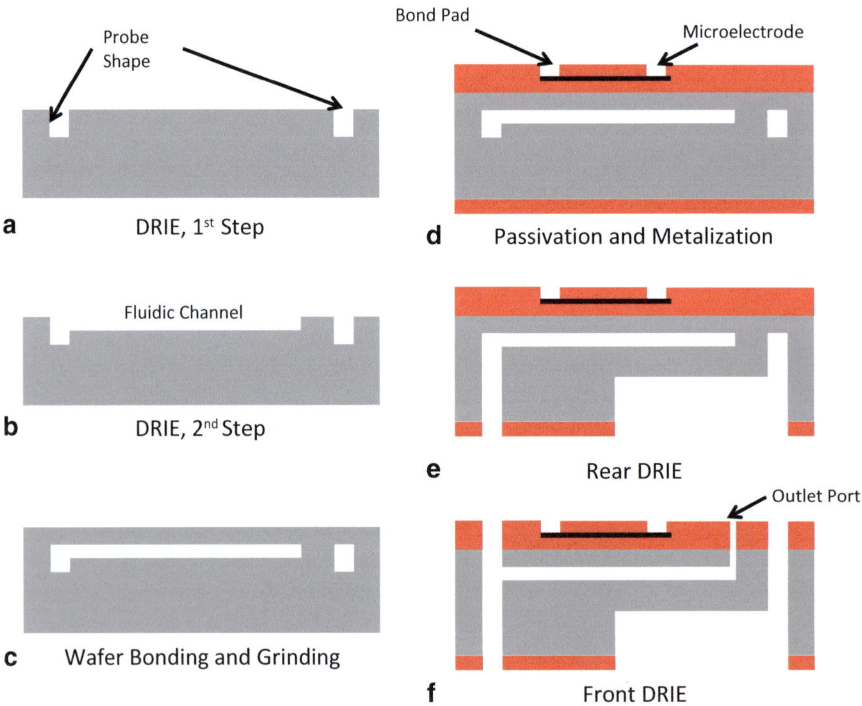

Fig. 5.19 Fabrication process flow to integrate fluidic ports into the EU NeuroProbes. Reproduced with permission from [62], © (2011) Springer

CMOS electronics have been integrated with the EU NeuroProbes as reported by the EU NeuroProbes research team [64, 65]. The *active probe* comprises a 100 μm thick, 4 mm long probe realized using DRIE etching of silicon. It comprises 188 electrodes (20 μm diameter with a pitch of 40 μm) arranged in two columns along

Fig. 5.20 SEM images of EU NeuroProbes with integrated fluidic ports. Reproduced with permission from [63],

the 4 mm probe. The CMOS circuitry is fabricated as the first step in the construction of these active probes. The post-CMOS fabrication is based on the standard EU NeuroProbes process. The four-mask process for the deposition and patterning of the electrodes, bonding pad metallization and structuring of the probe shafts are depicted in Fig. 5.21. In the first step the passivation layer for the CMOS substrate is opened up to connect to the electrodes and bonding pads realized in post-CMOS metallization. This metallization (300 nm of Pt sandwiched between 30 nm thick Ti adhesion layers) is sputter deposited and patterned using lift-off. Following the post-CMOS metallization, stress-compensated silicon oxide and silicon nitride layer stacks are deposited on both sides of the wafer utilizing a PECVD step. The topside passivation is patterned using RIE etching, and the top Ti layer is wet etched using 1 % Hydrofluoric Acid (HF) to expose the Platinum (Pt) electrodes. The backside passivation is additionally patterned using an RIE process, and the probes are released using DRIE to define a final thickness of 100 µm. A short DRIE step follows from the front side to complete the release of the probes. Figure 5.22 depicts optical micrographs of the *active probe* array.

The EU NeuroProbes project has extended the application platform of this technology for surface electromyography [28] and the integration of amperometric biosensors on the probe shafts for in vivo monitoring of choline and glutamate in the brain [66]. The former uses a suspended etch mask technology which utilizes ICP etching of silicon both anisotropically and isotropically to create microneedles

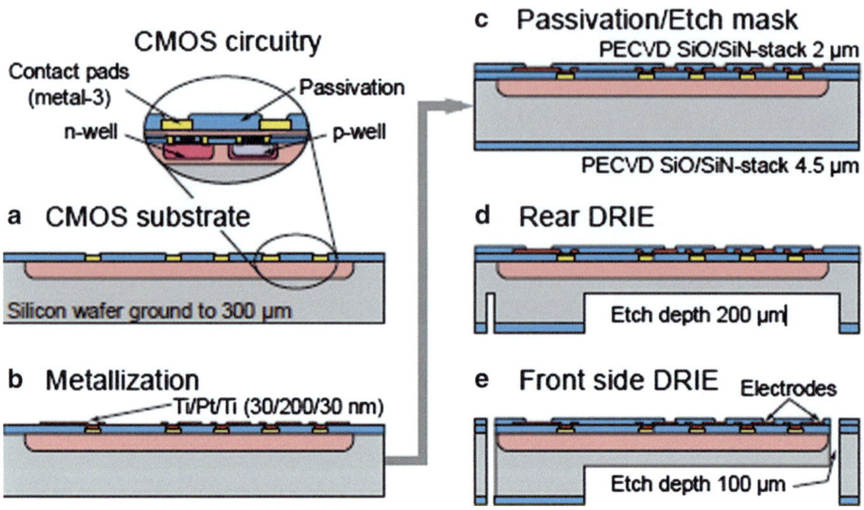

Fig. 5.21 Simplified four-mask process to build EU NeuroProbes on CMOS wafers, using post-CMOS MEMS processing [65]

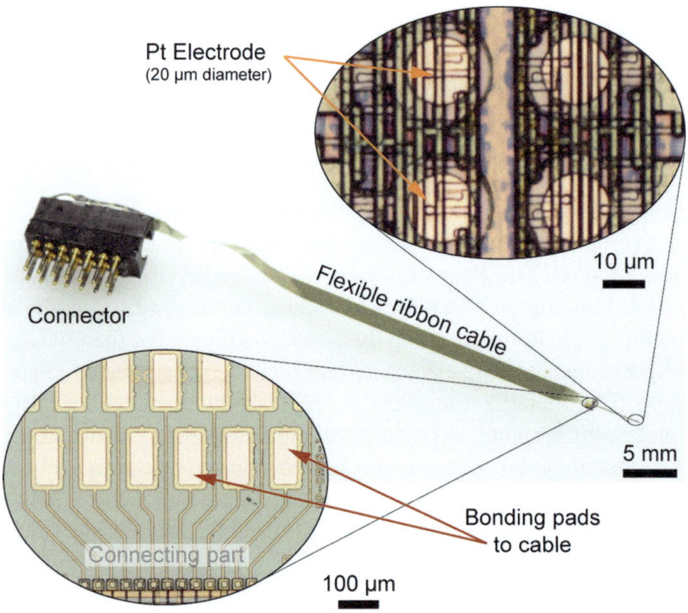

Fig. 5.22 Optical images of an active probe array fabricated using post-CMOS MEMS processing [65]

as depicted in Fig. 5.23. Interconnects are formed utilizing through wafer etching of vias from the backside and metallization (sputter deposition followed by electroplating). The application described by Frey et al. [66] involves the same fabrication process described by Aarts et al. [60] and Herwik et al. [57] which has been

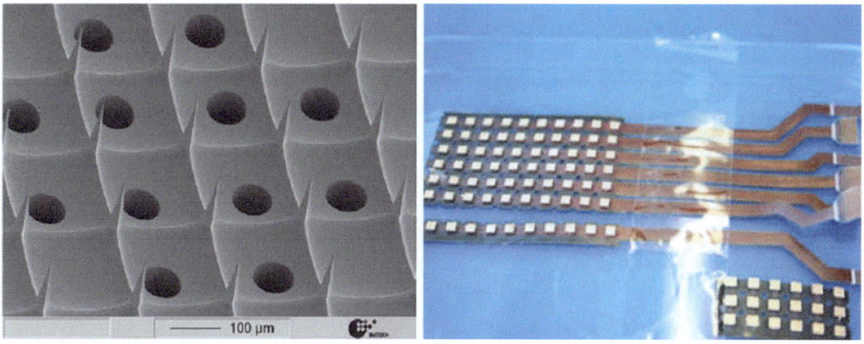

Fig. 5.23 SEM image of the utilization of suspended etch mask technology to create silicon microneedle electrodes (*left*). Packaging these microneedle electrode arrays on a flex circuit (*right*) [28]

described earlier but has a modified probe design. The probe consists of two shanks that are 8 mm in length and have a cross section of 250 μm × 250 μm (height × width). The dimensions of the probe were chosen with the application in mind. Each shank is connected to a common rectangular probe, which comprises the fluidic ports of 25 μm × 50 μm. The shanks comprise five platinum electrodes that are placed around the fluidic ports and have a pitch of 200 μm. The electrodes are 50 μm × 150 μm and are recessed by 10 μm to protect a biosensitive membrane subsequently immobilized on the electrodes. One of the electrodes acts as a reference.

Compared to the Utah arrays and the Michigan probes, both of which have been around for a lot longer than the EU NeuroProbes, the applications developed with these devices are limited but impressive. The in vivo micrograph of the EU NeuroProbes during experimentation and examples of recordings obtained from them are illustrated in Fig. 5.24.

5.2.4 Some Other Major Silicon-Based Approaches

Apart from the three major silicon-based three-dimensional microelectrode array fabrication approaches—Michigan probes, Utah arrays, and EU NeuroProbes—which have been widely published and utilized in a variety of applications, several other researchers have reported other novel silicon-based technologies to fabricate 3-D MEAs. Some of these approaches are briefly described below.

Koo et al. [67] describe a clever utilization of the crystalline structure of silicon to achieve sharp pyramidal tips that have been utilized as 3-D MEAs. In this process the electrode areas on a silicon wafer are first defined using a plasma-enhanced chemical vapor deposition (PECVD) layer of tetraethyl orthosilicate (TEOS) and silicon nitride. Multiple RIE and DRIE steps are carried out with the TEOS and photoresist acting as hard masks to define the vertical dimensions of the 3-D electrodes. Thermal oxidation is then performed to protect the sidewalls of the

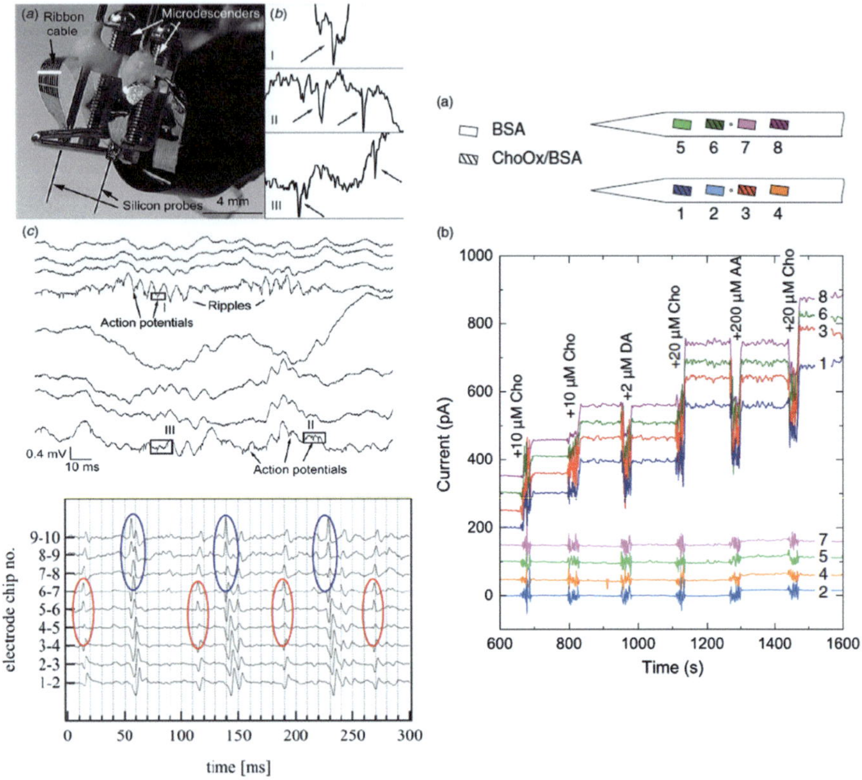

Fig. 5.24 Photograph of the NeuroProbes made ready for implantation and action potentials recorded from individual recording sites in an array (*top left*, reproduced with permission from [58], © (2009) IOP Publishing. All rights reserved); action potentials from regularly firing motor units in the biceps brachii muscle (*bottom* [28]); response of the microprobes to choline and dopamine (*top right*, reproduced with permission from [66], © (2011) IOP Publishing. All rights reserved)

electrodes followed by a potassium hydroxide (KOH) etch to define the 3-D pyramids taking advantage of the fact that the etching stops on the {1 1 1} plane as shown in Fig. 5.25. The electrical pathways are created in the last step after removal of the masking layers (with a 49 % hydrofluoric acid etch) by sputter deposition of a Ti/Au layer and photolithographic definition. The 3-D electrodes are constructed in an 8×8 array roughly 50 μm tall with a 120 μm pitch between electrodes.

Kusko et al. [68] report the fabrication of a 5×5 array of 3-D microelectrodes by the definition of a silicon dioxide/photoresist mask that is carefully undercut by silicon etching—a combination of wet and dry etching—DRIE process to define micropillars first followed by a KOH sharpening to achieve pyramidal tips and DRIE followed by an isotropic silicon etch to create conical tips. The contact metallization and recording sites are then defined by lift-off techniques on 3-D electrodes followed by the deposition of a PECVD silicon dioxide layer and definition of the

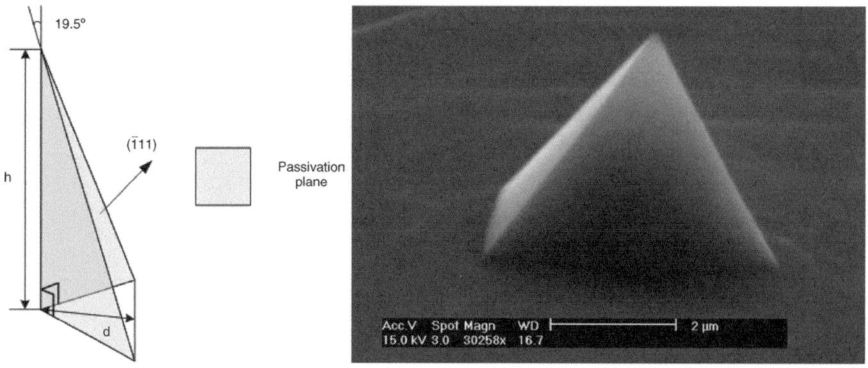

Fig. 5.25 Schematic of the passivation plane where the silicon etching stops (*left*) and SEM image of the defined single microneedle electrode (*right*). Reproduced with permission from [67], © (2006) Elsevier

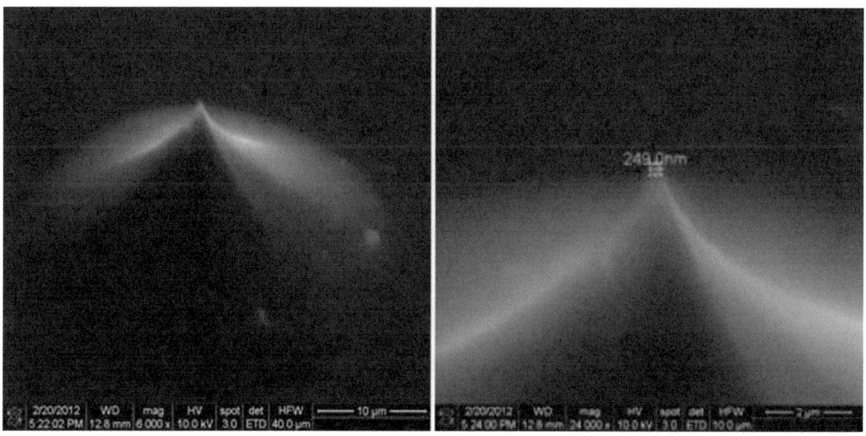

Fig. 5.26 Combination of wet and dry silicon etching to undercut a silicon dioxide/photoresist mask to achieve really sharp nanometer scale tips for 3-D microelectrodes [68]

recording sites lithographically. Three-dimensional microelectrodes that are ~35 µm tall with singular microelectrode tip sharpness of less than 500 nm are achieved using this process as shown in Fig. 5.26.

Du et al. [69] describe a silicon-based process where they utilize double-side thin silicon wafer processing and a carrier wafer to get around handling fragile wafers. So instead of an etch stop used in the Michigan probe approach, this approach uses a 25–50 µm silicon substrates as starting point for MEA fabrication. A 500 µm thick silicon or Pyrex or quartz wafer is used as the carrier in this process. The ultrathin silicon substrate is mounted on the carrier wafer, and metal is defined using a lift-off process. For insulation a 2 µm thick parylene layer is conformally deposited (room temperature vapor deposition process) and defined using

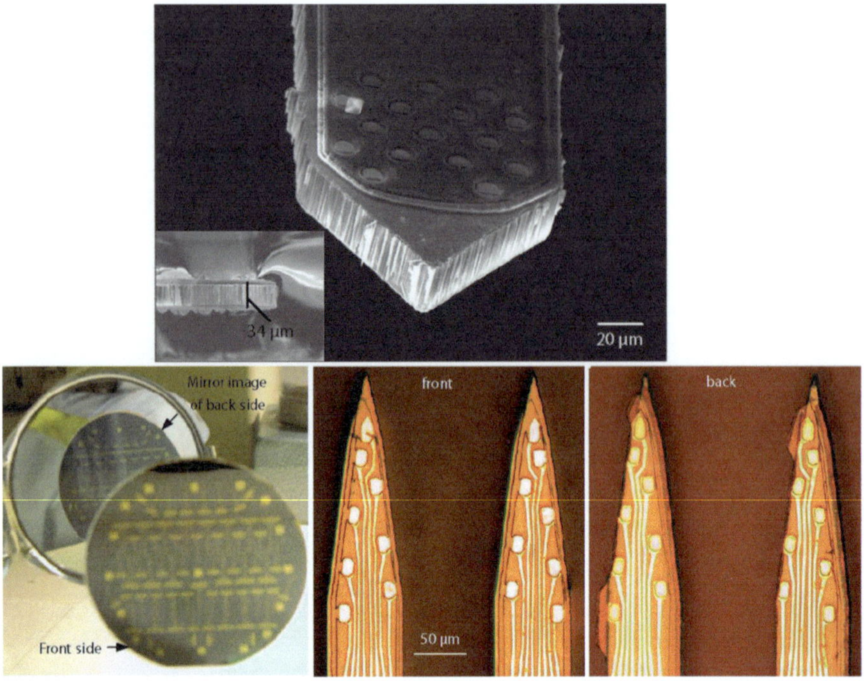

Fig. 5.27 Silicon probes fabricated with the utilization of ultrathin silicon wafers and performing double-side fabrication. Reproduced with permission from [69],

photolithography and RIE etching. Additionally this approach lends itself to the fabrication of identical probes on both sides of the ultrathin silicon wafers. Probes fabricated utilizing this approach are depicted in Fig. 5.27.

Hanein et al. [70] describe an interesting DRIE etching followed by RIE sharpening and interfacing with a polyimide flex circuit-based approach for the fabrication of intracellular 3-D MEAs. They utilize highly conductive p-type silicon wafers and fabricate protection pillars around 3-D electrodes in the DRIE step. These protection pillars completely etch away during the isotropic RIE sharpening process (utilizing a SF_6 plasma) leaving behind ~230 µm-tall 3-D microelectrodes that are electrically routed using a flex circuit. Figure 5.28 depicts SEM images of the protection pillars with the central 3-D electrode after the etching steps. These are completely etched away leaving behind sharp (~200 nm tips) 3-D microelectrode.

Chu et al. [71] create 3-D MEAs on SOI wafers of various thicknesses by defining the microelectrodes on the device layer first, followed by ICP etching of silicon to create micropillars which are then sharpened by either a dry, isotropic XeF_2 etch or a wet isotropic Hydrofluoric Acid/Nitric Acid/Acetic Acid (HNA) etch. The metal traces are protected during the entire process, which can prove to be difficult. Thermally grown SiO_2 acts as the insulation, and it is removed on the bond pads and the top of the 3-D electrodes by a carefully controlled wet etch.

Fig. 5.28 Silicon probes fabricated with the utilization of ultrathin silicon wafers and performing double-side fabrication. Scale bar is 100 μm on the *left* and 50 μm on the *right*. Reproduced with permission from [70], © (2003) IOP Publishing. All rights reserved

One of the more interesting fabrication approaches to integrate on-chip silicon probes and silicon dioxide microtubes is reported from Toyohashi University of Technology (Toyohashi, Japan). The researchers at this university have successfully developed a process by which they *grow* silicon and oxide microtubes on-chip utilizing a vapor–liquid–solid (VLS) method [72–74]. Interconnection materials such as tungsten, tungsten silicide, titanium nitride, and titanium are used to ensure that the Si and SiO_2 probes make contact with the Metal Oxide Semiconductor Field Effect Transistors (MOSFETs) that are fabricated in the first step of the process (Fig. 5.29). The MOSFETs are fabricated utilizing a standard 5 μm CMOS process. For the formation of the silicon probes, a 160 nm gold film was selectively deposited as a catalyst. VLS growth was carried out in a gas source molecular beam epitaxy (MBE) system with Si_2H_6 gas while heating the substrate to 680 °C. An Au–Si alloy is formed first by heating the substrate before the VLS growth. Phosphorous thermal diffusion can be carried out to reduce the resistivity of the probes. In order to create the microtubes, a silicon dioxide layer was grown on the tubes utilizing PECVD. This oxide is patterned utilizing standard photolithography, followed by wet etching, and once the gold–silicon alloy is removed, a XeF_2 etch is utilized to remove the silicon in the tube to create silicon dioxide-coated microtubes. The probes and microtubes are shown in Fig. 5.30. The probes and tubes are typically ~3 μm in diameter and ~30 μm in height.

5.3 Metal, Glass, and Polymer Probes as 3-D MEAs

Even though silicon has been the major material for the fabrication of three-dimensional MEAs, researchers in the last 15 years have looked to other materials due to several reasons: (a) costs associated with clean room microfabrication; (b) many universities may not have access to a high-class clean room; (c) end users of this tool (neuroscientists, physicians, cardiologists, toxicologists, pharmacologists, etc.) are very familiar with materials like polymers and biocompatible metals.

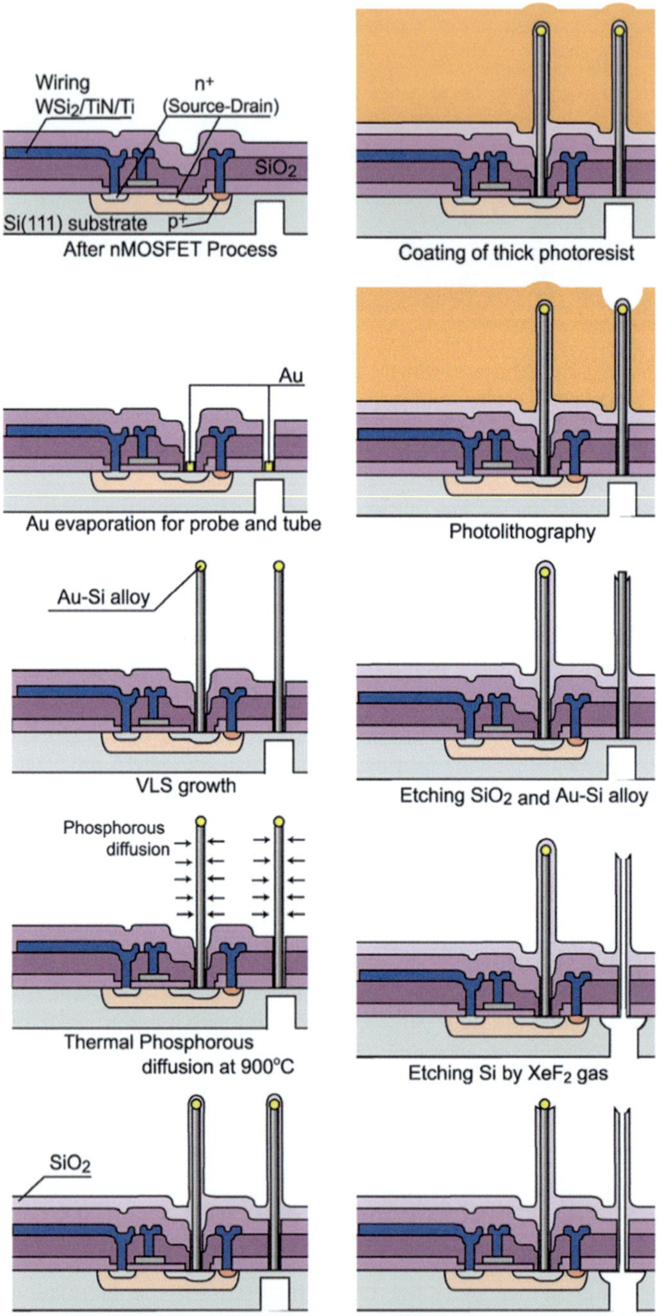

Fig. 5.29 Process flow for the fabrication of Si and SiO_2 probe arrays utilizing the VLS technique. Reproduced with permission from [74],

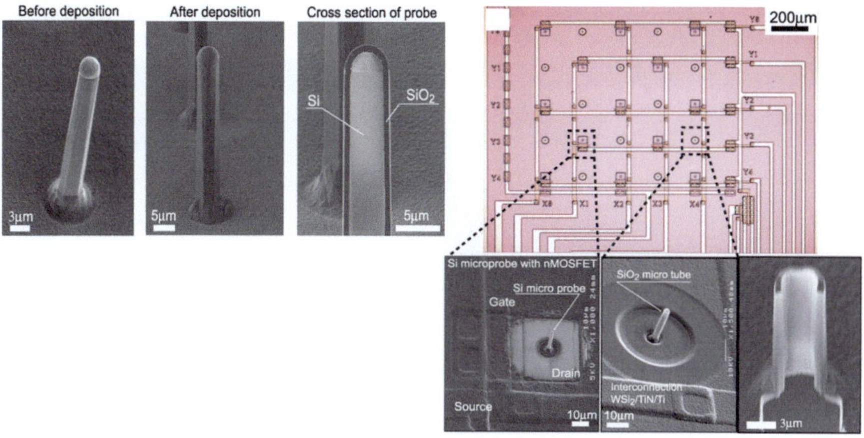

Fig. 5.30 Optical image of a MOSFET with regions of Si and SiO_2 nano-probe arrays (SEMs) on the *right*. SEM images of the microprobes on the *left*. Reproduced with permission from [74], © (2008) IOP Publishing. All rights reserved

Thus MEMS technologies developed on these materials are more likely to enjoy easy adoption and (d) many of the materials mentioned enjoy superior properties (such as mechanical stiffness comparable to tissue/cell networks and biocompatibility) to interface with cell cultures and tissue as compared to silicon.

Examples of such nonconventional approaches include glass micromachining, micromolding-based approaches, SU-8 micromachining, polyimide micromachining, electrical discharge micromachining, parylene-based approaches, etc. Some of these approaches are detailed below.

5.3.1 *EDM-Based 3-D MEAs*

Electrical discharge machining (EDM) is a process that makes use of computer-aided design (CAD) that runs under computer numerical control (CNC) and is capable of batch processing. It is used to generate intricate features with high aspect ratios and is capable of machining a large variety of conductive materials.

Fofonoff et al. [75–77] describe the use of EDM technology toward the microfabrication of 3-D MEAs. A zinc-coated brass wire and the work piece (made of suitable metals) form the basic materials in this process. An initial cut through one plane perpendicular to the EDM wire was first made followed by a cut in the orthogonal direction. This group from the Massachusetts Institute of Technology (Cambridge, MA, USA) along with collaborators at the University of Chicago (Chicago, IL, USA) and Brown University (Providence, RI, USA) report the fabrication of 3-D MEAs from a variety of conductive materials—titanium, titanium–aluminum–vanadium alloy, stainless steel, and tungsten carbide as shown in Fig. 5.31.

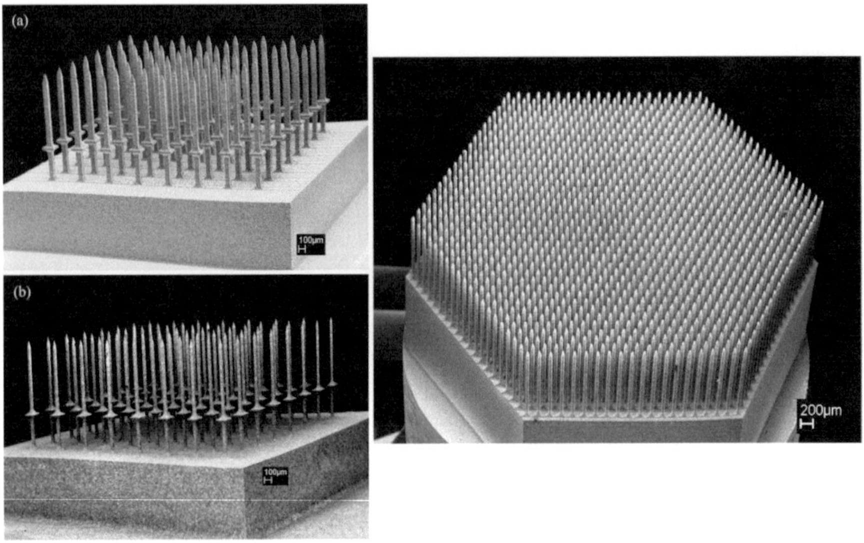

Fig. 5.31 SEM images electrical discharge machining (EDM)-based 3-D MEAs [77]

They report fabricating a variety of arrays with heights ranging from 1 to 5 mm and electrode widths ranging from 80 μm onward with interelectrode spacing ranging from 170 to 500 μm. They also report the fabrication of arrays with roughly 1,150 electrodes using this EDM technique. An etching step follows the EDM step where boiling hydrochloric acid (HCl) was utilized to not only prevent oxidation of the metals but also provide a suitable surface finish. The EDM wire 3-D MEA is subsequently electroplated with acid gold strike and platinum. The 3-D MEA was further insulated with parylene-C chosen due to its excellent biocompatibility [78] [Specialty Coating Systems Inc. 2013]. Parylene can additionally be deposited conformally at room temperature. The insulation was exposed at the tips of the microelectrodes utilizing laser micromachining. A Resonetics Excimer laser was utilized for this purpose. This laser micromachines polymers at 248 nm by chemical ablation (breaking of bonds).

5.3.2 *Glass-Based 3-D MEAs*

Metz et al. [79] and Heuschkel et al. [80] report what is probably one of the first approaches to batch fabricate integrated 3-D MEAs *on-chip* (3-D MEAs built monolithically on a substrate using surface/bulk micromachining processes) specifically targeted at in vitro neural applications. A simplified process flow for such a technique is depicted in Fig. 5.32, and it begins with the deposition of chromium on float glass substrates. The substrates are 700 μm thick, and the deposition is carried out utilizing a sputtering process at high temperatures (455 °C). The deposition is followed by the definition of the chromium layer (used as a mask for subsequent

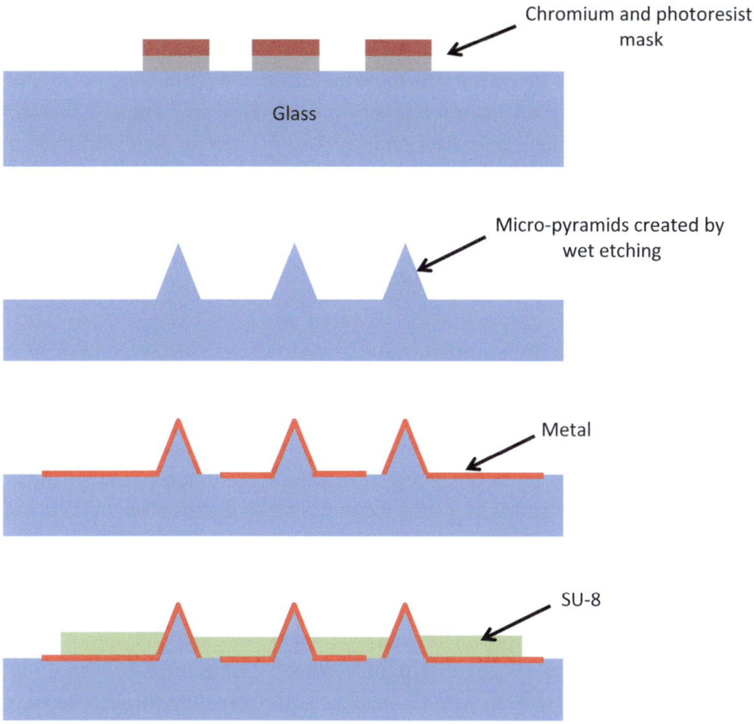

Fig. 5.32 Process flow for glass etching-/undercutting-based 3-D MEAs [79]

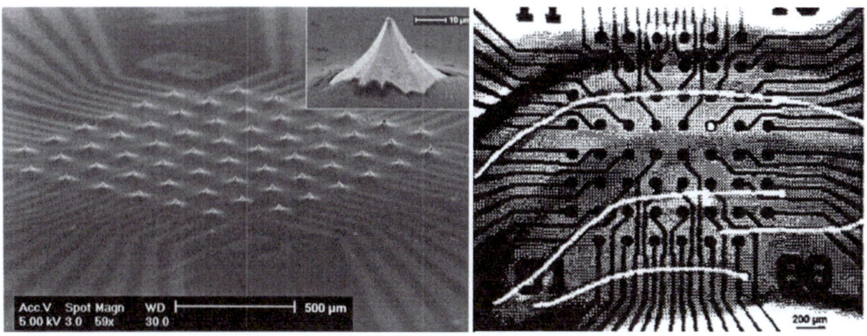

Fig. 5.33 SEM images of the microfabricated MEA (*left*) and a tissue slice placed on the MEA (*right*). Reproduced with permission from [80] © (2002) Elsevier

etching) using standard photolithography. The glass substrate was subsequently etched in a 10 % HF solution till the chromium mask was completely undercut and dethatched itself from the glass substrate. These researchers found out that chromium deposited at high temperatures was resistant enough to HF to carry out the etching. The HF etch step resulted in ~60 µm tall pyramidal structures that eventually will serve as the 3-D microelectrodes. Figure 5.33 depicts SEM images of such electrodes and an optical image of a tissue slice on a 3-D MEA. In the next step, metal tracks were deposited and patterned to define a layer of titanium/platinum on the microelectrodes. The final step was the deposition and patterning of a 5 μm layer of SU-8 which serves as the insulation layer for the 3-D MEA. These 3-D MEAs are developed in an 8×8 matrix and are typically 40×40 μm^2 in area and have a pitch of 200 μm.

5.3.3 Polyimide- or Kapton-Based 3-D MEAs

Polyimide or Kapton is a popular material to construct 3-D MEAs, and these MEAs are typically designed to be implantable. The properties of polyimide that make it attractive are (a) mechanical flexibility; (b) electrically insulating properties; (c) surface chemistry amenable to modification and preparation that allow a host of bioactive organic species to be absorbed or covalently bonded to the surface of the material; (d) acceptable low gas permeability; (e) acceptable vapor transmission rates; and (f) biocompatibility [81].

Rousche et al. [82] demonstrate one such polyimide probe where the probe fabrication involves surface micromachining of photosensitive polyimide and chromium/gold metal layers. The metal layers are sandwiched between two layers of polyimide. Both layers are in between 10 and 20 μm thick, which makes these probes suitable for flexible applications. A RIE step with O_2 plasma is performed before the deposition of metal and the spin coating of the insulation layer of polyimide. This step serves to roughen the underlying layer for successful adhesion of the layer deposited on top. Several 2-D and 3-D probes have been fabricated with electrode sizes ranging from 20–40 μm×20–40 μm with shaft lengths up to 1.5 mm. Silicon is used as the carrier substrate, and the first step is the growth of a thermal layer of SiO_2 (0.5 μm thick). This layer serves as the release layer after probe fabrication.

Owens et al. [83], Boppart et al. [84], and Stieglitz et al. [85] also report similar fabrication processes for the fabrication of implantable 3-D MEAs constructed out of polyimide. Figure 5.34 depicts probes fabricated using this approach.

Tekuchi et al. from the University of Tokyo [86] report a similar approach where the third dimension for the probe is attained by bending the in-plane metal out of plane using a manual motion. Even though the mechanical match to tissue is vastly improved as compared to silicon or glass due to lower modulus of elasticity of polyimide, stiffness for insertion still remains an issue. A couple of researchers have investigated the strengthening of these probes using either a 5–10 μm thick silicon backbone or a molybdenum backbone [87, 88].

5.3.4 Parylene-Based 3-D MEAs Probes

Parylene belongs to the thermoplastic polymer family and was developed by Union Carbide in 1965 and has found its way into several industries such as PCBs due to

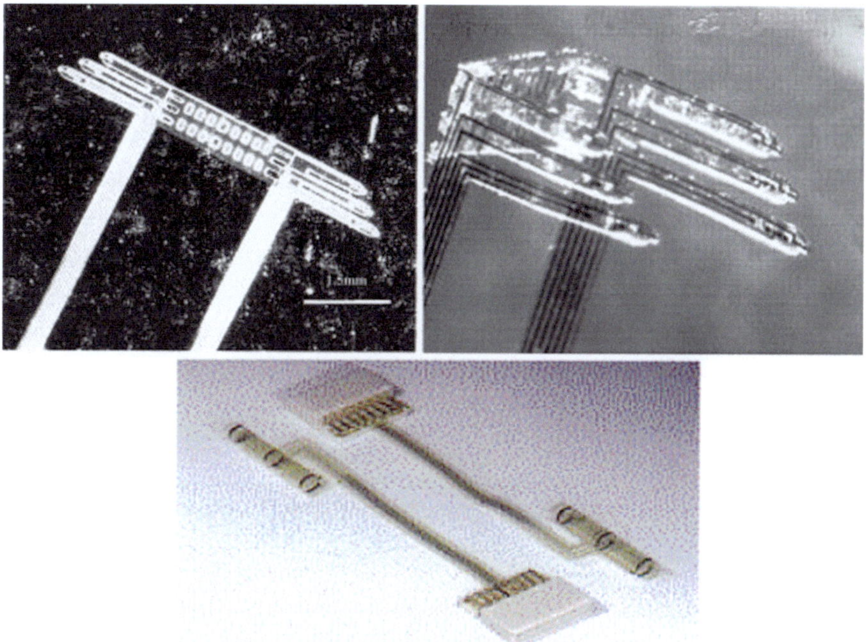

Fig. 5.34 Optical images of various approaches toward the microfabrication of polyimide-based 3-D MEA probes. *Top left* and *top right* [82]; *bottom*. Reproduced from [85], © (1999) The Author(s)

superior mechanical, chemical, biological, and thermal properties [78]. Additionally parylene can be deposited at room temperature making it ideal for low-temperature applications. Prof. Yu-Chong Tai's group at California Institute of Technology (Pasadena, CA, USA) has been leading the charge in the development of parylene-based 3-D MEAs.

Prof. Tai's group has been working in this area since 1999. Some of these efforts have been in the area of realization of neural cages [89, 90] that are fabricated with a multilayer parylene etching process utilizing multiple layers of photoresist as sacrificial layers. The neural cage acts as a "housing" structure for individual neurons with electrodes located right underneath the cage to stimulate and record from a neuron. Figure 5.35 depicts the SEM of one such cage.

Rodger et al. [91–93] report the development of a parylene MEMS technology for a retinal prosthetic system. High-density electrodes are fabricated on a parylene substrate with multiple layers of parylene deposition and etching with intermediate metallization processes. Additionally a parylene-based packaging technology has been developed to integrate the high-density array with an external CMOS chip for data processing [91, 94]. Figure 5.36 depicts a collage of parylene-based retinal MEAs and the packaging technologies associated with interfacing an ASIC with such a high-density probe.

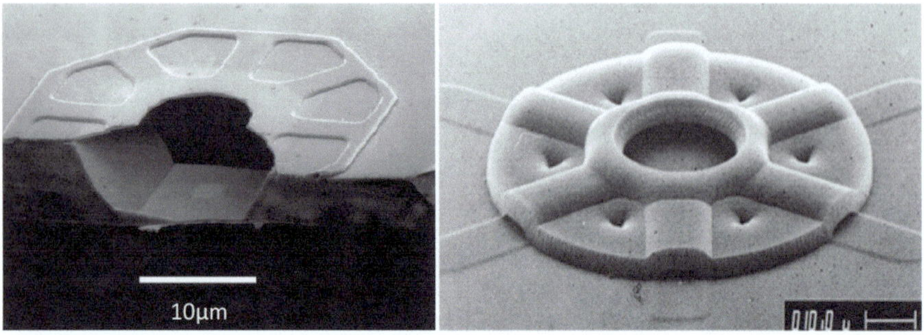

Fig. 5.35 SEM images of neural cages with microelectrodes located right below these cages fabricated by the MEMS group at the California Institute of Technology. *Left*, reproduced with permission from [89], © (1999) Elsevier; *right* [90]

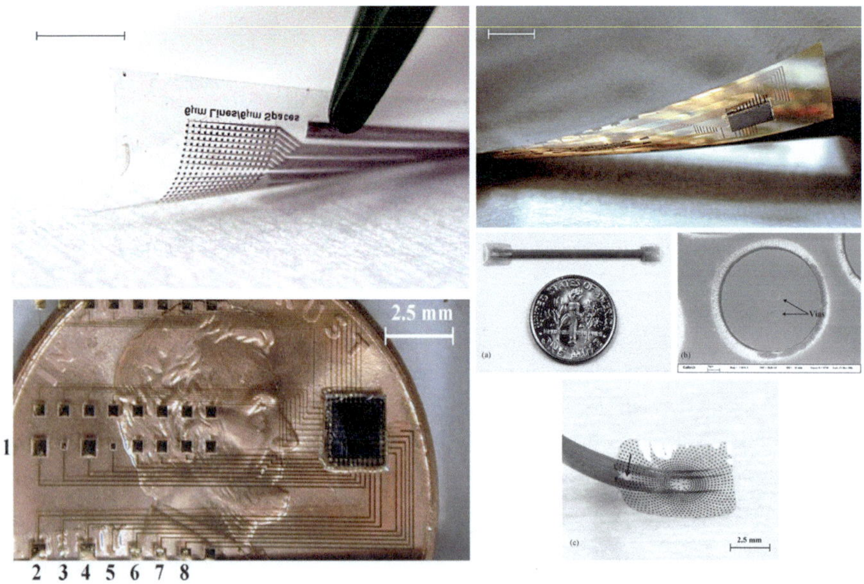

Fig. 5.36 Collage of parylene-based 3-D MEAs with and without a custom ASIC integrated for retinal prosthesis. *Top left*, *top right*, and *center left* [94]; *center right* and *bottom*, reproduced with permission from [83], © (2008) Elsevier

5.3.5 *Three-Dimensional MEA Work at the Georgia Institute of Technology*

Prof. Mark G. Allen's group at the Georgia Institute of Technology (Atlanta, GA, USA) has been working on the development of 3-D MEAs since the early 1990s. There are several reported technologies (all with metals or polymers or SU-8) developed both by Prof. Allen's group and Prof. Frazier's group from 1993 onward.

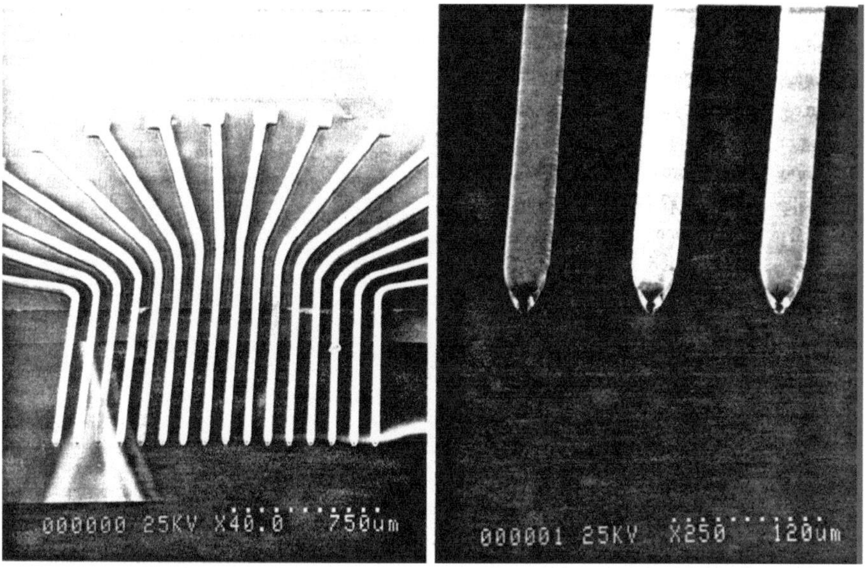

Fig. 5.37 One of the earliest demonstrations of metallic probes serving as 3-D MEAs fabricated on a silicon wafer using electrodeposition processes on a polyimide mold. The silicon wafer is etched in the final step to release the 3-D MEAs [95]

Frazier et al. [95, 96] report the fabrication of in-plane microelectrodes utilizing silicon as the substrate material onto which metallic probes are electroplated. The process begins with the doping of one side of the silicon wafer with boron to create a 4–6 μm thick p+ diffused layer. A 300 nm silicon nitride layer is then deposited on both sides of the wafer. The silicon nitride is patterned and etched on the undoped side to define the electrode area. This area is then etched anisotropically using a 20 % KOH solution, and the boron layer serves as an etch stop for the etching. The processing turns to the front side of the wafer now, and a seed layer of Ti/Cu is deposited on the silicon nitride layer. Polyimide is then spun coated and defined as an electroplating mold onto which several metallic probes can be electroformed. Gold, silver, nickel, copper, and combinations of these metals have been electroplated. The polyimide layer is then removed using a KOH solution, followed by the removal of the underlying seed layers used for electroplating. Silicon nitride insulation is then deposited and defined to open the recording sites. The probes are finally released by backside etching of the p+ membrane and the silicon nitride (Fig. 5.37). These probes are 15–20 μm in thickness, 25 μm in width, and roughly 1 mm long.

O'Brien et al. [97, 98] report a flexible microelectrode array (FMA) that can be sewn through a nerve for stimulation and recording from neural tissue applications. They utilize a glass substrate as the handle wafer on which the FMA is created. First a 10 μm × 10 mm long photolithographically defined gold metal traces are encapsulated within two layers of polyimide that are cured at 300 °C for 1 h to attain final mechanical strength. A sacrificial titanium mask is then defined. This Ti mask will be

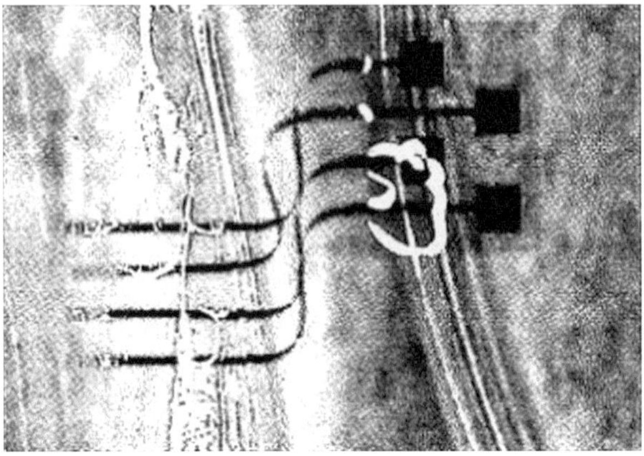

Fig. 5.38 Optical micrograph of a flexible microelectrode arrays (FMAs) fabricated on a polyimide substrate and inserted in a gelatin model [98]

used to define the FMA's outline, bond pads, and recording/stimulation sites. The rigid insertion piece of the device is then constructed by electroplating a metal layer using a thick photoresist mask. The photoresist layer is then removed followed by the removal of the metal seed layers. An RIE process is then utilized to define the recording sites and bond pads on the top polyimide layer. The FMA is released from the handle glass substrate by etching the Ti mask and immersing in DI water. Figure 5.38 depicts an optical micrograph of the released FMA inserted in a gelatin model.

Rowe et al. [99, 100] report an active 8×8 3-D microscaffold with integrated microelectrodes and microfluidic ports for culturing of neurons in vitro. The 3-D microscaffold consists of a silicon orifice plate above, which an 8×8 array of SU-8 micro-towers extends. The height of the micro-towers with integrated electrodes and fluidic channels extends to 1 or 1.5 mm from the silicon orifice plate as depicted in the schematic in Fig. 5.39. The SU-8 towers with the electrical and fluidic functionalities are fabricated in an *in-plane* fashion using a series of processes as described below. First a chromium release layer was deposited on a silicon wafer followed by the creation of the outer walls of the SU-8 towers utilizing a photolithographically defined layer of 5 μm SU-8. The outer wall layer houses the fluidic ports (20×20 μm^2) and electrodes (15×15 μm^2). Next 23 μm gold tabs were electrodeposited using a thick layer of AZ 4620 photoresist. The gold leads were then patterned and wet etched and subsequently insulated with a second 5 μm layer of SU-8. The hollow channel was defined in the next step utilizing a sacrificial layer of AZ 4620. A third SU-8 layer that is 75 μm thick was then defined to fabricate the outer wall of the SU-8 micro-tower. An intermediate layer of parylene-C is used to separate the SU-8 and the sacrificial resist layer. In the final release steps, the exposed parylene-C is etched using an RIE process, followed by the release of the towers by dissolving the sacrificial resist and lifting the towers off the substrate using chrome etch. The towers are then manually packaged into a DRIE-etched silicon orifice plate. Figure 5.40 depicts images of the microfabricated and assembled towers.

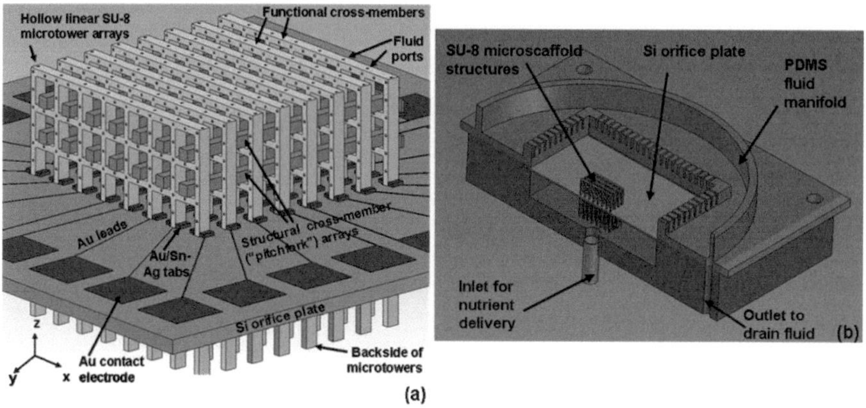

Fig. 5.39 Schematic of an 8×8 3-D SU-8 scaffold with integrated microelectrodes and microfluidic ports [100]

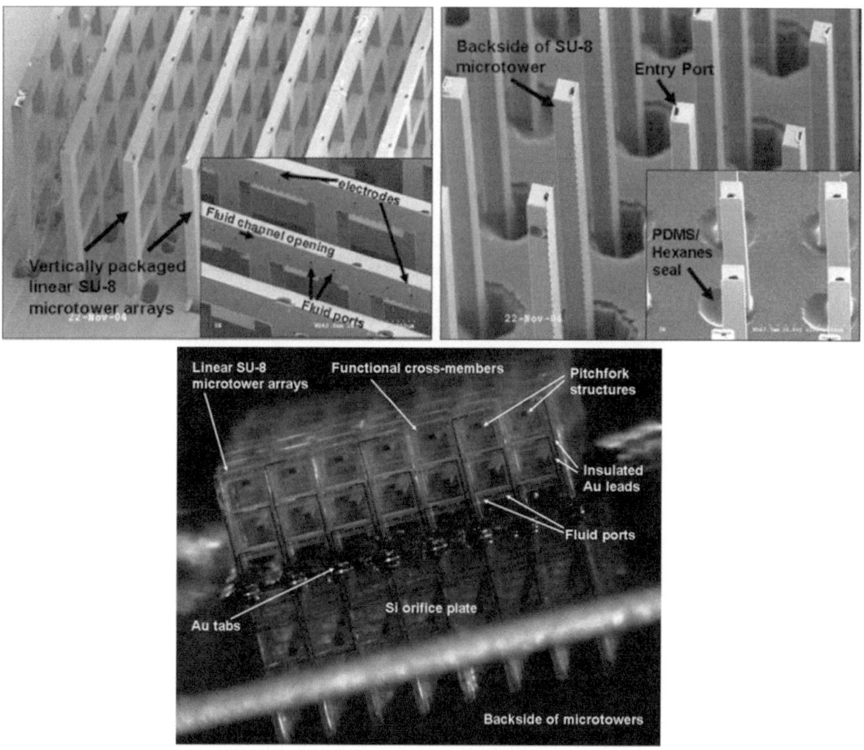

Fig. 5.40 SEM (*top*) and optical (*bottom*) images of the fabricated and assembled micro-towers with electrical and fluidic functionalities [100]

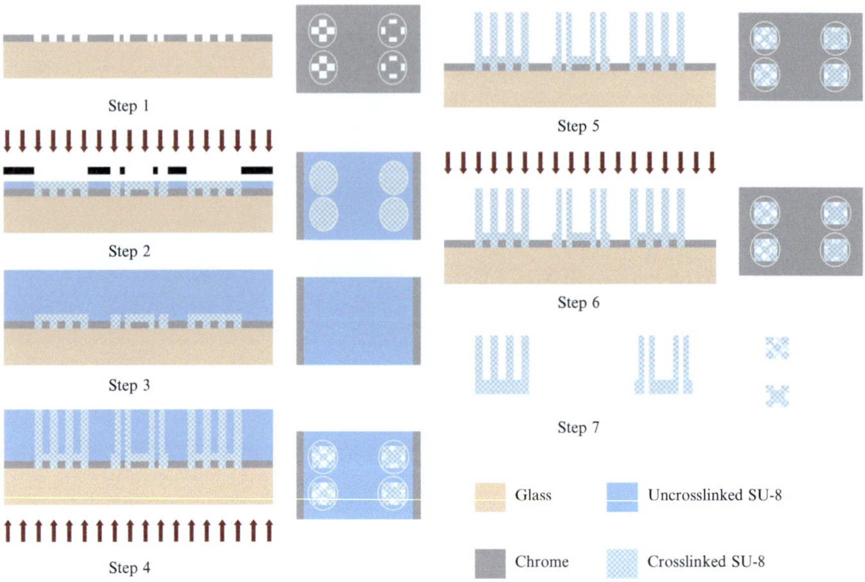

Fig. 5.41 Process flow for the fabrication of micro-towers utilizing a double-side exposure of SU-8 technology [101, 102]

Choi et al. [101] and Rajaraman et al. [102] report an SU-8-based active 3-D microscaffold technology with microelectrode and microfluidic functionalities. Their strategy is split into two different fabrication steps—(a) fabrication of micro-towers intended for electric or electric and fluidic functionalities and (b) addition of the 3-D microelectrode functionality in the second step. These 3-D coupled microfluidic, microelectrode MEAs have been utilized in a variety of in vitro network electrophysiological applications involving 2-D and 3-D cultures of neurons and tissue slices [103, 104].

Figures 5.41 and 5.42 depict the process flow for the fabrication of micro-towers. A novel "double-side" exposure technology of SU-8 is developed to fabricate the micro-towers in two steps. In the first step the base of the micro-towers is fabricated on a chrome plate (with features designed for bottom-side exposure) with a 100 μm thick layer of SU-8. A second layer of SU-8 (~500 μm thick) is then cast on the first layer without developing this layer. Bottom-side exposure of SU-8 is carried out with a larger UV energy dose. This exposure energy is coupled into SU-8 through the designed features in the chrome mask. Both layers are simultaneously developed thus defining ~500 μm tall SU-8 towers on a SU-8 substrate. These micro-towers are then individualized, and seed layers of Ti/Cu (30 nm/900 nm) are coated onto the towers. Parylene-C is then conformally coated onto the metallized micro-towers. This parylene layer is subsequently laser micromachined using a UV laser to define metal traces in 2-D and 3-D. Nickel is then electrodeposited onto the laser micromachined metal track areas to define the 2-D and 3-D microelectrodes.

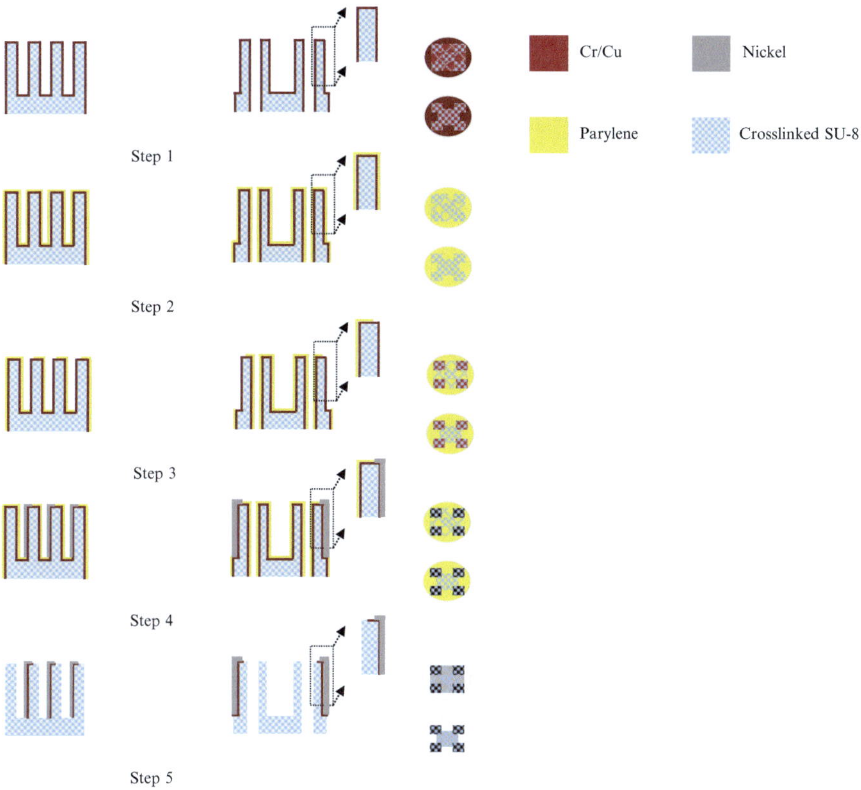

Fig. 5.42 Process flow for the fabrication of micro-towers utilizing a double-side exposure of SU-8 technology. These steps depict the fabrication of the microelectrodes on the micro-towers. [101, 102]

Figure 5.43 depicts images of the micro-towers and metal defined on the micro-towers. Thus, utilizing nonconventional microfabrication technologies such as "double-side" SU-8 exposure, laser micromachining, and electroplating, 3-D MEAs are created in a monolithic fashion with electrodes as tall as 500 μm and above. These devices are further packaged utilizing traditional approaches, and a thick insulation (parylene-C) is conformally deposited over 2-D and 3-D topographies to allow for cell cultures to seed onto a friendly material (parylene-C). Platinum is further electroplated in the last step to define the recording sites (process steps, images not shown).

Metal transfer micromolding (MTM) technology is an ideal technological platform for 3-D MEAs. It is a technology that has been pioneered by Prof. Allen's group [106] that has been used as a platform for the microfabrication of microneedles, RF MEMS components flow sensors, cellular scaffolds, and 3-D MEAs.

Rajaraman et al. [29, 105] report two different microfabrication approaches as shown in Figs. 5.44 and 5.45 for in vitro and in vivo 3-D MEAs respectively. The former starts with the exposure of pyramidal tips on 700 μm SU-8 utilizing inclined UV lithography. A silicon wafer is used as a carrier substrate on which the

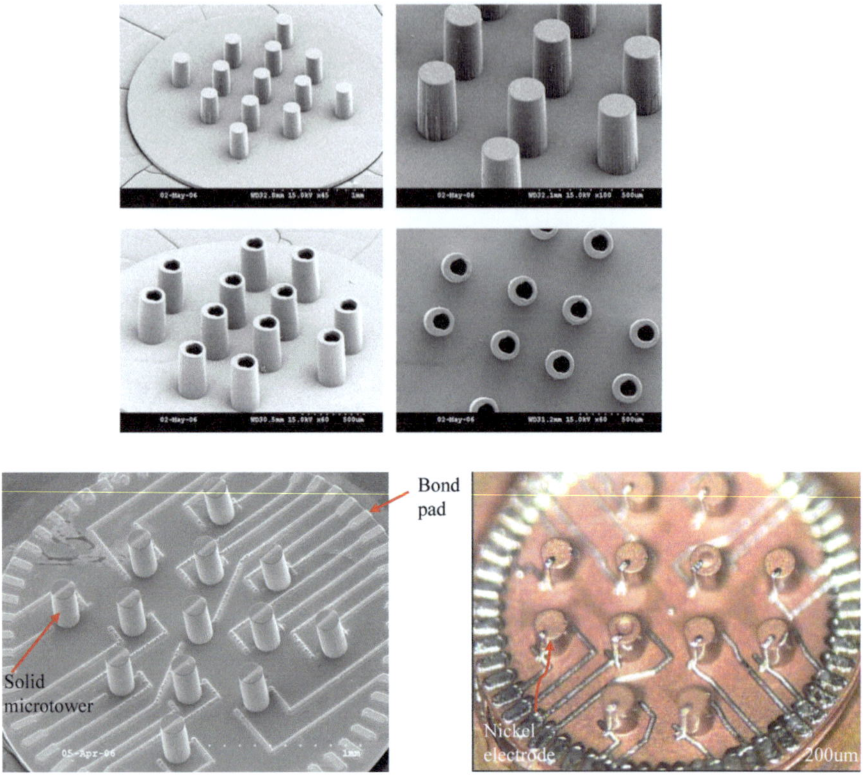

Fig. 5.43 SEM of the micro-towers fabricated utilizing double-side exposure technology (*top*). SEM and optical images of the 3-D microelectrodes defined on these towers (*bottom*). Reproduced with permission from [104], © (2007) IOP Publishing. All Rights Reserved

microfabrication is performed. A second layer of SU-8 (100 μm thick) is spin coated and exposed with the metal interconnect pattern without developing the first layer. Both layers are simultaneously developed after a postexposure bake in order to obtain a two-layer SU-8 rigid mold. PDMS micromolding is carried out twice to create a flexible mold with the same pattern as the rigid mold. A layer of Cr/Au is used as the intermediate release layer between the two materials. To define the microelectrodes, a Au/Cr layer is deposited on the PDMS mold using either an e-beam or filament evaporator. Metal transfer is achieved by bringing a high-surface energy plate in contact with the PDMS mold. Metal is patterned at this step to create 2-D and 3-D microelectrodes (up to a height of 500 μm is reported). Once the metal is patterned, the target polymer is cast into the patterned PDMS mold and processed. The final 3-D MEA structure is demolded in the last step with patterned microelectrodes. For the in vivo 3-D MEAs, the metal interconnect lithography step is eliminated, as each demolded array is a separate electrode in itself. Figure 5.46 depicts SEM and optical images of 3-D MEAs constructed with the metal transfer

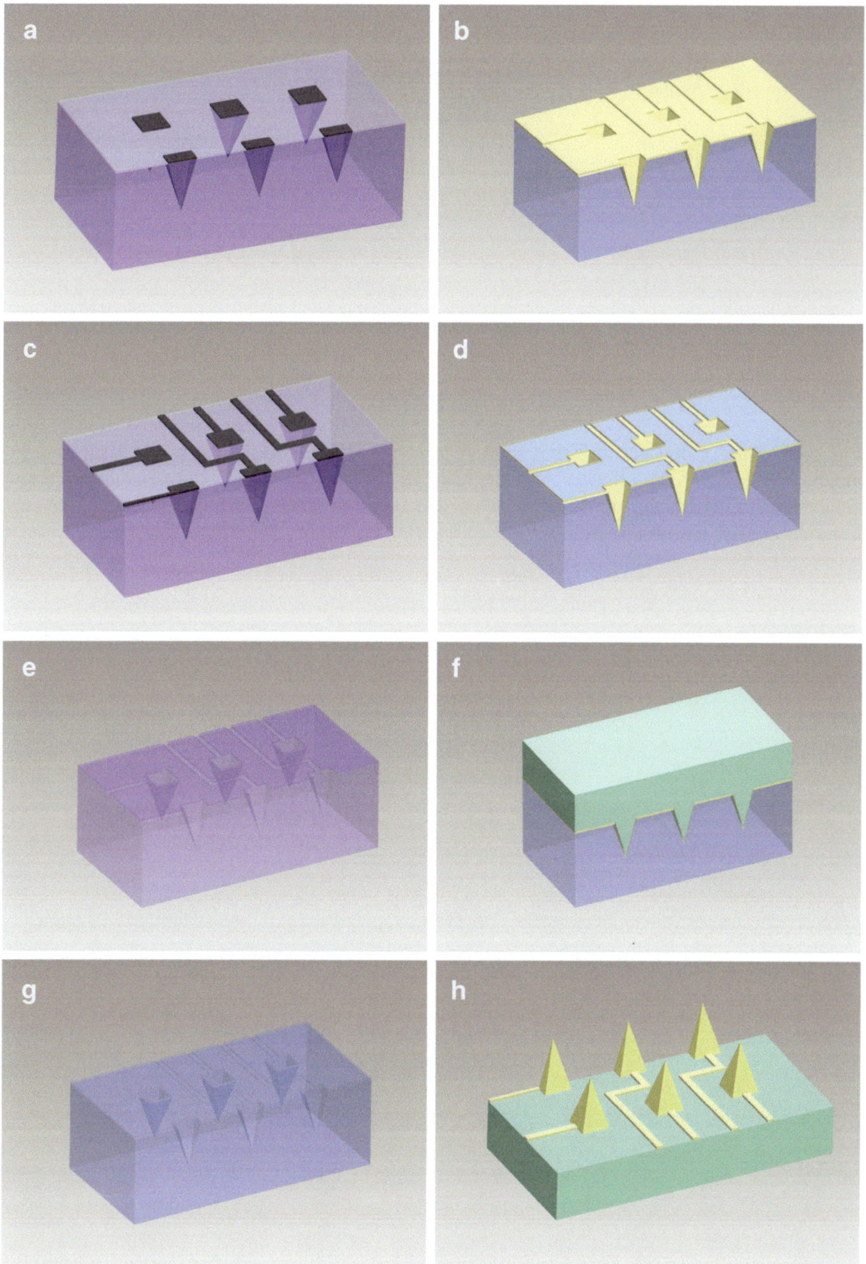

Fig. 5.44 Process flow for the fabrication of 3-D MEAs utilizing metal transfer micromolding technology [105]

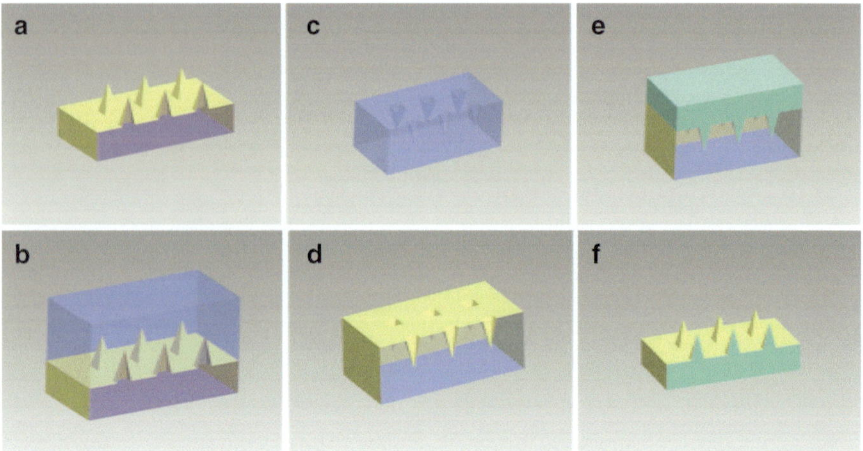

Fig. 5.45 Process flow for the fabrication of in vivo 3-D MEAs utilizing metal transfer micromolding technology. The master structure (**a**) can be fabricated using different techniques—inclined rotational exposure of SU-8 or double-side exposure of SU-8 followed by RIE sharpening. Reproduced with permission from [29], © (2011) IOP Publishing. All Rights Reserved

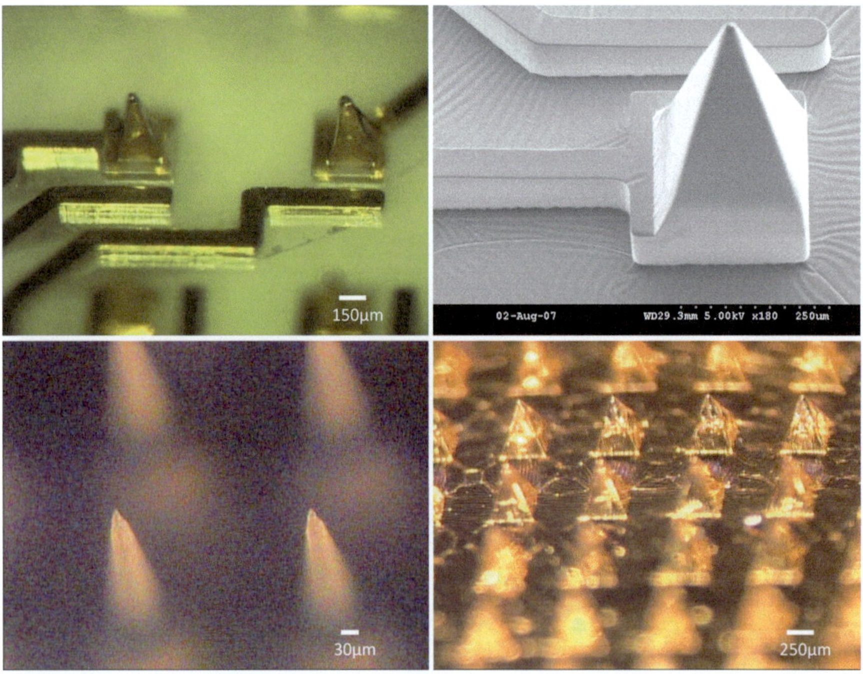

Fig. 5.46 SEM and optical micrographs of the 3-D MEAs fabricated utilizing the metal transfer micromolding technology for in vitro and in vivo applications. *Bottom left* and *bottom right*, reproduced with permission from [29], © (2011) IOP Publishing; (*top left* and *top right*) [105]

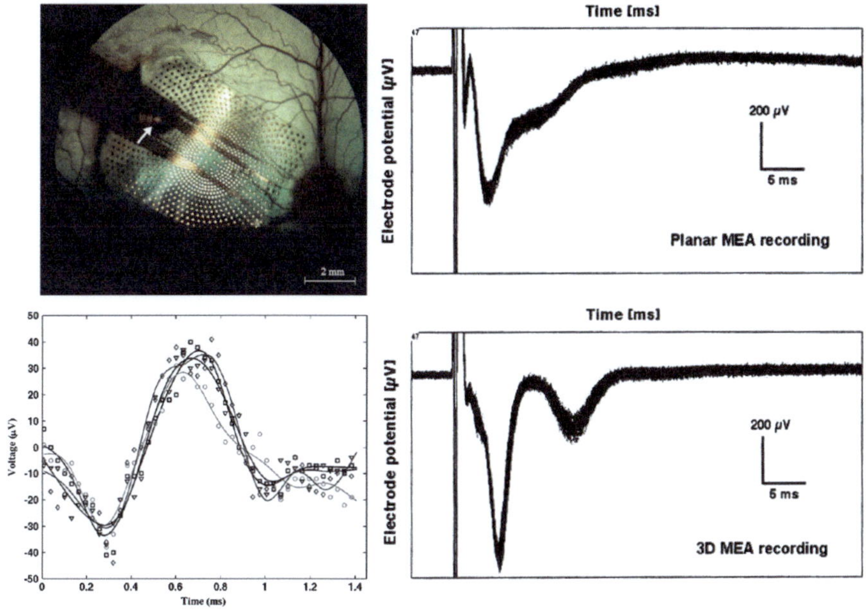

Fig. 5.47 Fundus photograph of a parylene 3-D MEAs tacked to a retina (*top left*, reproduced with permission from [92] © (2008) Elsevier); comparison between a single electrode in a planar and 3-D glass MEA showing much better spike signal acquisition in a rat hippocampal tissue slice (*top right*, reproduced with permission from [80] © (2002) Elsevier Heuschkel); spike recording from the cortex of a mouse using the EDM 3-D MEA (*bottom left* [77])

micromolding technology. For the in vitro 3-D MEAs, a parylene layer is conformally evaporated on a packaged device and patterned using laser micromachining to define the final electrode structure. The approach also reports the development of platinum electroplating on a packaged device to demonstrate low-impedance microelectrodes (results not shown).

Such metal-, polymer-, and SU-8-based technologies are becoming more popular with end users of these devices. Though not as thoroughly characterized as the silicon devices, several applications have been built on these devices. Examples of data acquisition from polymer-, glass-, SU-8-, and metal-based 3-D MEAs are depicted in Figs. 5.47 and 5.48. These images show the efficacy of these MEAs in biological experimentation.

5.4 Concluding Remarks

Microelectrode arrays (MEAs) are important tools for scientific discovery and medical advancement. Over the past 45 years, several three-dimensional MEA technologies have been developed and characterized worldwide. These technologies

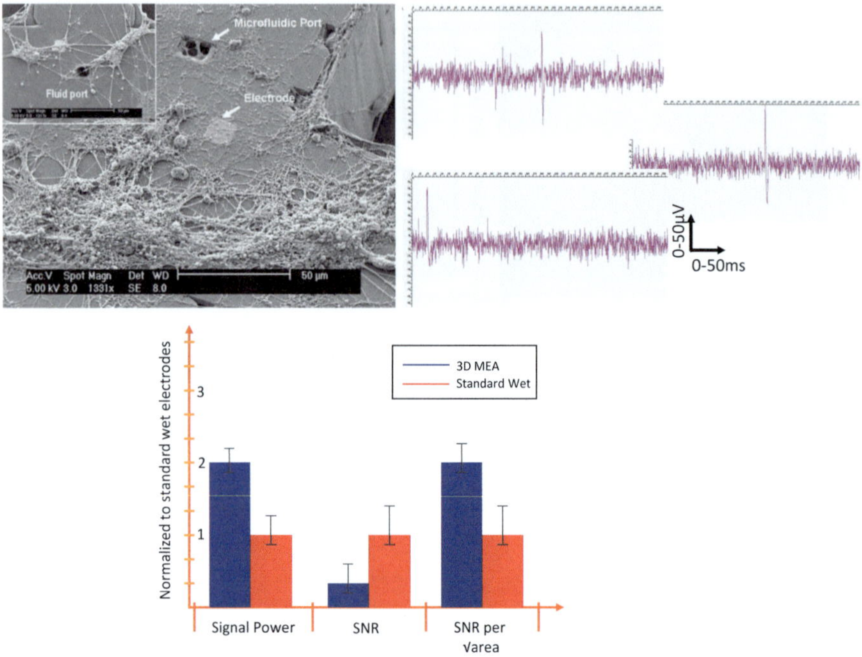

Fig. 5.48 Examples of applications developed on the Georgia Tech 3-D MEAs. An in vitro hippocampal neuronal culture on an SU-8 3-D MEA (*top left* [100]); action potential recordings from several electrodes utilizing MTM 3-D MEAs and tissue slice electrophysiology (*top right* [105]); and comparison of EMG data acquired on test human subjects between a MTM 3-D MEA and standard wet electrodes (*bottom*, reproduced with permission from [29], © (2011) IOP Publishing)

traditionally emanated from the IC fabrication industry, and hence many of the early processing techniques are based on using silicon as a substrate. To this day silicon-based 3-D MEAs remain the most well-characterized tools with extensive commercial and academic applications reported by several groups. As silicon technology continues to develop, so does the development of new and innovative fabrication methodologies for the construction of 3-D MEAs.

In the last 15 years, non-silicon-based approaches have gained a lot of traction. One of the big attractions toward moving away from silicon is familiarity of materials such as polymers for the end users of this technology. Several 3-D MEA microfabrication technologies have been reported on glass, metal, polyimide, parylene, SU-8, PDMS, and other polymer substrates. Though not as well characterized as silicon-based 3-D MEAs, the application development on these tools has increased in the last 5 years.

This chapter summarizes advances in both traditional silicon based and non-traditional Micro-Electro-Mechanical Systems (MEMS) approaches to the fabrication of 3-D MEAs. These MEAs will continue to serve as very important tools that have led to dramatic advances in the areas of neuroscience, prosthetics, pharmacology, diagnostics and implantable devices.

References

1. Galvani, L.: De Viribus Electricitatis in Motu Musculari Commentarius. In: De Bononiensi Scientiarum et Artium Instituto atque Academia Commentarii, vol. VII, Bononiae, Ex Typographia Instituti Scientiarum (1791)
2. Chowdhury, T.K.: Fabrication of extremely fine glass Micropipette Electrodes. Journal of Scientific Instruments **2**(12), 1087–1090 (1969)
3. Gray, C.M., Maldonado, P.E., Wilson, M. et al.: Tetrodes markedly improve the reliability and yield of multiple single-unit isolation from multi-unit recordings in cat striate cortex. J. Neurosc. Meth. **63**(1–2), 43–54 (1995)
4. Fishman, H.M.: Patch Voltage Clamp of a Squid Axon Membrane. Journal of Membrane Biology **24**(3–4), 265–279 (1975)
5. Chiappalone, M., Casagrande, S., Tedesco, M., Valtorta, F., Baldelli, P., Martinoia, S., Benfenati, F.: Opposite Changes in Glutamatergic and GABAergic Transmission Underlie the Diffuse Hyperexcitability of Synapsin I-Deficient Cortical Networks. Cerebral Cortex **19**(6), 1422–1439 (2009)
6. Cao, Z., Hulsizer, S., Tassone, F., Tang, H.T., Hagerman, R.J., Rogawski, M.A., Hagerman, P.J., Pessah, I.N.: Clustered burst firing in FMR1 premutation hippocampal neurons: amelioration with allopregnanolone. Hum. Mol. Genet. **21**(13), 2923–2935 (2012)
7. MacLaren, E.J., Charlesworth, P., Coba M.P., Grant, S.G.: Knockdown of mental disorder susceptibility genes disrupts neuronal network physiology in vitro. Mol. Cell. Neurosci. **47**(2), 93–99 (2011)
8. Morin, F.O., Takamura, Y., Tamiya, E.: Investigating neuronal activity with planar microelectrode arrays: Achievements and new perspectives. Journal of Bioscience and Bioengineering **100**, 131–143 (2005)
9. McConnell, E.R., McClain, M.A., Ross, J., LeFew, W.R. and Shafer, T.J.: Evaluation of multi-well microelectrode arrays for neurotoxicity screening using a chemical training set. NeuroToxicology **33**(5), 1048–1057 (2012)
10. Lignani, G., Raimondi, A., Ferrea, E., Rocchi, A., Paonessa, F., Cesca, F., Orlando, M., Tkatch, T., Valtorta F., Cossette, P., Baldelli, P., and Benfanati, F.: Epileptogenic Q555X SYN1 Mutant Triggers Imbalances in Release Dynamics and Short-Term Plasticity. Hum. Mol. Genet. **22**, (11), 2186–9219 (2013)
11. Wong, R.O.L., Meister, M., Shatz, C.J.: Transient Period of Correlated Bursting Activity During Development of the Mammalian Retina. Neuron **11**, 923–938 (1993)
12. Wagenaar, D.A., Madhavan, R., Pine, J., Potter, S.M.: Controlling bursting in cortical cultures with closed-loop multi-electrode stimulation. Journal of Neuroscience **25**, 680–688 (2005)
13. Haustein, M.D., Reinert, T., Warnatsch, A.: Synaptic transmission and short-term plasticity at the calyx of Held synapse revealed by multielectrode array recordings. Journal of Neuroscience Methods **174**(2), 227–236 (2008)
14. Bastrikova N., Gardner, G.A., Reece, J.M.: Synapse elimination accompanies functional plasticity in hippocampal neurons. Proceedings of the National Academy of Sciences **105**(8), 3123–3127 (2008)
15. Jimbo, Y., Tateno, T., Robinson, H.P.C.: Simultaneous induction of pathway-specific potentiation and depression in networks of cortical neurons. Biophysical Journal **76**, 670–678 (1999)
16. Dunlap, J., Bowlby, M., Peri, R.: High-throughput electrophysiology: an emerging paradigm for ion-channel screening and physiology. Nature Reviews Drug Discovery **7**(4), 358–368 (2008)
17. Chiappalone, M., Vato, A., Tedesco, M.B., Marcoli, M., Davide, F., Martinoia, S.: Networks of neurons coupled to microelectrode arrays: a neuronal sensory system for pharmacological applications. Biosensors & Bioelectronics **18**, 627–634 (2003)
18. Easter, A., Bell, M.E., Damewood, J.R., Redfern, W.S., Valentin, J-P., Winter, M.J., Fonck, C., Bialecki, R.A.: Approaches to Seizure Risk Assessment in Preclinical Drug Discovery. Drug Discovery Today **14**(17–18), 876–884 (2009)

19. Stett, A., Egert, U., Guenther, E., Hofmann, F., Meyer, T., Nisch, W., Haemmerle, H.: Biological application of microelectrode arrays in drug discovery and basic research. Analytical and Bioanalytical Chemistry **377**, 486–495 (2003)
20. Meyer, D.A., Carter, J.M., Johnstone, A.F., Shafer, T.J.: Pyrethroid Modulation of Spontaneous Neuronal Excitability and Neurotransmission in Hippocampal Neurons in Culture. Neurotoxicology **29**(2): 213–225 (2008)
21. Martinoia, S., Bonzano, L., Chiappalone, A.: In vitro cortical neuronal networks as a new high-sensitive system for biosensing applications. Biosensors and Bioelectronics **20**(10), 2071–2078 (2005)
22. Gholmieh, G., Courellis, S., Fakheri, S., Cheung, E., Marmarelis, V., Baudry, M., Berger, T.: Detection and classification of neurotoxins using a novel short-term plasticity quantification method. Biosensors & Bioelectronics **18**, 1467–1478 (2003)
23. Kovacs, G.T.A.: Electronic sensors with living cellular components. Proceedings of the IEEE **91**(6), 915–929 (2003)
24. DeBusschere, B.D., and Kovacs, G.T.A.: Portable cell-based biosensor system using integrated CMOS cell-cartridges. Biosensors and Bioelectronics **16**, 543–556 (2001)
25. Wise, K.D.: Silicon microsystems for neuroscience and neural prostheses. Engineering in Medicine and Biology Magazine, IEEE **24**(5), 22–29 (2005)
26. Normann, R.A.: Technology Insight: future neuroprosthetic therapies for disorders of the nervous system. Nature Clinical Practice Neurology **3**(8), 444–452 (2007)
27. Griss, P., Enoksson, P., Tolvanen-Laakso, H.K., Merilainen, P., Ollmar, S., and Stemme, G.: Micromachined Electrodes for Biopotential Measurements. IEEE Journal of Microelectromechanical Systems **10**(1), 10–16 (2001)
28. Ruther, P., Lapatki, B., Trautmann, A., and Paul, O.: Micro Needle Based Electrode Arrays for Surface Electromyography. Mikrosystemtechnik Kongress, October, Dresden, Germany. VDE Verlag, Berlin, Offenbach, Germany (2007)
29. Rajaraman, S., Bragg, J.A., Ross, J.D., and Allen, M.G.: Micromachined Three-Dimensional Electrode Arrays for Transcutaneous Nerve Tracking. Journal of Micromechanics and Microengineering **21**(8), 085014 (2011)
30. Plummer, J.D., and Meindl, J.D.: Micropower Electronics for a Reading Aid for the Blind. IEEE J. Solid State Circuits **7**(2), 111–119 (1972)
31. Allen, H.V., Knutti, J.W., and Meindl, J.D.: Integrated circuits for an Implantable Pulsed Doppler Ultrasonic Blood Flowmeter. IEEE J. Solid State Circuits **13**(6), 853–863 (1978)
32. Samaun, Wise, K.D., and Angell, J.B.: An IC Piezoresistive Pressure Sensor for Biomedical Instrumentation. IEEE Trans. Biomed. Engr. **20**(3), 101–109 (1973)
33. Wise, K.D., Angell, J.B., and Starr, A.: An Integrated Circuit Approach to Extracellular Microelectrodes. Proc. of Intrl Conf. on Med and Bio Engr., 14–15 (1969)
34. Wise K.D., and Angell, J.B.: A Low-Capacitance Multielectrode Probe for Use in Extracellular Neurophysiology. Biomedical Engineering, IEEE Transactions on **BME-22**(3), 212–219 (1975)
35. Najafi, K., Wise, K.D., and Mochizuki, T.: A High Yield IC-Compatible Multichannel Recording Array. IEEE Transactions on Electron Devices **32**(7), 1206–1211 (1985)
36. Hoogerwerf, A.C., and Wise, K.D.: A Three-Dimensional Microelectrode Array for Chronic Neural Recording. IEEE Trans. Biomed. Engr. **41**(12), 1136–1146 (1994)
37. Wise, K.D., Anderson, D.J., Hetke, J.F., Kipke, D.R., and Najafi, K.: Wireless implantable microsystems: high-density electronic interfaces to the nervous system. Proceedings of the IEEE **92**(1), 76–97 (2004)
38. Bell T.E., Wise, K.D., and Anderson, D.J.: A flexible micromachined electrode array for a cochlear prosthesis. Solid State Sensors and Actuators, 1997. TRANSDUCERS '97 Chicago., 1997 International Conference on **2**, 1315–1318 (1997)
39. Bell T.E., and Wise, K.D.: A dissolved wafer process using a porous silicon sacrificial layer and a lightly-doped bulk silicon etch-stop. Paper presented at the Eleventh Annual International Workshop on Micro Electro Mechanical Systems. MEMS 98. pp. 251–256, 25–29 January 1998

40. Huang, C., and Najafi, K.: Fabrication of Ultrathin p++ Silicon Microstructures using Ion Implantation and Boron Etch Stop. IEEE Journal of MicroElectroMechanical Systems (JMEMS) **10**, 532–537 (2001)
41. Ghovanloo, M., Beach, K., Wise, K.D., and Najafi, K.: A BiCMOS Wireless Interface Chip for Micromachined Stimulating Microprobes. IEEE EMBS Conference on Microtechnologies in Medicine and Biology, pp. 277–282, 2–4 May 2002
42. Bai Q., Wise, K.D., and Anderson, D.J.: A high-yield microassembly structure for three-dimensional microelectrode arrays. Biomedical Engineering, IEEE Transactions on **47**(3), 281–289 (2000)
43. Ghovanloo, M., and Najafi, K.: A Three-Dimensional Microassembly Structure for Micromachined Planar Microelectrode Arrays. Proc. IEEE EMBS Conf., pp. 112–115 (2005)
44. Merriam, M.E., Srivannavit, O., Gulari, M.N., Wise, K.D.: A Three-Dimensional 64-Site Folded Electrode Array using Planar Fabrication. IEEE Journal of Microelectromechanical Systems. **20**(3), 594–600 (2011)
45. Papageorgiou, D.P., Shore, S.E., Bledsoe, S.C., Wise, K.D.: A Shuttered Neural Probe with On-Chip Flowmeters for Chronic in-vivo Drug Delivery. IEEE Journal of Microelectromechanical Systems **15**(4), 1025–1033 (2006)
46. NeuroNexus (2013) http://www.neuronexus.com
47. Campbell, P.K., Jones, K.E., Huber, R.J., Horch, K.W., and Normann, R.A.: A Silicon-based, 3-Dimensional Neural Interface - Manufacturing Processes for an Intracortical Electrode Array. IEEE Transactions on Biomedical Engineering **38**(8), 758–768 (1991)
48. Jones, K.E., Campbell, P.K., and Normann, R.A.: A Glass Silicon Composite Introcortical Electrode Array. Annals of Biomedical Engineering **20**(4), 423–437 (1992)
49. Sharma, A., Rieth, L., Tathireddy, P., Harrison, R., Oppermann, H., Klein, M., Topper, M., Jung, E., Normann, R., Clark, G., and Solzbacher, F.: Long-term in-vitro Functional Stability and Recording Longevity of Fully Integrated Wireless Neural Interfaces based on the Utah Slant Electrode Array. Journal of Neural Engineering **8**, 045005 (2011.)
50. Rousche, P.J., Normann, R.A.: Chronic Recording Capability of the Utah Intracortical Electrode Array in a cat sensory cortex. Journal of Neuroscience Methods **82**, 1–15 (1998)
51. Bhandari, R., Negi, S., Rieth, L., Normann, R.A., and Solzbacher, F.: A novel masking method for high aspect ratio penetrating microelectrode arrays. J. Micromech. Microeng. **19**, 035004 (8pp) (2009)
52. Nordhausen, C.T., Maynard, E.M., and Normann, R.A.: Single unit Recording Capabilities of a 100 Microelectrode Array. Brain Research **726**, 129–140 (1996)
53. Blackrock Microsystems Inc. (2013) http://www.blackrockmicro.com
54. Brenner, A., Stein R.B., and Normann, R.A.: Selective Stimulation of Cat Sciatic Nerve using an Array of Varying Length Microelectrodes. Journal of Neurophysiology **85**, 1585–1594 (2001)
55. Neves, H.P., and Ruther, P.: The NeuroProbes Project. NeuroProbes Consortium Report, (2007)
56. Stieglitz, T., Rubehn, B., Henle, C., Kisban, S., Hervik, S., Ruther, P., and Schuettler M.: Brain-Computer Interfaces: an Overview of the Hardware to Record Neural Signals from the Cortex. Prog. In Brain Research **175**(20), 297–315 (2009)
57. Kisban, S., Herwik, S., Seidl, K., Rubehn, B., Jezzini, A., Umiltà, M.A., Fogassi, L., Stieglitz, T., Paul, O., and Ruther, P.: Microprobe Array with Low Impedance Electrodes and Highly Flexible Polyimide Cables for Acute Neural Recording. IEEE EMBS, pp. 175–178, Lyon, 22–26 August 2007
58. Herwik, S., Kisban, S., Aarts A.A.A., Seidl, K., Girardeau, G., Benchenane, K., Zugaro, M.B., Wiener, S.I., Paul, O., Neves, H.P., and Ruther, P.: Fabrication Technology for Silicon-based Microprobe Arrays used in Acute and Sub-Chronic Neural Recording. Journal of Micromechanics and Microengineering **19**, 074008 (2009)
59. Neves, H., and Ruther, P.: NeuroProbes - Development of Multifunctional Probe Arrays for Cerebral Applications. R&D Fact Sheet (2008)
60. Aarts, A.A.A., Neves, H.P., Ulbert, I., Wittner, L., Grand, L., Fontes, M.B.A., Herwik, S., Kisban S., Paul, O., Puers, R.P., Ruther, P., and Van Hoof, C.: A 3D Silm-base Probe Array for In Vivo Recorded Neuron Activity. Proc. IEEE EMBS Conf., pp. 5798–5801 (2008)

61. Aarts A.A.A.: 3D Interconnect Technology for out-of-plane Biomedical probe arrays - a modular approach with slim-base solutions. Dissertation, Katholieke Universiteit Leuven (2011)
62. Spieth, S., Schumacher, A., Seidl, K., Hiltmann, K., Haeberle, S., McNamara, R., Dalley, J.W., Edgley, S.A., Ruther, P., and Zengerle, R.: Robust Microprobe Systems for Simultaneous Neural Recordings and Drug Delivery. Proc. IFMBE, pp. 2426–2430 (2008)
63. Spieth, S., Brett, O., Seidl, K., Aarts, AAA, Erismis, M.A., Herwik, S., Trenkle, F., Tatzner, S., Auber, J., Daub, M., Neves, H.P., Puers, R., Paul, O., Ruther, P., and Zengerle, R.: A Floating 3D Silicon Microprobe Array for Neural Drug Delivery Compatible with Electrical Recording. Journal of Micromechanics and Microengineering **21**, 125001 (16pp) (2011)
64. Aarts, A.A.A., Neves, H.P., Puers, R.P., Herwik, S., Seidl, K., Ruther, P., and Van Hoof, C.: A Slim out-of-plane 3D Implantable CMOS based Probe Array. Proc. Smart Systems Integration, pp. 258–263 (2009)
65. Seidl, K., Herwik, S., Nurcahyo, Y., Torfs, T., Keller, M., Schuttler, M., Neves, H., Stieglitz, T., Paul, O. and Ruther, P.: CMOS-Based High Density Silicon Microprobe Array for Electronic Depth Control in Neural Recording. Proc. IEEE (2009)
66. Frey, O., van der Wal, P.D., Spieth, S., Brett, O., Seidl, K., Paul, O., Ruther, P., Zengerle, R., and de Rooij, N.F.: Biosensor Microprobes with Integrated Microfluidic Channels for bi-directional Neurochemical Interaction. Journal of Neural Engineering, **8**, 066001 (2011)
67. Koo, K., Chung, H., Yu, Y., Seo, J., Park, J., Lim, J-M., Paik, S-J., Park, S., Choi, H.M., Jeong M-J., Kim, G.S., and Cho, D-I.: Fabrication of Pyramid Shaped Three-Dimensional 8x8 Electrodes for Artificial Retina. Sensors and Actuators A **130–131**, 609–615 (2006)
68. Kusko, M., Craciunoiu, F., Amuzescu, B., Halitzchi, F., Selescu, T., Radoi, A., Popescu, M., Simion, M., Bragaru, A., and Ignat, T.: Design, Fabrication and Characterization of a Low-Impedance 3-D Electrode Array System for Neuro-Electrophysiology. Sensors Journal **12**, 16571–16590 (2012)
69. Du, J., Roukes, M.L., and Masmanidis, S.C.: Dual-side and Three-Dimensional Microelectrode Arrays Fabricated from ultra-thin Silicon Substrates. Journal of Micromechanics and Microengineering **19**, 075008 (2009)
70. Hanein, Y., Schabmueller, C.G.J., Holman, G., Lucke, P., Denton, D.D., and Bohringer, K.F.: High Aspect Ration Submicrometer Needles for Intracellular Applications. Journal of Micromechanics and Microengineering **13**, S91–95 (2003)
71. Chu, H-Y., Kuo, T-Y. Chang, B., Chiao, C-C., and Fang, W.: Development of the Three-Dimensional Multi-Electrode Array for Neural Recording. Proc. IEEE Transducers Conf., pp. 1804–1807 (2005)
72. Harimoto, T., Takei, K., Kawano, T., Ishihara A., Kawashima, T., Kaneko H., Ishida, M., and Usui, S.: Enlarged Gold-Tipped Silicon Microprobe Arrays and Signal Compensation for Multi-site Electroretinogram Recordings in the Isolated Carp Retina. Biosensors and Bioelectronics **26**, 2368–7523 (2011)
73. Takei K., Kawashima, T., Takao, H., Sawada, K., and Ishida, M.: Si Micro Probe and SiO2 Micro Tube Array with NMOSFETs. Transducers 2007: The 14th International Conference on Solid-State Sensors, Actuators and Microsystems IEEE, pp. 1381–1384 (2007)
74. Takei, K., Kawashima, T., Kawano, T., Takao, H., Sawada, K., and Ishida, M.: Integration of out-of-plane Silicon Dioxide Microtubes, Silicon Microprobes and on-chip NMOSFETs by Selective Vapor-Liquid-Solid Growth. Journal of Micromechanics and Microengineering **18**, 035033 (2008)
75. Fofonoff, T.A., Martel, S., Wiseman, C., Dyer, R., Hunter, I.W., Hatsopoulos, N.G., and Donoghue, J.P.: A highly flexible manufacturing technique for microelectrode array fabrication. Engineering in Medicine and Biology, 2002. 24th Annual Conference and the Annual Fall Meeting of the Biomedical Engineering Society EMBS/BMES Conference, 2002. Proceedings of the Second Joint **3**, 2107–2108, Houston, Tx, USA, 23–26 October 2002
76. Fofonoff, T.A., Wiseman, C., Dyer, R., Malasek, J., Burgert, J., Martel, S., Hunter, I.W., Hatsopoulos, N.G., and Donoghue, J.P.: Mechanical assembly of a microelectrode array for use in a wireless intracortical recording device. 2nd Annual International IEEE EMBS

Special Topic Conf. on Microtechnologies in Medicine and Biology, pp. 269–72, Madison, Wisconsin USA, 24 May 2002

77. Fofonoff, T.A., Martel, S.M., Hatsopoulos, N.G., Donoghue, J.P., and Hunter, I.W.: Microelectrode array fabrication by electrical discharge machining and chemical etching. IEEE Transactions on Biomedical Engineering **51**(6), 890–895 (2004)
78. Specialty Coating Systems Inc. (2013) http://scscoatings.com
79. Metz, S., Heuschkel, M.O., Valencia Avila, B., Holzer, R., Bertrand, D., and Renaud, P.: Microelectrodes with Three-Dimensional Structures for Improved Neural Interfacing. Proc. of 23rd Annual EMBS Intr. Conf., IEEE, 765–768 (2001)
80. Heuschkel, M.O., Fejtl, M., Raggenbass, M., Bertrand, D., and Renaud, P.: A three-dimensional multi-electrode array for multi-site stimulation and recording in acute brain slices. Journal of Neuroscience Methods **114**(2), 135–148 (2002)
81. DuPont Inc. (2013) http://www2.dupont.com/Kapton/en_US/assets/downloads/pdf/summaryofprop.pdf
82. Rousche, P.J., Pellinen, D.S., Pivin, D.P., Williams, J.C., Vetter, R.J., and Kipke, D.R.: Flexible Polyimide-based Intracortical Electrode Arrays with Bioactive Capability. IEEE Trans. In Biomedical Engineering **48**(3), 361–371 (2001)
83. Owens, A.L., Denison, T.J., Versnel, H., Rebbert, M., Peckerar, M., and Shamma, S.A.: Multielectrode Array for Measuring Evoked Potentials from Surface of Ferret Primary Auditory Cortex. J. Neuroscience Methods **58**(1–2), 209–220 (1995)
84. Boppart, S.A., Wheeler, B.C., and Wallace, C.S.: A Flexible Perforated Microelectrode Array for Extended Neural Recordings. IEEE Transactions on Biomedical Engineering, **39**(1), 37–42 (1992)
85. Stieglitz, T., and Meyer, J.U.: Implantable Microsystems: Polyimide-based Neuroprostheses for Interfacing Nerves. Med. Dev. Technol., **10**(6), 28–30 (1999)
86. Takeuchi, S., Suzuki, T., Mabuchi, K., and Fujita, H.: 3D Flexible Multichannel Neural Probe Array. Journal of Micromechanics and Microengineering **14**, 104–107 (2004)
87. Lee, K-K., He, J., Singh, A., Massia, S., Ehteshami, G., Kim, B., and Raupp, G.: Polyimide-based Intracortical Neural Implant with Improved Structural Stiffness. Journal of Micromechanics and Microengineering **14**, 32–37 (2004)
88. Blum, N.A., Carkhuff, B.G., Charles, Jr. H.K., Edwards, R.L., and Meyer, R.A.: Multisite Microprobes for Neural Recordings. Biomedical Engineering, IEEE Transactions on **38**, 68–74 (1991)
89. Maher, M.P., Pine, J., Wright, J., and Tai, Y.C.: The Neurochip: A New Multielectrode Device for Stimulating and Recording from Cultures Neurons. Journal of Neuroscience Methods **87**, 45–56 (1999)
90. Tooker, A., Meng, E., Ericson, J., Tai, Y.C., and Pine, J.: Biocompatible Parylene Neurocages. IEEE Engineering in Medicine and Biology Magazine **24**, 30–33 (2005)
91. Rodger, D.C., Weiland, J.D., Humayun, M.S., and Tai, Y.C.: Scalable high lead-count parylene package for retinal prostheses. Sensors and Actuators B **117**, 107–114 (2006)
92. Rodger, D.C., Fong, A.J., Li, W., Ameri, H., Lavrov, I., Zhong, H., Saati, S., Menon, P., Meng, E., Burdick, J.W., Roy, R.R., Edgerton, V.R., Weiland, J.D., Humayun, M.S. and Tai, Y.C.: High-Density Flexible Parylene-Based Multielectrode Arrays for Retinal and Spinal Cord Stimulation. Solid-State Sensors, Actuators and Microsystems Conference, 2007. TRANSDUCERS 2007. International, pp. 1385–1388 (2007)
93. Rodger, D.C., Fong, A.J., Li, W., Ameri, H., Ahuja, A.K., Guiterrez, C., Lavrov, I., Zhong, H., Menon, P., Meng, E., Burdick, J.W., Roy, R.R., Edgerton, V.R., Weiland, J.D., Humayun, M.S., and Tai, Y.C.: Flexible Parylene-Based Multielectrode Arrays for High-Density Neural Stimulation and Recording. Sensors and Actuators B **132**, 449–460 (2008)
94. Rodger, D.C., and Tai, Y.C.: Microelectronic Packaging for Retinal Prosthesis. IEEE Engineering in Medicine and Biology Magazine **24**(5), 52–57 (2005)
95. Frazier, A.B., O'Brien, D.P., and Allen, M.G.: Two dimensional metallic microelectrode arrays for extracellular stimulation and recording of neurons. Micro Electro Mechanical Systems, MEMS '93, IEEE, pp. 195–200 (1993)

96. Frazier, A.B., Ahn, C.H., and Allen, M.G.: Development of Micromachined Devices using Polyimide-based Processes. Sensors and Actuators A **45**, 47–53 (1994)
97. O'Brien, D.P., Allen, M.G., and Nichols, T.R.: Flexible Microelectrode Arrays with Integrated Insertion Devices. Annals of Biomedical Engineering **25**(sup. 1), 58 (1997)
98. O'Brien, D.P., Nichols, T.R., and Allen, M.G.: Flexible microelectrode arrays with integrated insertion devices. Micro Electro Mechanical Systems, 2001. MEMS 2001. The 14th IEEE International Conference on, pp. 216–219 (2001)
99. Rowe L., Almasri, M., Fogleman, N., Frazier, A.B., and Brewer, G.J.: An active microscaffold for culturing 3-D neuronal networks. Solid-State Sensors, Actuators and Microsystems, 2005 Digest of Technical Papers. TRANSDUCERS '05. The 13th International Conference on, pp. 948–951 (2005)
100. Rowe L., Almasri, M., Lee, K., Fogleman, N., Brewer, G.J., Nam, Y., Wheeler, B.C., Vukasinovic, J., Glezer, A., and Frazier, A.B.: Active 3-D microscaffold system with fluid perfusion for culturing in vitro neuronal networks. Lab on a Chip **7**(4), 475–482 (2007)
101. Choi, Y., Choi, S-O., Shafer, R.H., and Allen, M.G.: Highly Inclined Electrodeposited Metal Lines Using an Excimer Laser Patterning Technique. Proc. IEEE Transducers Conf. (2005). doi: 10.1109/SENSOR.2005.1497360
102. Rajaraman, S., Choi, S.O., Shafer, R.H., Ross, J.D., Vukasinovic, J., Choi, Y., DeWeerth, S.P., Glezer, A., and Allen, M.G.: Microfabrication technologies for a coupled three-dimensional microelectrode, microfluidic array. Journal of Micromechanics and Microengineering **17**(1), 163–171 (2007)
103. Choi, Y., Powers, R., Vernekar, V., Frazier, A.B., LaPlaca, M.C., and Allen, M.G.: High Aspect Ration SU-8 Structures for 3-D Culturing of Neurons. ASME International Mechanical Engineering Congress and Expo, Microelectromechanical Systems, Washington, DC, USA, 15–21 November 2003.
104. Choi Y., McClain, M.A., LaPlaca, M.C., Frazier, A.B., and Allen M.G.: Three-Dimensional MEMS Microfluidic Perfusion System for Thick Brain Slice Cultures. Biomedical Microdevices **9**(1), 15–23 (2007)
105. Rajaraman, S., Choi, S-O., McClain, M.A., Ross, J.D., LaPlaca, M.C., and Allen, M.G.: Metal Transfer Micromolded Three-Dimensional Microelectrode Arrays for in-vitro Brain Slice Recordings, IEEE Journal of Microelectromechanical Systems. **20**(2), 396–409 (2011)
106. Allen, M.G., Choi, S-O., Park, J-H., Wu, X., Zhao, Y., Yoon, Y-K., and Rajaraman, S.: Method for Making Electrically Conductive Three-Dimensional Structures. U.S. Patent Application No. 2008/0063866 A1 (2008)

Chapter 6
Focused Ion Beam Technology as a Fabrication and Inspection Tool in Neuron Interfacing

Leonardo Sileo, Ferruccio Pisanello, Luigi Martiradonna, and Massimo De Vittorio

Abstract Recent trends in the development of devices for electrophysiology involve the fabrication of electrodes with three-dimensional micro- and nanoprotrusions. These devices take advantage of the natural capacity of cells to actively interact with nanostructured substrates in order to realize a more intimate cell-to-electrode coupling. In this chapter, we review the use of focused ion beam (FIB) technology as a versatile tool for fabricating nanostructures of different shape and size on top of freely chosen substrates. This approach allows custom design and fabrication of nanoprotrusions to optimize cell-to-electrode electrical coupling, while at the same time allowing leeway to optimize the microelectronic substrate. Examples of enhanced interaction of cells with nanostructures are reviewed, with respect to nanoprotrusion geometry and surface functionalization, to illustrate the potential of FIB-based deposition as a tool for realizing new types of nanostructures for neurophysiological measurements. Finally, the combined focused ion beam/scanning electron microscope is presented as a tool for investigating the physical basis for interactions between neuronal cell membranes and nanostructured surfaces.

6.1 Introduction

The elucidation of the biophysical mechanisms of cell excitability and the encoding of information in neural networks is constrained by the spatiotemporal resolution at which the neural tissue can be studied. Accordingly, advances in neurotechnology

L. Sileo (✉) • F. Pisanello • L. Martiradonna
Center for Biomolecular Nanotechnologies @UNILE, IIT, Arnesano 73010, Italy
e-mail: leonardo.sileo@iit.it

M. De Vittorio
Center for Biomolecular Nanotechnologies @UNILE, IIT, Arnesano 73010, Italy

NNL-National Nanotechnology Laboratory, CNR-NANO, Università del Salento, Lecce 73100, Italy

M. De Vittorio et al. (eds.), *Nanotechnology and Neuroscience: Nano-electronic, Photonic and Mechanical Neuronal Interfacing*, DOI 10.1007/978-1-4899-8038-0_6,

have driven new insights in neuroscience. For example, the development of sharp and patch electrodes for electrically accessing the interior of single cells allowed the investigation of electrical activity in electrogenic cells, including action potentials and subthreshold potentials [1, 2]. The development of Multi Electrode Arrays (MEA) for recording the activity of many single neurons simultaneously facilitated the study of dynamics, computation, and learning in neural networks [3–6] and provided a platform for screening the effects of drugs on neural networks [7–10]. The realization of silicon-based implantable needle chips with multiple recording sites ("Michigan electrodes") [11] and of 3D arrays of microwires ("Utah arrays") [12] opened the way for long-term recording and stimulation of multiple neurons in vivo by means of intracortical chronic implants. Other techniques such as electroencephalography (EEG), electrocorticography (ECoG), magnetoencephalography (MEG), and functional magnetic resonance imaging (fMRI) provide signatures of the collective activity of large groups of neurons for studying brain function or for medical diagnostics [13, 14]. Despite the rich toolkit available to neuroscientists, there is still the need for new tools for electrophysiology that could enable real-time, highly spatially resolved recordings for long periods of time. Ideally, such tools should be able to record the electrical activity of a great number of neurons with high temporal fidelity and with low invasiveness. Such an approach could reveal how neural computations emerge from the coordinated activity of hundreds or thousands of neurons [13].

The sensitivity of electrophysiology tools is a direct consequence of the nature of the interface between the electrode and the neural tissue. Patch-clamping produces high-quality recordings of transmembrane potential or currents due to the direct contact between the electrolyte filling the pipette and the cytosol and to the gigaohm seal between the pipette and the cell membrane [2]. However, such intimate access requires high mechanical stability and affects the viability of the cell. Metal extracellular electrodes capacitively transduce the voltage drop caused by extracellular currents originating from the nearby neurons. Extracellular recordings are less invasive but come at the price of low signal-to-noise ratio (SNR) and low bandwidth, which preclude detection of subthreshold synaptic potentials. Another limitation is that the extracellular signal results from the superposition of the activity of a number of neurons [15] and thus requires spike-sorting methods that can be somewhat subjective [16]. A one-to-one correspondence between cells and electrodes would be desirable to reduce or eliminate the uncertainty of reconstructing single unit activity. Reducing electrode size increases the spatial resolution of extracellular recordings, although the consequent increase of the electrode impedance negatively affects the SNR. This drawback can be in part overcome by increasing the effective surface area of the electrodes with nanoporous platinum black or Ti_3N_4, gold nanoflakes, and carbon nanotubes [17–23] or with conductive polymer coatings such as poly(pyrrole) (PPy) and poly(3,4-ethylenedioxythiophene) (PEDOT) [24–27]. For stimulating electrodes, charge-injection limit is an additional constraint to miniaturization. In this case, noble metals such as Ti, TiN, Pt, and iridium oxide (IrOx) enable higher charge-density delivery without electrode degradation or electrolysis of water [28]. More generally, the SNR of extracellular recordings can be increased

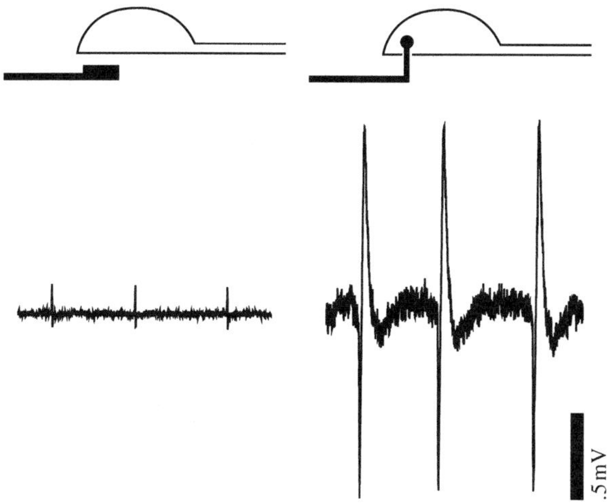

Fig. 6.1 Spine-shaped electrodes (*right*) enhance the neuron/electrode electrical coupling enabling higher amplitude recordings in comparison to flat electrodes (*left*) [37]

by (1) achieving a tight cell adhesion onto the electrode and by (2) increasing the surface area of the cell/electrode junction. Tighter adhesion increases the resistance to the flow of ionic currents in the reduced cleft between the electrode and the cell, thus increasing the extracellular potential (by Ohm's law), while a more extended junction also leads to a higher resistance of the cleft, as well as a higher total current in the cleft. Previous approaches for enhancing the cell-to-substrate adhesion include protein coatings [29, 30] or topographical patterning of the electrode surface [31, 32]. The latter can also increase the effective surface area of the junction. Both approaches strengthen cell adhesion and facilitate cell retention from suspension [33]. However, ultrastructural analysis of the cell-to-substrate junction reveals that the junction is invariably characterized by points of tight adhesion and other parts of the junction with higher cell-to-substrate separation [67–71] (see Sect. 6.4). A recent method to achieve a tight adhesion on an extended portion of membrane, thus producing a substantially narrower cleft, is based on the employment of mushroom-shaped electrodes that are effectively engulfed by cells [34, 38], provided they are functionalized with a specific "engulfment-promoting peptide" (for a review see Chap. 3). The method exploits the tendency of cells to wrap the three-dimensional microprotrusions in a phagocytosis-like fashion. This was demonstrated for cultured Aplysia neurons [36–39], cardiomyocytes [36, 42], and mammalian neurons [34, 35, 40]. The analysis of the interface revealed a reduced average distance of the cell membrane from the electrode surface in comparison to flat electrode surfaces (see Sect. 6.4). In large Aplysia neurons, this increased electrical coupling revealed larger extracellular potentials (Fig. 6.1) [37] and even attenuated "intracellular" recordings [39]; however, these improvements in recording quality have not yet been demonstrated with smaller mammalian neurons.

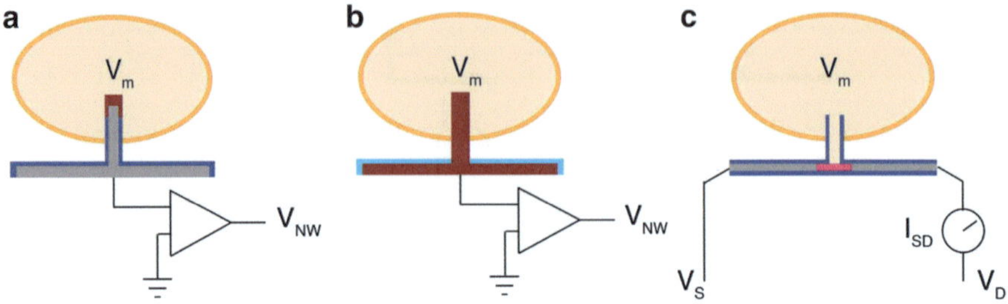

Fig. 6.2 Nanoscale electrodes for minimally invasive intracellular recordings. (**a**) Silicon vertical nanowires with insulated sidewalls penetrating the cell membrane. (**b**) Platinum nanopillars (on an insulated platinum electrode) engulfed by the cell. (**c**) Silicon dioxide nanotube penetrating the cell membrane for intracellular recording with an NwFET [43]

A more recent breakthrough was achieved by using new nanotechnology approaches to fabricate nanoscale devices for minimally invasive *intracellular* recordings, which allow for much higher SNR than extracellular recordings [43].

In one case (Fig. 6.2a), nine Silicon vertical nanowires (diameter 150 nm, height 3 μm), arranged in a 3 × 3 array (pitch 2 μm) on each of a 16 electrodes MEA, were realized by advanced silicon nanofabrication processes [44]. The sharpness of the nanowires is used to induce membrane penetration (directly or after the application of a voltage cycle), allowing intracellular recording from embryonic rat cortical neurons (Fig. 6.2a). In another case (Fig. 6.2b), similar-sized platinum nanopillars were grown on standard MEA electrodes by ion beam-induced deposition and used to record from cardiomyocytes extracellularly and then following transient membrane electroporation intracellularly [45]. A third solution was based on scaling the hollow pipet electrode down to nanometer size (Fig. 6.2c). A silicon dioxide nanotube (diameter 55 nm) was placed on top of the gate of a nanowire field effect transistor (NwFET) [46]. After penetration into the plasma membrane of a cardiomyocyte, the nanotube was filled by a minimal quantity of intracellular liquid, thus connecting the NwFET with the cell interior.

These examples make clear that "nanoprotrusions" offer the prospect of highly localized interfacing with cells, to enhance electrophysiological signals and reduce invasiveness. Among the approaches for fabricating three-dimensional micro- and nano-electrodes, focused ion beam (FIB) technology is a promising candidate. In the next sections, we review recent examples of FIB technology applied to nanoprotrusion-based electrodes, highlighting the versatility of FIB for fast prototyping of complex nanostructures and for realizing platforms for high-density, high-resolution neuronal recordings. We then summarize recent reports of enhanced interactions of cells with protruding electrodes, as related to the combination of protrusion geometry and surface functionalization, and we consider the use of FIB as a tool to probe the basis for enhanced electrical coupling at the cell-to-nanoprotrusion interface. In particular, we highlight the potential use of the focused ion beam/scanning electron microscope (FIB/SEM) to resolve the ultrastructural details of the cell-to-substrate interface.

6.2 Focused Ion Beam Technology

The focused ion beam (FIB) microscope is nowadays broadly used in both material science and technological applications, spanning from microstructural analysis to nanomachining and nanofabrication [47–50]. It is similar to a scanning electrode microscope (SEM), except that an ion beam from a liquid metal ion source is focused and rastered on the target sample in place of a standard field-emission electron beam. Like the SEM, FIB can be used for high-resolution imaging, although the peculiar interaction of the heavier ions with the sample material gives rise to a different material-related contrast [48]. Because the ions impinge on the sample with high kinetic energy, a certain degree of material removal is obtained during FIB action. This sputtering effect can be controlled with a resolution down to 5–10 nm [49] and has been exploited for micromachining applications [50]. In addition, FIB systems equipped with injectors of gas precursors are capable of highly localized material deposition at the beam–substrate interaction site. Depending on the type of (mostly organic) precursor, both metals (W, Pt, Au, Mo, Al, Cu) and insulators (TEOS, TEOS/O_2) can be deposited with a resolution of a few tens of nanometers [49]. Specific scanning strategies can realize complex, submicron-sized three-dimensional structures, in particular structures with overhanging features, which cannot be obtained by etching from the top down [47]. Additional functionality has been afforded by the development of FIB/SEM dual beam systems, with both ion and electron beams convergent at the sample site. With FIB/SEM systems it is possible to perform high-resolution, non-damaging SEM microscopy in situ directly after FIB milling and/or deposition. For example, SEM inspection of FIB-cut cross sections is commonly employed for visualization of embedded structures and chip-fabrication defects [47]. In Sect. 6.5, we will describe FIB/SEM as a tool to inspect the cell-to-substrate interface. Moreover, the electron beam is able to deposit material with higher resolution (but lower material purity) than FIB, extending down the range of achievable nanostructure dimensions.

6.3 Neurons Interfacing with Nanostructures Fabricated by Focused Ion Beam Deposition

In recent years, FIB-induced deposition has been employed to fabricate three-dimensional nanostructures on top of substrates for in vitro cell cultures, to study the effect of nanoprotrusions on cell viability and mobility, and on top of the electrodes of MEA substrates for investigating the influence of the nanostructures on electrical coupling to electrogenic cells. Xie et al. [51] fabricated platinum nanopillars (diameter 150 nm, height 1 μm) on a custom MEA substrate to assess the influence of nanoprotrusions on cell motility during the first few days in culture. The nanopillars were arranged in rings (10 μm in diameter) or in 5×5 square arrays (2 μm pitch), both on the electrode sites and on transparent open areas of the MEA to facilitate

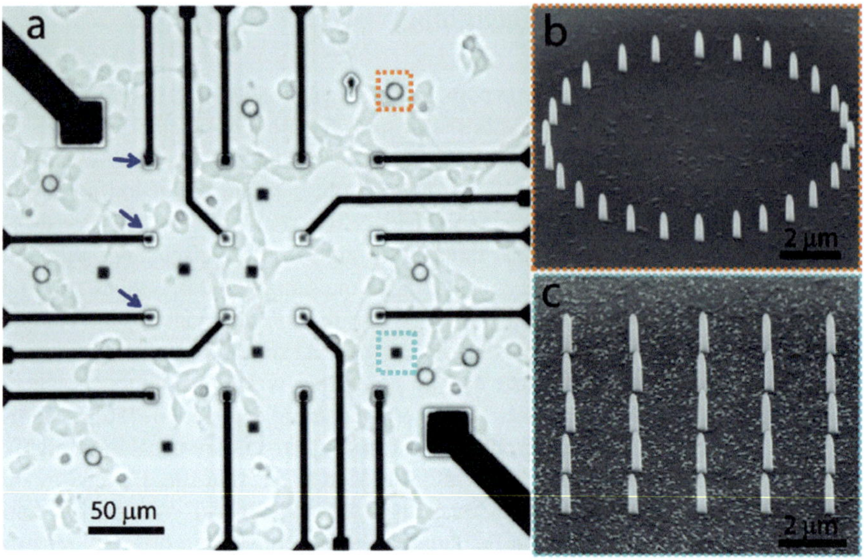

Fig. 6.3 Focused ion beam deposition of arrays of nanopillars on top of an MEA substrate to study the effect nanoprotrusions on cell mobility. (**a**) The arrays, located both in open areas (*cyan* and *orange squares*) and on the microelectrode sites (*blue arrows*), were either (**b**) ring shaped or (**c**) square (5 × 5) shaped. (**a**) Bright field image; (**b** and **c**) SEM images. Reproduced with permission from [51], © (2010) American Chemical Society

optical inspection (Fig. 6.3). The substrate was treated with poly-l-lysine (PLL) as an adhesion promoter before plating embryonic cortical neurons from E18 rats.

The authors observed that cells that had not been initially plated on nanopillars could migrate normally, whereas cells that were initially plated on nanopillars or that encountered nanopillars along their path were effectively pinned at the pillars. In all the reported cases, neurons showed similar viability. It was also observed that pinned neurons extended their projections preferentially toward nanopillars (Fig. 6.4a). Sometimes, cells increased the surface of contact by developing a thin membrane wrapping the pillars (Fig. 6.4b). In other cases, neurons interacted with nanopillars only by means of their neurites (Fig. 6.4c). Immunostaining of actin filaments followed by confocal fluorescence microscopy revealed an intimate interaction between nanopillars and the cytoskeletal network (Fig. 6.4d). There was even apparent bending of some nanopillars (Fig. 6.4b), possibly due to mechanical tension generated by the cytoskeleton-guided process of engulfment.

Xie et al. obtained similar results with Si and SiO_2 nanopillars fabricated with other techniques, indicating that the interaction of the neurons with the nanopillars is related to the nanopillar geometry, rather than being a material effect [51].

An important advantage of FIB technology is the ability to realize several three-dimensional nanostructures of different shapes and sizes on the same substrate, to compare their interactions with cells. For example, Martiradonna et al. [52] fabricated straight nanopillars, nail-headed pillars, and sphere-headed pillars together on top of the recording electrodes of an MEA substrate (Fig. 6.5).

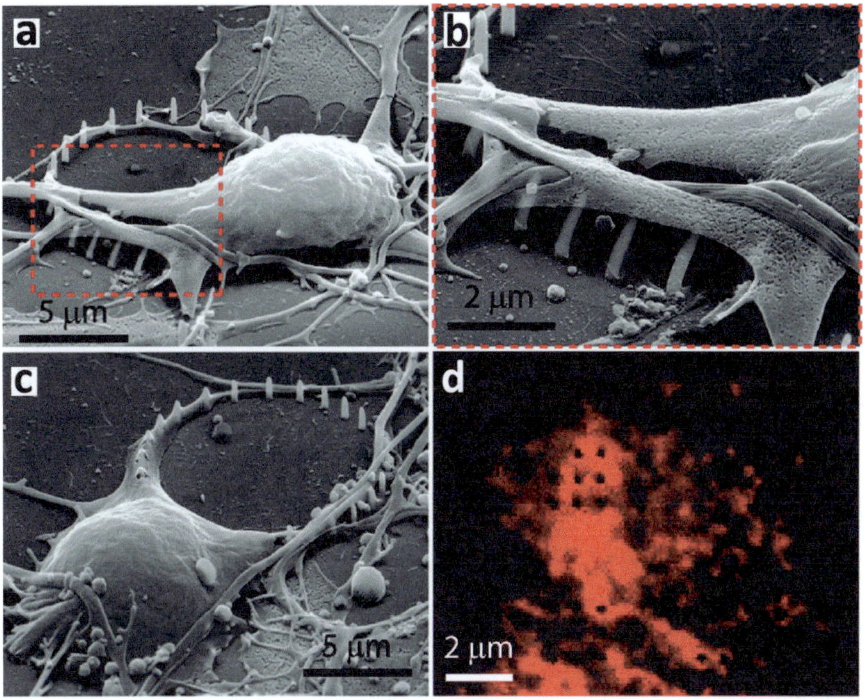

Fig. 6.4 Interaction of neurons with nanopillars. (**a**) A pinned neuron extending neuritis toward the nanopillars and (**b**) wrapping the nanopillars with the cell membrane. (**c**) A different neuron extending one of its neuritis along a ring-shaped nanopillar array. (**d**) Nanopillars (*black dots*) are embedded in the immunostained actin cytoskeletal network. (**a–c**) SEM images; (**d**) confocal microscope image. Reproduced with permission from [51], © (2010) American Chemical Society

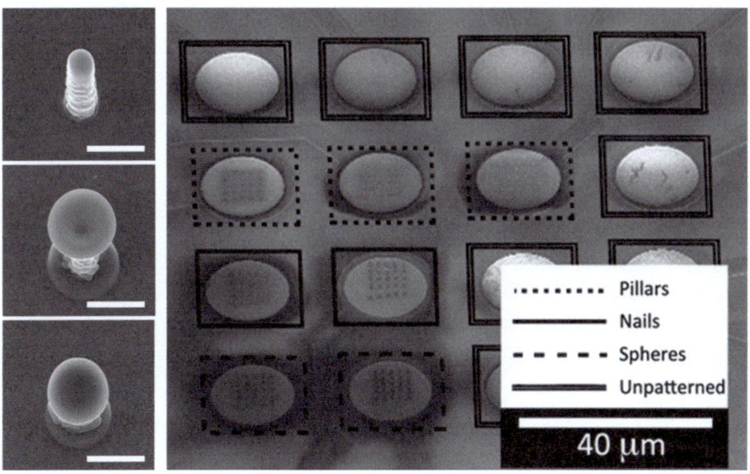

Fig. 6.5 Focused ion beam deposition of nanoprotrusions with different shapes and sizes on top of an MEA substrate for comparative analysis of the interaction with cultured neurons. On the *left-hand side*, from *top* to *bottom*, SEM pictures of nanopillar, nail-headed, and sphere-headed nanoprotrusions (scale bar 1 μm). On the *right-hand side*, 5×5 arrays of different nanoprotrusions as fabricated on targeted pads of the MEA. Adapted with permission from [52], © (2013) Elsevier

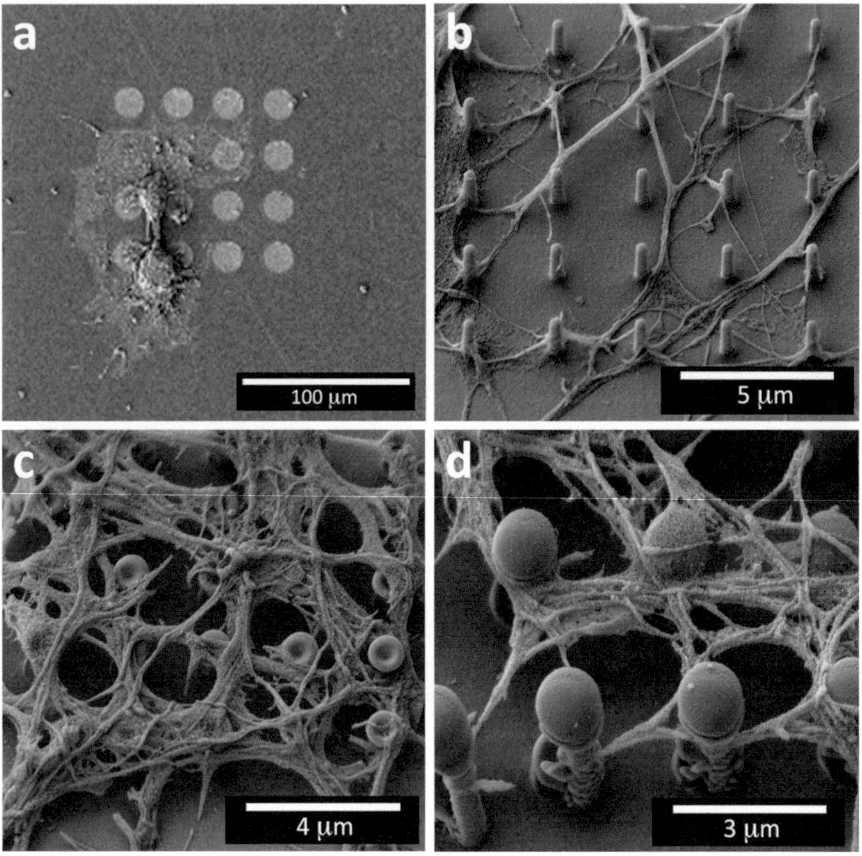

Fig. 6.6 Results of cell culturing on top of the MEA substrate processed as shown in Fig. 6.5. (**a**) Top view showing the neurons agglomerating in the flat region between the processed pads and extending neurites toward the nanoprotrusions. (**b**) Straight nanopillars induced a guiding effect on neuritis. Both nail-headed (**c**) and sphere-headed (**d**) pillars supported the formation of denser networks detached from the substrate. Reproduced with permission from [52], © (2013) Elsevier

Hippocampal neurons from E18 rat embryos were cultured on the modified chip after functionalization with PLL. After 7 days in culture, cells were found to adhere next to the nanostructured electrodes (Fig. 6.6a). Pillars were wrapped by neurites extending from neurons attached to the flat insulating layer between the electrodes, and all three kinds of nanostructures readily supported cellular projections. In particular, a guiding effect was observed in the case of straight nanopillars (Fig. 6.6b). Sometimes neurites were seen to form suspended bridges between adjacent pillars, while in other cases they contacted the underlying gold pad and were only partially guided by the nanostructures. In contrast, nail-headed and sphere-headed pillars promoted the formation of a denser network, with the neurites completely detached from the substrate (Fig. 6.6c, d). These observations demonstrate that FIB technology

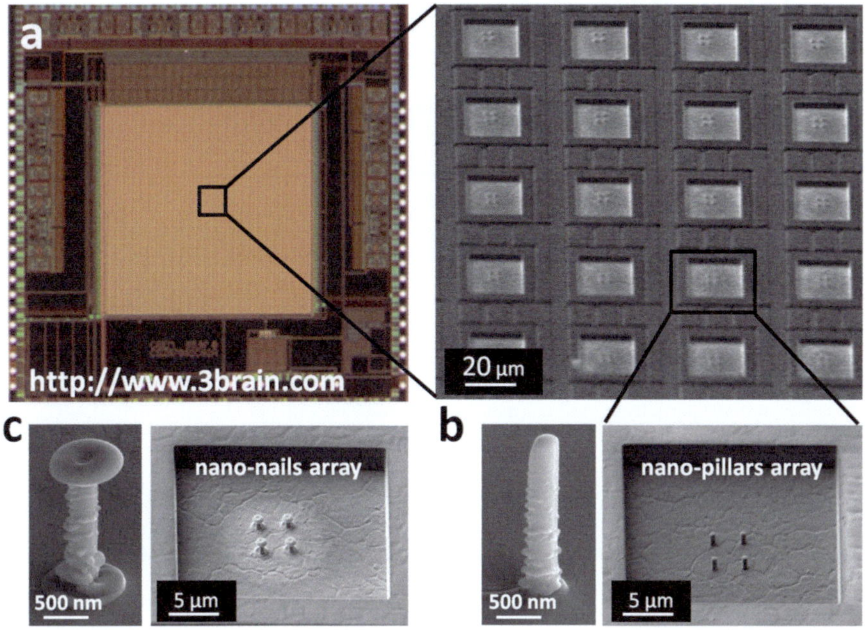

Fig. 6.7 2×2 Arrays of nanoprotrusions fabricated by FIB-induced deposition on top of the electrodes of a high-density APS-MEA chip (**a**). Both pillars (**b**) and nail-like nanoprotrusions (**c**) were deposited on the same substrate. Adapted with permission from [53], © (2013) Elsevier

can be used to examine the interactions of neurons with different nanostructures on the same substrate.

In [52], the versatility of FIB technology was exploited to grow different structures on the microelectrodes of an active pixel sensor microelectrode array (APS-MEA). Straight or nail-headed nanopillars were fabricated as 2×2 arrays (2.5 µm pitch) (Fig. 6.7). Nanostructures were made both of conductive platinum and low-conducting silicon dioxide to distinguish the contributions of the tight attachment and of the reduced electrode impedance to the electrical coupling between neuron and electrode.

Again, rat hippocampal neurons were cultured on the substrate after functionalization with PLL. As exemplified in Fig. 6.8a, the presence of nanoprotrusions often induced the placement of a single neuron on a single electrode. The engulfment process, shown in Fig. 6.8b, c in the case of platinum nanonails, was observed for all types of fabricated nanostructures. However, the authors could not correlate the engulfment of the nanostructures to a substantial enhancement of the recorrding sensitivity, differently on what observed with Aplysia neurons on mushroom-shaped electrodes [37, 39]. In fact, in this case, the enhanced electrical coupling was observed only for electrodes functionalized with a specific adhesion-promoting peptide [37, 39]. This result

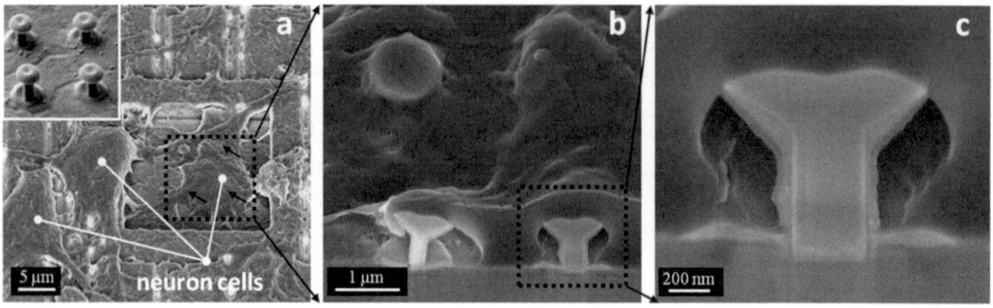

Fig. 6.8 (**a**) SEM image of neurons attached on the surface of an APS-MEA microelectrode with a 2×2 array of nanonails (*inset* shows an as fabricated one), the *arrows* indicating the hidden nanonails. One of the visible neurons is pinned by the nanoprotrusions on the electrode site. (**b**) Section of the same cell, realized by ion milling, showing the attachment sites. (**c**) Detail showing the cell membrane partially engulfing a nanonail. Reproduced with permission from [53], © (2013) Elsevier

suggests that the enhancement of the neuron-to-electrode coupling can be related to the combination of geometric and chemical properties of the electrode.

6.4 Geometric and Chemical Cues for the Enhanced Cell-to-Nanoprotrusion Interaction

As discussed in the previous sections, the fabrication of electrodes with nanoprotrusions can substantially improve neuron/electrode electrical coupling, due to the intimate contact resulting from the cells' engulfing the nanoprotrusions. The mechanisms responsible for the engulfing are still unclear, although the engulfing process is related to the cellular regulation of adhesion, shape, and motility and is likely regulated by the chemical composition of the extracellular environment and the nano-to-micrometer-level topography of the substrate [54]. In particular, the relative roles of geometry (shape and dimensions) vs. surface functionalization in the engulfment of protruding electrodes are unclear.

6.4.1 The Role of Coatings to Promote Adhesion

Cells likely use the same sensory apparatus to sense both surface adhesiveness and topographic features of the surface [55]. In particular, the interaction between cells and an artificial substrate is regulated by the same mechanisms with which cells interact with the extracellular matrix (ECM) in tissues. Indeed, a number of ECM molecules have been employed for cell attachment and efficient neurite outgrowth, such as laminin [56, 57], fibronectin [58, 59], and collagen [60].

Also, adhesion-promoting peptides derived from laminin, containing the sequence Arg–Gly–Asp (RDG) [61, 62], have been employed for effective culturing of neurons. Neuronal cell adhesion is also promoted by the employment of positively charged polyamino acids polyornithine and polylysine [64, 65]. Cells also adhere to substrate topography on nanostructured materials. Both carbon nanotubes (CNT) and conductive polymers such as poly(pyrrole) (PPy) and poly(3,4-ethylenedioxythiophene) (PEDOT) have been employed to enhance neuronal interfacing [22–27, 73, 74]. The ability of these materials to promote neuronal adhesion and proliferation has been related to their ECM-like fibrous structure [75, 76]. In the context of neuro-electrode interfaces, many studies have characterized the neuron-to-substrate distance on planar electrodes, with the aim of achieving a tight junction for enhanced electrical coupling. Measurements of the cleft thickness of neurons or fibroblasts grown on laminin or fibronectin [66–69], or of HEK293 cells on fibronectin [70], to cite a few examples, invariably reveal that the distance between the cell membrane and the substrate is not homogeneous. Instead, points with close spacing (0–10 nm) alternate with regions with larger spacing (tens to hundreds of nanometers) and the frequency of tight contacts are generally independent on the type of adhesion proteins. It has been suggested that the total cleft width is determined both by the thickness of the adhesion protein layer and by cellular components such as the glycocalyx, the carbohydrate-rich layer which coats the cell surface [71, 72]. The glycocalyx may prevent a closer contact of the cell membrane to the surface by steric hindrance [63, 73], which varies along the membrane surface. Therefore, the characteristics of the neuron-to-substrate interface cannot be precisely predicted, and biochemical treatments alone have not substantially improved electrical coupling at neuron/electrode interfaces.

6.4.2 The Role of Combined Geometric and Chemical Cues

It is conceivable that the enhanced interfacing of micro- and nanoprotrusions could result from the combination of the type of ligands presented by the substrate and the shape that the cell membrane is forced to assume to engulf the nanoprotrusion. In the first reports on neuronal culturing on substrates with mushroom- or spine-shaped protrusions, substrate functionalization with specific peptides appeared to be critical for cells to effectively engulf the nanoprotrusions [34–38]. PA22-2 and RGD synthetic peptides, featuring two different functional sites on the laminin molecule, were employed to induce a "phagocytic"-like activity of mouse hippocampal neurons [35], Aplysia neurons [36], and cardiomyocytes [36] (Fig. 6.9).

Hai et al. [38] cultured Aplysia neurons expressing fluorescent GFP-actin on gold-spine substrates coated either with an RGD-based peptide (EPP) or with standard PLL. Live confocal microscopy revealed that hotspots of actin, corresponding to focal adhesions, were localized around the stems of engulfed spines, forming an "actin ring." However, RGD-functionalized spines where engulfed more frequently than PLL-coated spines, and focal adhesions were more sparsely distributed in the

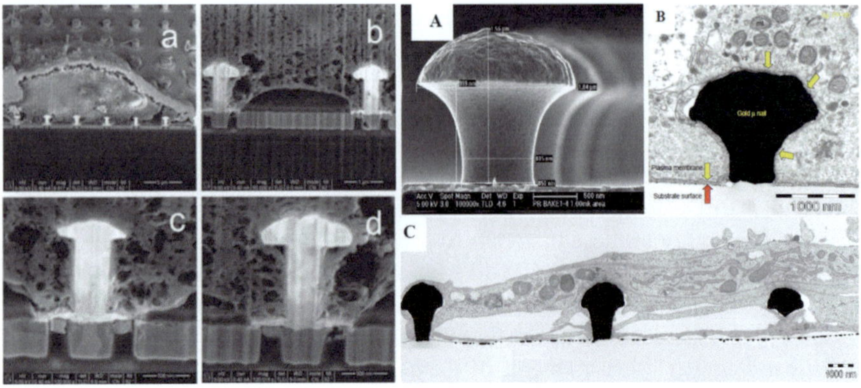

Fig. 6.9 Mushroom- or spine-shaped protrusions engulfed by mammalian neurons (*left*) and cardiomyocytes (*right*) upon substrate functionalization with laminin-derived PA22-2 or RGD-based peptides, respectively. *Left* Reproduced with permission from [35], © (2008) IEEE. *Right* Reproduced with permission from [36], © (2007) IEEE

case of PLL. This result further demonstrates the importance of presenting specific ligands to the cell membrane to stimulate engulfment of the protrusions. The stronger cytoskeleton remodelling around engulfed RGD-coated spines might also explain the reduced cleft width compared to flat substrates, as observed with TEM [37]. In general, estimating the exact width of the cleft is complicated by artifacts related to fixation, dehydration, embedding, and sectioning [37, 72]. Nonetheless, estimations of the cleft electrical resistance per unit area based on the measured cleft thickness agree with the measured improvement in extracellular recording quality [37]. In other cases, attenuated intracellular-like recordings were reported with the same geometry–ligand combination [39]. In those cases, an increased membrane conductance in the membrane around the protrusion was hypothesized to explain the observed resistive electrical coupling. The formation of such a "leaky" membrane was related either to a redistribution or extra activation of ion channels, phenomena that are known to participate in the mechano-response of cells to membrane deformation [54]. Another important factor could be deformation-induced membrane tension, similar to experiments with gateless field effect transistors (FETs) [77], where the application of force on the cell was able to switch the electrical coupling from extracellular to intracellular.

The need of a specific peptide to stimulate the engulfment of nanoprotrusions is less evident in the case of mammalian neurons, as seen in Sect. 6.3 for FIB-fabricated nanoprotrusions (Figs. 6.4 and 6.8). In [40], two different peptides or PLL were tested for functionalization of mushroom-shaped electrodes, and all three coatings gave similar results in terms of distribution of cleft thickness along sectors of the protrusions. Moreover, a recent systematic study of the interaction of rat hippocampal neurons with nanopillars of varying dimensions coated with standard PLL found that cells readily engulf the nanoprotrusions and that the average thickness of the cell–protrusion junction was reduced in comparison to cells cultured on flat substrates [78]. That study included nanopillars with sub-micrometer diameter.

However, it is possible that very thin (<200–300 nm in diameter) vertical nanopillars or nanowires penetrate through the cell membrane. A number of studies have reported on the ability of nanopillars and similar structures to facilitate the delivering of biomolecules into cells [79–81], supporting the hypothesis that thin nanopillars can impale cells. In [46, 82], nanowire penetration was promoted by a phospholipid coating. In [44], PLL-coated silicon nanowires were argued to penetrate the cell membrane. Membrane penetration was either spontaneous or it was induced by a voltage pulse for membrane electroporation. The sign of nanowires penetrating the membrane was the possibility to change the membrane potential by current injection through the nanowires themselves. Also, membrane penetration was indirectly indicated by the shape of signals recorded with the nanowires. The intracellular configuration of the nanowires was successively confirmed by three-dimensional confocal reconstruction [44]. In the case of PLL-coated platinum nanopillars, electroporation of the cell membrane induced a temporary (several minutes) change in the signature of the electric signal recorded with the nanowires from low-level extracellular to higher-level intracellular recordings [45]. Such change was attributed to the formation of pores in the membrane, which allowed the nanopillars to contact the intracellular medium while remaining outside the cell. Sealing of the pores eventually restored the extracellular condition [45]. In the last two examples, electroporation takes advantage of the small size of nanoprotrusions and of their intimate contact with neurons in order to induce a large electric field across the membrane with a small applied voltage and cause poration of the membrane, although it is not established if this causes nanoprotrusion penetration. Nonetheless, it is worth mentioning that also in the case of larger mushroom-shaped microelectrodes, which had been functionalized with PLL, it was possible to temporarily switch from an initial extracellular coupling to an intracellular-like one after the application of an electroporation voltage pulse [41]. In this case, the protrusions are tightly engulfed by the cell membrane, but retain the extracellular condition.

In conclusion, the precise nature of the interaction of neurons with micro- and nanostructured electrodes is still unclear, but it likely involves geometric as well as chemical factors at the cell-to-nanostructure interface. To elucidate the mechanisms of the interaction, it would be useful to test nanoprotrusions with shapes other than straight pillars or spine-shaped structures, for example, protrusions with systematically varied radius of curvature. For this purpose, the versatility of FIB nanofabrication can be exploited to produce different structures on the same substrate, to examine the mechanisms of cells-nanoprotrusion interaction under highly controlled conditions.

6.5 FIB/SEM Techniques for the Analysis of the Neuron-to-Nanostructures Interface

The interaction of living cells with the surface of artificial materials is of general interest for the development of new substrates for cell culture, neuroprosthetics, and tissue engineering [31, 75]. In the context of the development of cell–sensor hybrids for electrical recording from electrogenic cells, a number of techniques have been established to characterize this interaction at the cell-to-substrate interface, which ultimately determines the quality of the electrical coupling. In the course of the development of recording devices for in vitro electrophysiology such as MEAs and FET-based sensors, many studies have attempted to measure the distance between the cell membrane and the surface materials of the recording devices. These studies were conducted with optical methods such as total internal reflection fluorescence (TIRF) microscopy [83, 84], fluorescence interference-contrast (FLIC) microscopy [66–68, 71], and confocal microscopy [79, 84]. Although these techniques have excellent resolution in depth, their lateral resolution is limited to a few hundred nanometers by the wavelength of light. Analyzing the details of the cell–substrate interaction instead requires electron microscopy. Cells cultured on specific substrates can be subjected to critical-point drying (CPD) and then imaged by scanning electron microscopy (SEM), with high surface detail [79, 84–86]. However, the cell–substrate interface is mostly hidden by the cell itself, and high tilt angle can reveal only the peripheral portion of the interface. Instead, the cellular ultrastructure and the details of the cell–substrate interface can be revealed with nanometer resolution using transmission electron microscopy (TEM) on thin sections cut on an ultramicrotome, stained with heavy metals and embedded in resin [37, 78, 87, 88]. However, cutting sections from samples formed by soft biological matter on hard substrates can cause distortions and delamination [89–91]. For this reason, the substrate is usually removed by etching [37, 78] or by temperature-induced cleavage [30, 72] and then replaced by an epoxy replica, but these procedures have been applied only with glass or silicon substrates. An alternative is the use of a soft substrate such as PMMA [92], which can be sectioned by an ultramicrotome. Nevertheless, this does not solve an important drawback of TEM-based investigation—that direct targeting of the region of interest requires time-consuming serial sectioning at different sample sites.

The FIB/SEM hybrid technique represents a valid alternative to these methods and offers specific advantages, depending on the histological preparation method. In contrast to TEM, dual beam systems allow in situ screening of large sample areas and precise selection of the region of interest for FIB cross-sectioning and subsequent SEM imaging. Slice-by-slice parallel cross sections (Fig. 6.10a) can be performed with resolution of <10 nm, and the acquired stack of images can be post-processed to generate a 3D reconstruction of the observed structure [93, 94]. This stepwise process can reveal the fine features of the cell–substrate interface, such as the distribution of focal adhesions, over an extended portion of the surface contact—or even over the entire surface. In addition, it is possible to change the direction of cutting during the process with the so-called pie-slice technique (Fig. 6.10b), to unravel hidden structures and to better analyze complex interactions of the cell membrane with the substrate structures [93].

Both the cell structure and the substrate morphology can be affected by the procedure used for sample preparation, which can induce shrinkage, fracture, and/or detachment. Preparation of samples for FIB/SEM investigation is typically performed based on standard methods for either SEM or TEM microscopy. In standard

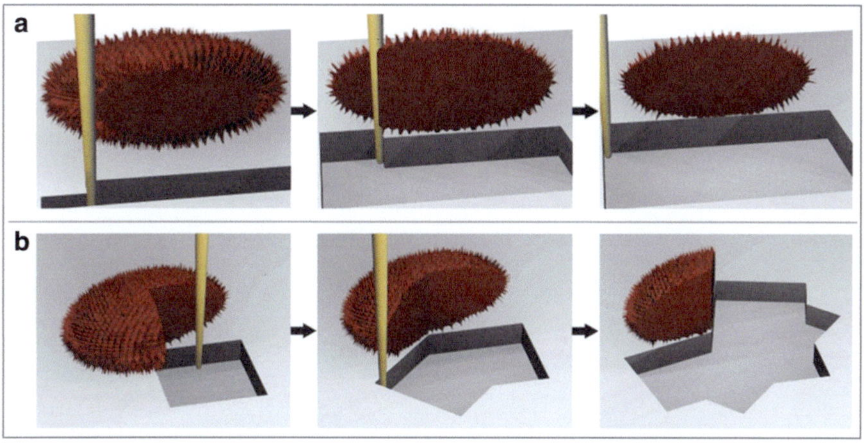

Fig. 6.10 Schematic representation of the (**a**) slice-by-slice and (**b**) pie-slice FIB preparation techniques. Reproduced with permission from [93], © (2011) Elsevier

SEM preparation, cells are first fixed with glutaraldehyde (or a combination of glutaraldehyde and formaldehyde) to preserve cellular ultrastructure. Then, the sample is either dehydrated and critical-point dried or freeze-fixed and subsequently freeze-dried in high vacuum [95]. Finally, the sample is coated with a conductive metal or carbon layer. Alternatively, since surface structure can be masked by the coating layer, samples can be made conductive by staining with reagents based on osmium tetroxide [96]. In standard TEM preparation, after chemical fixation and dehydration, the sample is embedded with resin to facilitate cutting sections. FIB/SEM visualization of subcellular structures has been demonstrated with cells cultured on different substrates and then chemically fixed, dehydrated, and then either CPD-dried [72, 93, 95] or freeze-dried [90, 97, 98]. The advantage of these approaches is that cell-to-substrate interface can be analyzed in situ, independently of the substrate material and without removing the substrate itself [72, 87, 95]. Multiple regions of interest can be selected from a relatively large field of view and then targeted by cross-sectioning, as exemplified in Fig. 6.11 for the case of a CPD-dried neuron on glass. However, the CPD procedure causes significant alteration of cellular structures [96]. Instead, the cryogenic approach has been shown to better preserve cellular structures and the ECM [90, 96]. Modern ion beam systems implementing cryostages have been developed to preserve ultrastructure of biological preparations such as bacterial samples [99, 100] as well as eukaryotic cells and tissues [98]. Concerning the interface of cells with nanopatterned surfaces, cryogenic (Cryo-) FIB/SEM has been shown to better resolve extracellular details at the interface compared to traditional FIB/SEM approaches [90]. Cryo-FIB/SEM has also been used to prepare TEM samples, thus avoiding damage typically induced by drying and microtomy [97]. Cryo-FIB/SEM is a time-consuming process, but it provides the advantages of cryogenic preparation with the versatility of FIB and the

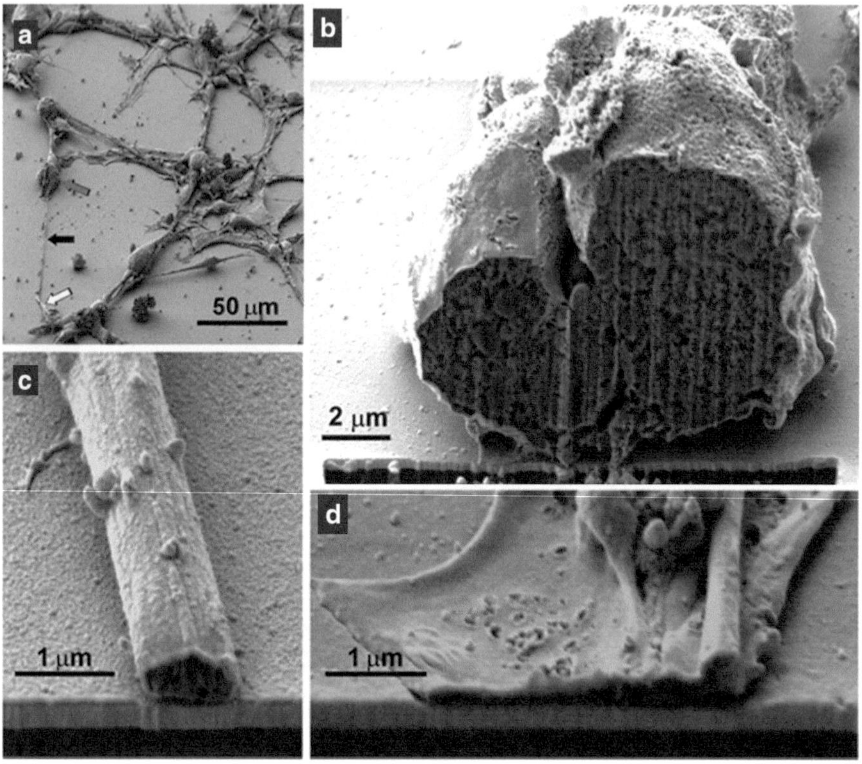

Fig. 6.11 Example of region-of-interest selection with the FIB/SEM dual beam microscope, in the case of CPD-dried neurons cultured on glass. (**a**) Overview of the neural network. (**b**) Cross section of the neuronal cell body (*open arrow* in **a**). (**c**) Cross section of a neurite (*black arrow* in **a**). (**d**) Cross section of a neuronal growth cone (*white arrow* in **a**). Reproduced with permission from [87], © (2009) Wiley

high resolution of TEM. Difficulties in employing cryo-techniques are related to platinum deposition at low temperature and to frost buildup on sample surfaces [97, 101].

Compared to samples prepared by CPD or by freezing, FIB/SEM with resin-embedded samples has been found to provide higher-resolution images of subcellular structures [87, 96, 98, 102, 103]. However, during embedding procedures, natural structures can be altered by sample heating and shrinkage. Except a few cases [94, 104], substrate removal is usually needed in order to visualize the cells from below, due to the large volume of resin left above the cells after the embedding process. FIB/SEM also has the advantage of eliminating further damage produced by ultramicrotomy. However, image resolution and contrast are lower for FIB/SEM than for TEM; in particular, the cell-to-substrate interface is difficult to visualize [87]. This is shown in Fig. 6.12, where a conventional TEM sample preparation is compared with in situ cross sections performed by FIB milling.

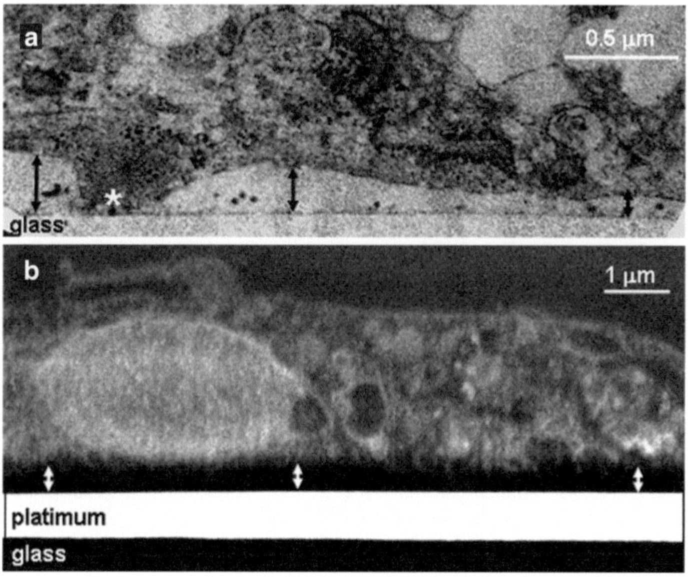

Fig. 6.12 Comparison of the cell-to-substrate interface visualized (**a**) by TEM and (**b**) by SEM after in situ FIB milling of resin-embedded samples. Even if the FIB/SEM image reveals a number of details on the intracellular structure, the cell-to-substrate interface is better resolved by TEM analysis. Reproduced with permission from [87], © (2009) John Wiley and Sons

In conclusion, FIB/SEM represents a versatile approach for the investigation of the interaction of cells cultured on nanostructured substrates, although with a resolution not as good as TEM. It remains the only method available when substrate removal is not possible, as is the case of cells cultured on recording devices.

6.6 Summary

Noninvasive neuronal electrical interfacing has been improved by the use of electrodes with three-dimensional micro- and nanoprotrusions. FIB-induced deposition allows the versatile fabrication of three-dimensional structures on various substrates, including passive substrates for studying cell–nanostructure interactions and MEA substrates for electrophysiological recording. By facilitating the development of novel three-dimensional structures, FIB can facilitate the study of how nanostructure geometry and surface functionalization affect mechanical and electrical coupling to neurons. Finally, in situ SEM analysis of FIB-milled cross sections, realized with FIB/SEM dual beam microscopes, presents a promising alternative to other methods, such as TEM, for the study of the interface between neurons and nanostructured substrates.

References

1. Hodgkin, A., Huxley, A.: A quantitative description of membrane current and its application to conduction and excitation in nerve. J. Physiol. **117**, 500–544 (1952).
2. Purves, R.D. Microelectrode Methods for Intracellular Recording and Ionophoresis. Academic press, London (1981)
3. Shahaf, G., and Marom, S.: Learning in networks of cortical neurons, J. Neurosci. **21**, 8782–8788 (2001).
4. Shahaf, G., Eytan, D., Gal, A., Kermany, E., Lyakhov, V., Zrenner, C., Marom, S.: Order based representation in random networks of cortical neurons. PLoS Comput. Biol. **4**, e1000228 (2008)
5. Bonifazi, P., Ruaro, M.E., Torre, V.: Statistical properties of information processing in neuronal networks. Eur. J. Neurosci. **22**, 2953–2964 (2005)
6. Chiappalone, M., Bove, M., Vato, A., Tedesco, M., and Martinioia, S.: Dissociated cortical networks show spontaneously correlated activity patterns during in vitro development. Brain Res. **1093**, 41–53 (2006)
7. Chiappalone, M., Vato, A., Tedesco, M., Marcoli, M., Davide, F., and Martinoia, S.: Networks of neurons coupled to microelectrode array: a neuronal sensory system for pharmacological applications. Biosens. Bioelectron. **18**, 627–634 (2003)
8. Scelfo, B., Politi, M., Reniero, F., Palosaari, T., Whelan, M., Zaldívar, J-M.: Application of multielectrode array (MEA) chips for the evaluation of mixtures neurotoxicity. Toxicology **299**, 172–183 (2012)
9. Gopal, K.V., Gross, G.W. Emerging histiotypic properties of cultured neuronal networks. In: Baudry, M., Taketani, M. (eds.) Advances in Network Electrophysiology Using Multi-Electrode Arrays, pp. 193–214. Springer, New York (2006)
10. Xia, Y., and Gross, G.W.: Histotypic electrophysiological responses of cultured neuronal networks to ethanol. Alcohol **30**, 167–174 (2003)
11. Wise, K.D.: Silicon microsystems for neuroscience and neural prostheses. IEEE Eng. Med. Bio. Magazine **24**, 22–29 (2005)
12. Normann, R.A.: Technology insight: future neuroprosthetic therapies for disorder of the nervous system. Nat. Clin. Pract. Neuro. **3**, 444–452 (2007)
13. Buzsáki, G.: Large-scale recording of neuronal ensembles. Nat. Neurosc. **7**, 446–451 (2004)
14. Huettel, S.A., Song, A.W., McCarthy, G.: Functional Magnetic Resonance Imaging, 2nd edn. Sinauer Associates, Massachusetts (2008)
15. Gold, C., Henze, D.A., Koch, C., and Buzsáki, G.: On the origin of the extracellular action potential waveform: a modeling study. J. Neurophysiol. **95**, 3113–3128 (2006)
16. Buzsáki, G., Anastassiou, C. A., and Koch, C.: The origin of extracellular fields and currents – EEG, ECoG, LFP and spikes. Nat. Rev. Neurosc. **13**, 407–420 (2012)
17. Pine, J.: Recording action potentials from cultured neurons with extracellular microcircuit electrodes. J. Neurosci. Methods **2**, 19–31 (1980)
18. Oka, H., Shimono, K., Ogawa, R., Sugihara, H., and Taketani, M.: A new planar multielectrode array for extracellular recording: application to hippocampal acute slice. J. Neurosci. Methods **93**, 61–67 (1999)
19. Grumet, A.E., Wyatt, J.L., and Rizzo, J.F.: Multi-electrode stimulation and recording in the isolated retina. J. Neurosci. Methods **101,** 31–42 (2000)
20. Kim, J.H., Kang, G., Nam, Y., and Choi, Y.K.: Surface-modified microelectrode array with flake nanostructure for neural recording and stimulation. Nanotechnology **21**, 85303 (2010)
21. Bruggemann, D., Wolfrum, B., Maybeck, V., Mourzina, Y., Jansen, M., Offenähusser, A.: Nanostructured gold microelectrodes for extracellular recording from electrogenic cells. Nanotechnology **22**, 265104 (2011)
22. Shein, M., Greenbaum, A., Gabay, T., Sorkin, R., David-Pur, M., Ben-Jacob, E., Hanein, Y.: Engineered neuronal circuits shaped and interfaced with carbon nanotube microelectrode arrays. Biomed. Microdevices **11**, 495–501 (2009)

23. Keefer, E.W., Botterman, B.R., Romero, M.I., Rossi, A.F., and Gross, G.W.: Carbon nanotube coating improves neuronal recordings. Nat. Nanotech. **3**, 434–439 (2008)
24. Kim, D.H., Richardson-Burns, S.M., Hendricks, J.L., Sequera, C., Martin, D.C.: Effect of immobilized nerve growth factor on conductive polymers: electrical properties and cellular response. Adv. Funct. Mater. **17**, 79–86 (2007)
25. Kim, Y.T., Haftel, V.K., Kumar, S., Bellamkonda, R.V.: The role of aligned polymer fiber-based constructs in the bridging of long peripheral nerve gaps. Biomaterials **29**, 3117–3127 (2008)
26. Abidian, M.R., Kim, D.H., Martin, D.C.: Conducting-polymer nanotubes for controlled drug release, Adv. Mater. **18**, 405–409 (2006)
27. Cui, X., Wiler, J., Dzaman, M., Altschuler, R. A., Martin, D. C.: In vivo studies of polypyrrole/peptide coated neural probes. Biomaterials **24**, 777–787 (2003)
28. Cogan, S.F.: Neural stimulation and recording electrodes. Annu. Rev. Biomed. Eng. **10**, 275–309 (2008)
29. Cai, N., Gong, Y., Chian, K.S., Chan, V., and Liao, K.: Adhesion dynamics of porcine esophageal fibroblasts on extracellular matrix protein-functionalized poly(lactic acid). Biomed. Mater. **3**, 15014 (2008)
30. Wrobel, G., Höller, M., Ingebrandt, S., Dieluweit, S., Sommerhage, F., Bochem, H.P., and Offenhäusser, A.: Transmission electron microscopy study of the cell–sensor interface. J. R. Soc. Interface **5**, 213–222 (2008)
31. Sniadecki, N.J., Desai, R.A., Ruiz, S.A., and Chen, C.S.: Nanotechnology for cell–substrate interactions. Ann. Biomed. Eng. **34**, 59–74 (2006)
32. Spatz, J.P., and Geiger, B.: Molecular engineering of cellular environments: cell adhesion to nano-digital surfaces. Methods Cell Biol. **83**, 89–111 (2007)
33. Nikkhah, M., Edalat, F., Manoucheri, S., Khademhosseini, A.: Engineering microscale topographies to control the cell-substrate interface. Biomaterials **33**, 5230–5246 (2012)
34. Van Meerbergen, B., Raemaekers, T., Winters, K., Braeken, D., Bartic, C., Engelborghs, Y., Annaert, W., and Borghs, G.: On chip induced phagocytosis for improved neuronal cell adhesion. NSTI-Nanotech. **2**, 107–110 (2006)
35. Van Meerbergen, B., Jans, K., Loo, J., Reekmans, G.G., Braeken, D., Seon-Ah, C., Bonroy, K., Maes, G., Borghs, G., Engelborghs, Y., Annaert, W., and Bartic, C.: Peptide-functionalized microfabricated structures for improved on-chip neuronal adhesion. Paper presented at the 30th annual international IEEE EMBS conference, Vancouver, British Columbia, Canada, 20–24 August 2008
36. Spira, M.E., Kamber, D., Dormann, A., Cohen, A., Bartic, C., Borghs, G., Langedijk, J.P.M., Yitzchaik, S., Shabthai, K., and Shappir, J.: Improved neuronal adhesion to the surface of electronic device by engulfment of protruding micro-nails fabricated on the chip surface. Paper presented at the 14th international conference on solid-state sensors, actuators and microsystems, Lyon, France, 10–14 June 2007
37. Hai, A., Dormann, A., Shappir, J., Yitzchaik, S., Bartic, C., Borghs, G., Langedijk, J.P.M., and Spira, M.E.: Spine-shaped gold protrusions improve the adherence and electrical coupling of neurons with the surface of micro-electronic devices. J. R. Soc. Interface **6**, 1153–1165 (2009)
38. Hai, A., Kamber, D., Malkinson, G., Erez, H., Mazurski, N., Shappir, J. and Spira, M.E.: Changing gears from chemical adhesion of cells to flat substrata toward engulfment of micro-protrusions by active mechanisms. J. Neural Eng. **6**, 066009 (2009)
39. Hai, A., Shappir, J., and Spira, M.E.: Long-term, multisite, parallel, in-cell recording and stimulation by an array of extracellular microelectrodes. J. Neurophysiol. **104**, 559–568 (2010)
40. Fendyur, A., Mazurski, N., Shappir, J., and Spira, M.E.: Formation of essential ultrastructural interface between cultured hippocampal cells and gold mushroom-shaped MEA - toward "IN-CELL" recordings from vertebrate neurons. Front. Neuroeng. (2011). doi: 10.3389/fneng.2011.00014
41. Hai, A. and Spira, M.E.: On-chip electroporation, membrane repair dynamics and transient in-cell recordings by arrays of gold mushroom-shaped microelectrodes. Lab. Chip **12**, 2865–2873 (2012)

42. Fendyur, A., and Spira, M.E.: Toward on-chip, in cell recordings from cultured cardiomyocytes by arrays of gold mushroom-shaped microelectrodes. Front. Neuroeng. (2012) doi: 10.3389/fneng.2012.00021
43. Robinson, J.T., Jorgolli, M., and Park, H.: Nanowire electrodes for high stimulation and measurements of neural circuits. Front. Neuroeng. (2013) doi: 10.3389/fncir.2013.00038
44. Robinson, J.T., Jorgolli, M., Shalek, A.K., Yoon, M-H., Gertner, R.S., Park, H.: Vertical nanowire electrode arrays as a scalable platform for intracellular interfacing to neuronal circuits. Nat Nanotechnol **7**,180–184 (2012)
45. Xie, C., Lin, Z., Hanson, L., Cui, Y., Cui, B.: Intracellular recording of action potentials by nanopillar electroporation. Nat. Nanotechnol. **7**, 185–190 (2012)
46. Duan, X., Gao, R., Xie, P., Cohen-Karni, T., Quing, Q., Choe, H.S., Tian, B., Jiang, X. and Lieber, C.M.: Intracellular recordings of action potentials by an extracellular nanoscale field-effect transistor. Nat. Nanotechnol. **7**, 174–179 (2012)
47. Giannuzzi, L.A., Stevie, F.A.: Introduction to Focused Ion Beams: Instrumentation, Theory, Techniques, and Practice. Springer, New York (2005)
48. Volkert, C.A., and Minor, A.M.: Focused ion beam microscopy and micromachining. MRS Bulletin **32**, 389–399 (2007)
49. Langford, R.M., Nellen, P.M., Gierak, J., and Fu, Y.: Focused ion beam micro- and nanoengineering. MRS Bulletin **32**, 417–423 (2007)
50. Ali, M.Y., Hung, W., Yongqi, F.: A review of focused ion beam sputtering. Int. J. Precis. Eng. Man. **11**, 157–170 (2010)
51. Xie, C., Hanson, L., Xie, W., Lin, Z., Cui, B., Cui, Y.: Noninvasive neuron pinning with nanopillar arrays. Nano Lett. **10**, 4020–4024 (2010)
52. Martiradonna, L., Quarta, L., Sileo, L., Schertel, A., Maccione, A., Simi, A., Dante, S., Scarpellini, A., Berdondini, L., De Vittorio, M.: Beam induced deposition of 3D electrodes to improve coupling to cells Microelectron. Eng. **97**, 365–368 (2013)
53. Sileo, L., Pisanello, F., Quarta, L., Maccione, A., Simi, A., Berdondini, L., De Vittorio, M., Martiradonna, L.: Electrical coupling of mammalian neurons to microelectrodes with 3D nanoprotrusions. Microelectron. Eng. (2013). doi:10.1016/j.mee.2013.03.152.
54. Vogel, V. and Sheetz, M.: Local force and geometry sensing regulate cell functions. Nat. Rev. Mol. Cell. Biol. **7**, 265–275 (2006)
55. Sniadecki, N.J., Desai, R.A., Ruiz, S.A., Chen, C.S.: Nanotechnology for cell-substrate interactions. Ann. Biomed. Eng. **34**, 59–74 (2006)
56. Previtali, S.C., Nodari, A., Taveggia, C., Pardini, C., Dina, G., Villa, A., Wrabetz, L., Quattrini, A., Feltri, M.L.: Expression of laminin receptors in Schwann cell differentiation: evidence for distinct roles. J. Neurosci. **23**, 5520–5530 (2003)
57. Clegg, D.O., Wingerd, K.L., Hikita, S.T., Tolhurst, E.C.: Integrins in the development, function and dysfunction of the nervous system. Front. Biosci. **8**, 723–750 (2003)
58. Qian, L., Saltzman, W.M.: Improving the expansion and neuronal differentiation of mesenchymal stem cells through culture surface modification. Biomaterials **25**, 1330–1337 (2004)
59. Mao, Y., Schwarzbauer, J.E.: Fibronectin fibrillogenesis, a cell-mediated matrix assembly process. Matrix Biol. **24**, 389–399 (2005)
60. Rohr, S., Flückiger-Labrada, R., Kucera, J.P.: Photolithographically defined deposition of attachment factors as a versatile method for patterning the growth of different cell types in culture. Eur. J. Physiol. **446**, 125–132 (2003)
61. Glass, J., Blevitt, J., Dickerson, K., Pierschbacher, M.D., Craig, W.S.: Cell attachment and motility on materials modified by surface-active RGD-containing peptides. Ann. N. Y. Acad. Sci. **745**, 177–186 (1994)
62. Heiduschka, P., Gopel, W., Beck, W., Kraas, W., Kienle, S., Jung, G.: Microstructured peptide-functionalized surfaces by electrochemical polymerization. Chem. Eur. J. **2**, 667–672 (1996)
63. Sorribas, H., Braun, D., Leder, L., Sonderegger, P., Tiefenauer, L.: Adhesion proteins for a tight neuron-electrode contact. J. Neurosci. Meth. **104**, 122–141 (2001)

64. Romanova, E.V., Oxley, S.P., Rubakhin, S.S., Bohn, P.W., Sweedler, J.V.: Self-assembled monolayers of alkanethiols on gold modulate electrophysiological parameters and cellular morphology of cultured neurons. Biomaterials **27**, 1665–1669 (2006)
65. Richert, L., Schneider, A., Vautier, D., Vodouhe, C., Jessel, N., Payan, E., Schaaf, P., Voegel, J.C., Picart, C.: Imaging cell interactions with native and cross-linked polyelectrolyte multilayers. Cell Biochem. Biophys. **44**, 273–285 (2006)
66. Lambacher, A., and Fromherz, P.: Fluorescence interference contrast microscopy on oxidized silicon using a monomolecular dye layer. Appl. Phys. A **63**, 207–216 (1996)
67. Braun, D., and Fromherz, P.: Fluorescence interference contrast microscopy of cell adhesion on oxidized silicon. Appl. Phys. A **65**, 341–348 (1997)
68. Iwanaga, Y., Braun, D., and Fromherz, P.: No correlation of focal contacts and close adhesion by comparing GFP vinculin and fluorescence interference of Dil. Eur. Biophys. J. **30**, 17–26 (2001)
69. Gleixner, R., Fromherz, P.: The extracellular electrical resistivity in cell adhesion. Biophys. J. **90**, 2600–2611 (2006)
70. Brittinger, M., and Fromherz, P.: Field-effect transistor with recombinant potassium channels: fast and slow response by electrical and chemical interactions. Appl. Phys. A **81**, 439–447 (2005)
71. Zeck, G., Fromherz, P.: Repulsion and attraction by extracellular matrix protein in cell adhesion studied with nerve cells and lipid vesicles on silicon chips. Langmuir **19**, 1580–1585 (2003)
72. Greve, F., Frerker, S., Bittermann, A.G., Burkhardt, C., Hierlemann, A., Hall, H.: Molecular design and characterization of the neuron-microelectrode array interface. Biomaterials **28**, 5246–5258 (2007)
73. Martinelli, V., Cellot, G., Toma, F.M., Long, C.S., Caldwell, J.H., Zentilin, L., Giacca, M., Turco, A., Prato, M., Ballerini, L., Mestroni, L.: Carbon nanotubes promote growth and spontaneous electrical activity in cultured cardiac myocytes. Nano Lett. **12**, 1831–1838 (2012)
74. Bareket-Keren, L., and Hanein, Y.: Carbon nanotube-based multielectrode arrays for neuronal interfacing: progress and prospects. Front. Neuroeng. (2013) doi: 10.3389/fncir.2012.00122
75. Kotov, N.A., Winter, J.O., Clements, I.P., Jan, E., Timko, B.P., Campidelli, S., Pathak, S., Mazzatenta, A., Lieber, C.M., Prato, M., Bellamkonda, R.V., Silva, G.A., Wong Shi Kam, N., Patolsky, F., and Ballerini, L. (2009) Nanomaterials for neural interfaces. Adv. Mater. **21**, 3970–4004
76. Heim, M., Yvert, B., Kuhn, A.: Nanostructuration strategies to enhance microelectrode array (MEA) performance for neuronal recording and stimulation. J. Physiol.-Paris **106**, 137–145 (2012)
77. M. Jenkner, and P. Fromherz, Bistability of membrane conductance in cell adhesion observed in a neuron transistor. Phys. Rev. Lett. **79**, 4705–4708 (1997)
78. Hanson, L., Lin, Z.C., Xie, C., Cui, Y. and Cui, B.: Characterization of the cell–nanopillar interface by transmission electron microscopy. Nano Lett. **12**, 5815–5820 (2012)
79. Shalek, A.K., Robinson, J.T., Karp, E.S., Lee, J.S., Ahn, D.R., Yoon, M.H., Sutton, A., Jorgolli, M., Gertner, R.S., Gujral, T.S., MacBeath, G., Yang, E.G., Park, H.: Vertical silicon nanowires as a universal platform for delivering biomolecules into living cells. Proc. Natl. Acad. Sci. U. S. A., **107**, 1870–1875 (2010)
80. Vandersarl, J.J., Xu, A.M., Melosh, N.A.: Nanostraws for direct fluidic intracellular access. Nano Lett. **12**, 3881–3886 (2012)
81. Mann, D.G.J., McKnight, T.E., McPherson, J.T., Hoyt, P.R., Melechko, A.V., Simpson, M.L., Sayler, G.S.: Inducible RNA interference-mediated gene silencing using nanostructured gene delivery arrays. ACS Nano **2**, 69–76 (2008)
82. Tian, B., Cohen-Karni, T., Qing, Q., Duan, X., Xie, P., Lieber, C.M.: Three-dimensional, flexible nanoscale field-effect transistors as localized bioprobes. Science **329**, 831–834 (2010)
83. Gingell, D., Todd, I., and Bailey, J.: Topography of cell glass apposition revealed by total internal-reflection fluorescence of volume markers. J. Cell Biol. **100**, 1334–1338 (1985)

84. Berthing, T., Bonde, S., Sørensen, C.B., Utko, P., Nygård, J., Martinez, K.L.: Intact mammalian cell function on semiconductor nanowire arrays: new perspectives for cell-based biosensing. Small **7**, 640–647 (2011)
85. Yang, P., Yan, R., Fardy, M.: Semiconductor nanowire: what's next? Nano Lett. **10**, 1529–1536 (2010)
86. Hällström, W., Mårtensson, T., Prinz, C., Gustavsson, P., Montelius, L., Samuelson, L., and Kanje, M.: Gallium phosphide nanowires as a substrate for cultured neurons. Nano Lett. **7**, 2960–2965 (2007)
87. Bittermann, A. G., Burkhardt, C. and Hall, H.: Imaging of cell-to-material interfaces by SEM after in situ focused ion beam milling on flat surfaces and complex 3D-fibrous structures. Adv. Eng. Mater. **11**, B182–B188 (2009)
88. Murphy, G.E., Narayan, K., Lowekamp, B.C., Hartnell, L.M., Heymann, J.A.W., Fu, J., Subramniam, S.: Correlative 3D imaging of whole mammalian cells with light and electron microscopy. J. Struct. Biol. **176**, 268–278 (2011)
89. Edwards, R.L., Coles, G., Sharpe, W.N.: Comparison of tensile and bulge tests for thin-film silicon nitride. Exp. Mech. **44**, 49–54 (2004)
90. Lamers E., Walboomers, X.F., Domanski, M., McKerr, G., O'Hagan, B.M., Barnes, C.A., Peto, L., Luttge, R., Winnubst, L.A.J.A., Gardeniers, H.J.G.E., and Jansen, J.A.: Cryo DualBeam focused ion beam–scanning electron microscopy to evaluate the interface between cells and nanopatterned scaffolds. Tissue Eng. Part C Methods **17**, 1–7 (2011)
91. Volkert, C.A., Busch, S., Heiland, B., and Dehm, G.: Transmission electron microscopy of fluorapatite-gelatine composite particles prepared using focused ion beam milling. J. Microscopy **214**, 208–212 (2004)
92. Dalby, M.J., Riehle, M.O., Sutherland, D.S., Agheli, H., Curtis, A.S.G.: Changes in fibroblast morphology in response to nano-columns produced by colloidal lithography. Biomaterials **25**, 5415–5422 (2004)
93. Friedmann, A., Hoess, A., Cismak, A., Heilmann, A.: Investigation of cell-substrate interactions by focused ion beam preparation and scanning electron microscopy. Acta Biomater. **7**, 2499–2507 (2011).
94. Wierzbicki, R., Købler, C., Jensen, M.R.B., Łopacińska, J., Schmidt, M., Skolimowski, M., Abeille, F., Qvortrup, K., Mølhave, K.: Mapping the complex morphology of cell interactions with nanowire substrates using FIB-SEM. PLoS ONE **8**, e53307 (2013)
95. Martinez, E., Engel, E., Lopez-Iglesias, C., Mills, C.A., Planell, J.A., Samitier, J.: Focused ion beam/scanning electron microscopy characterization of cell behavior on polymer micro-nanopatterned substrates: a study of cell–substrate interactions. Micron **39**, 11–116 (2008)
96. Lešer, V., Drobne, D., Pipan, Ž., Milani, M., Tatti, F.: Comparison of different preparation methods of biological samples for milling and SEM investigation. J. Microsc. **233**, 309–319 (2009)
97. Edwards, H.K., Fay, M.W., Anderson, S.I., Scotchford, C.A., Grant, D.M., and Brown, P.D.: An appraisal of ultramicrotomy, FIBSEM and cryogenic FIBSEM techniques for the sectioning of biological cells on titanium substrates for TEM investigation. J. Microsc. **234**, 16–25 (2009)
98. Heymann, J.A.W., Hayles, M., Gestmann, I., Giannuzzi, L.A., Lich, B., and Subramniam, S.: Site-specific 3D imaging of cells and tissues with a dual beam microscope. J. Struct. Biol. **155**, 63–73 (2006)
99. Nordestgaard, B.G., Rostgaard, J.: Critical-point drying versus freeze drying for scanning electron microscopy: a quantitative and qualitative study on isolated hepatocytes. J. Microsc. **137**, 189–207 (1985)
100. Marko, M., Hsieh, C., Shalek, R., Frank, J., and Mannella, C.: Focused-ion-beam thinning of frozen hydrated biological specimens for cryo-electron microscopy. Nat. Methods **4**, 215–217 (2007)
101. Hayles, M.F., Stokes, D.J., Phifer, D., and Findlay, K.C.: A technique for improved focused ion beam milling of cryoprepared life science specimens. J. Microsc. **226**, 263–269 (2007)

102. JiméNez, N., Van Donselaar, E.G., De Winter, D.A.M., Vocking, K., Verkleij, A.J., Post, J.A., Gridded Aclar: preparation methods and use for correlative light and electron microscopy of cell monolayers, by TEM and FIB-SEM. J. Microsc. **237**, 208–220 (2010)
103. Knott, G., Marchman, H., Wall, D., Lich, B.: Serial section scanning electron microscopy of adult brain tissue using focused ion beam milling. J. Neurosc. **28**, 2959–2964 (2008)
104. Gnauck, P., Burkhardt, C., Wolburg, H., Nisch, W.: Investigation of the interface between biological cell tissue and hard substrate materials using crossbeam technology. Microsc. Microanal. **11**, 65–66 (2005)

Chapter 7
Active Pixel Sensor Multielectrode Array for High Spatiotemporal Resolution

L. Berdondini, A. Bosca, T. Nieus, and A. Maccione

7.1 Introduction

Among the different methodologies used for electrophysiological measures in the brain, electrodes have played an undisputed role in high-quality intracellular signal recordings from a few neurons and in chronic extracellular measures with electrode-array probes implanted in the brain. Electrode arrays providing multisite extracellular measures have become a key methodology in neuroscience for studying coding and transmission of information by neuronal ensembles [1] and for the development of Brain–Machine Interfaces (BMIs) and neural prosthetics [2–8]. This is mainly because electrode arrays combine the unique features of bidirectionality (i.e., recording and stimulation), long-term stability (up to years), and of a large signal bandwidth that enables recordings of action potentials from multiple neurons as well as low-frequency field potentials (LFPs).

The development of multielectrode arrays (MEAs) started in the 1970s and progressed in parallel with advances in microfabrication processes and microelectronics [9–12]. Since the end of the 1970s, these developments were mainly based on silicon machining and thin-film technologies used to realize a wide range of brain-implantable probes as well as chips for in vitro studies. This was a period in which the microscale was at the center of attention. R&D efforts contributed to the optimization of the electrode–electrolyte–neuron interface, to the development of improved microstructuring processes for MEAs and to the exploitation of microfluidics for neural probes [13]. More recently, the advent of nanotechnologies and nanostructuring capabilities has opened novel perspectives for MEAs. Indeed, the biophysical interactions at the nanoscale occurring among neurons, electrodes, and substrates

L. Berdondini (✉) • A. Bosca • T. Nieus • A. Maccione
NetS3 Laboratory, Neuroscience and Brain Technologies, Istituto Italiano di Tecnologia, Genoa, Italy
e-mail: luca.berdondini@iit.it

M. De Vittorio et al. (eds.), *Nanotechnology and Neuroscience: Nano-electronic, Photonic and Mechanical Neuronal Interfacing*, DOI 10.1007/978-1-4899-8038-0_7,

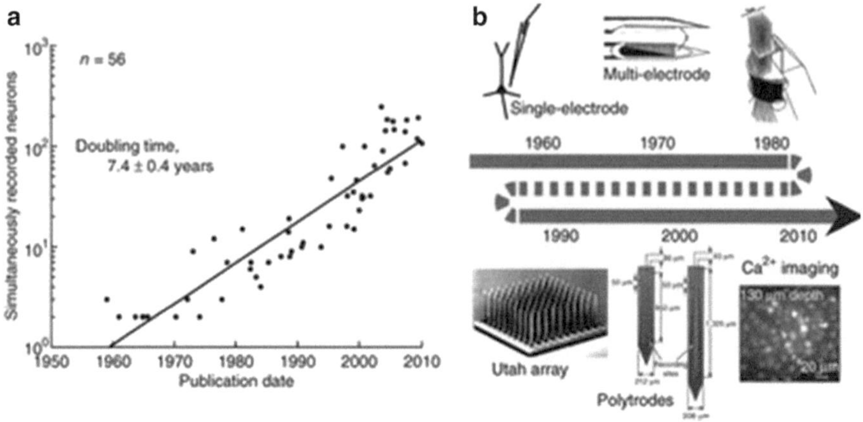

Fig. 7.1 State of the art in the number of simultaneously measured single neural units. (**a**) Examining 56 studies published over the last five decades, it emerges that the number of simultaneously recorded neurons in vivo doubled approximately every 7 years. (**b**) This analysis considers both electrical and optical recording neurotechnologies [25]

have become an exciting field of research [14] that constantly provides novel ways [15] and nanomaterials (e.g., carbon nanotubes [16, 17]) to sense and actuate electrophysiological signals more efficiently [18]. Among these novel neural interfacing modalities [19], "in-cell" neuroelectronic coupling was recently demonstrated by the group of M. Spira [20], showing that peptide-functionalized microstructured "mushroom-shaped" electrodes [21] are able to record subthreshold synaptic activity in *Aplysia* neurons. Furthermore, the use of nanopillars [22] and nanowires [23] was shown to enable intracellular recordings for a short time (~1 h) in vertebrate neurons, while electrodes remain in the cell or pores on the membrane remain open.

However, even though the signal quality of electrode arrays can be advanced by these novel neurotechnologies that are able to reach submicron sizes, the simultaneous measure at millisecond temporal resolution of the electrical activity of a very large number of single neurons in the brain [24] remains a major challenge. Interestingly, a recent review on the number of single neurons that can be measured in vivo (Fig. 7.1) shows that the number of simultaneously recorded neurons has been doubling approximately every 7 years [25] and this evolution appears following Moore's law (i.e., an empirical law that has driven the scalability of the number of integrated transistors in microelectronic devices). However, a way to practically increase by one order of magnitude the current range of 10^2 single units in vivo is still not established. Furthermore, an increase in recordable single neurons poses not only a technological issue for the design and implementation of adapted sensing and actuating devices but, as pointed out by Stevenson and Kording, also computational challenges that need to be solved. Indeed, while analysis techniques to study interactions between neurons have become more precise over the last decades, their computational cost has increased with the number of recorded neurons. Therefore, to face this challenge of large-scale single-neuron recordings, novel methodologies able to manage the readout and also the analysis of electrophysiological signals are needed.

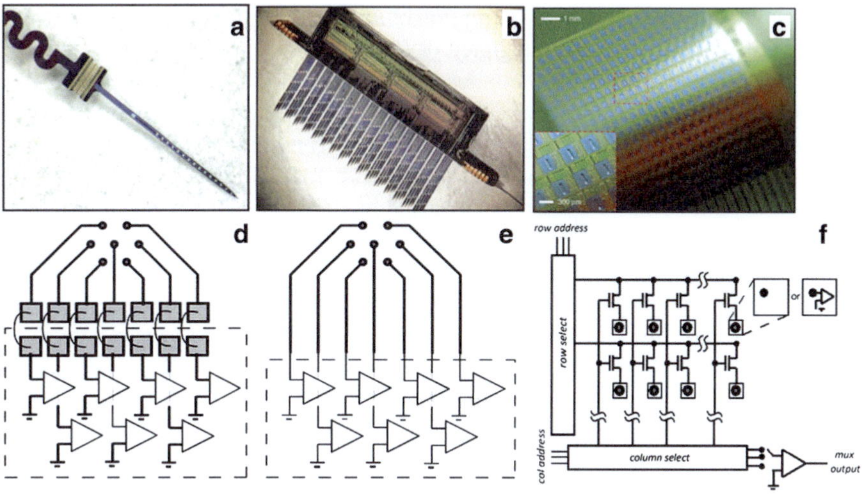

Fig. 7.2 Examples and concepts of *passive*, *hybrid*, and *active electrode arrays*. (**a**) Passive 16-site implantable probe with flexible polymeric interconnection [34]. (**b**) Hybrid implantable probe with 64-site probe (8 electrodes per shaft) and on-chip CMOS circuits for amplification, multiplexing, and signal processing [10]. (**c**) Active electrode array for epicortical recordings (360 electrodes). Circuits on thin silicon substrates (300 × 300 μm) are assembled on polymeric interconnecting substrates with a spacing of 500 μm [25]. (**d**) In *passive MEAs* electrodes are individually wired to an instrument amplifier. (**e**) In *hybrid MEAs* electrodes are wired to on-probe CMOS circuits or to external ASICs. (**f**) *Active MEAs* integrate an active matrix for wiring the electrodes. (**a**) Reproduced with permission from [34], © (2007) Elsevier; (**b**) [10] © (2005) IEEE; (**c**) [25]

In the R&D of neurotechnologies targeting large-scale neuronal recordings, electrode arrays are nowadays competing with advanced optical recording technologies that have remarkably progressed toward single-neuron resolution and enable accessing ~100 cells in local brain circuits [26, 27]. Electrode arrays have also progressed toward the ambitious objective of massively parallel single-unit recordings, taking advantage of low-power microelectronics for signal conditioning, wireless transreceiving, and multiplexing [28]. In particular, with advances in novel *active* multielectrode arrays, it is becoming realistic goal to record from several thousands of micro-/nano-electrodes simultaneously. This emerging generation of electrode arrays is realized with standard complementary metal–oxide–semiconductor (CMOS) technologies to implement active circuits for addressing large and dense matrices composed of several thousands of electrodes. This approach radically differs from *passive* MEAs that rely on the individual metal wiring of each microelectrode. While passive arrays of m columns and n rows need $m \times n$ interconnecting wires, active arrays require only $m + n$ wires and can also integrate signal conditioning and addressing circuits very close to each electrode. Another emerging trend in neuroelectronic R&D is to combine innovative fabrication processes of flexible and biodegradable materials with the functionalities provided by CMOS microelectronic circuits (Fig. 7.2). As an example, Viventi and his collaborators have recently proposed an epicortical probe [29] with CMOS silicon circuits assembled on a flexible electrode-array substrate.

In this chapter, we will use the definition of *hybrid MEAs* to refer to those devices that consist of passive MEAs integrated or assembled with CMOS circuits that are not used to modify the passive connectivity of the microelectrodes in the array. This hybrid approach was developed during the 1980s [30] to realize neural probes with a short distance among microelectrodes and front-end circuits and to reduce the number of output wires. In contrast, *active MEAs* for addressing and multiplexing a large number of closely spaced (tens of microns) microelectrodes were proposed very recently and have been only validated in vitro [31]. These devices, which provide an image-like view of extracellular potential distributions in brain tissue, are becoming an interesting tool to study synaptic plasticity in cultured networks formed by several thousands of neurons [32]. The availability of these devices is also driving development of computational tools adapted to large-scale recordings, for example, algorithms for estimating functional connectivity based on simultaneously recorded spike trains from nearby sites [33].

The aim of this chapter is to review ongoing progress in the design and fabrication and performance of microelectrode arrays, brought about by recent advances in low-power microelectronics, nanotechnology, and computational power. The first section of this chapter provides a short review on developments that brought MEAs from passive to active devices. Next we will present the implementation of the first platform based on a monolithic CMOS multielectrode array, enabling simultaneous record from 4,096 microelectrodes [35]. Developed over the last 13 years, this active MEA chip is based on the Active Pixel Sensor (APS) concept that was originally developed for high-speed light imaging devices [36, 37]. The design of this system was a significant engineering effort oriented toward the development of a highly scalable circuit architecture that can be a springboard for next-generation large-scale MEAs. Finally, we will discuss the experimental performance of the current generations of APS-MEA.

The goal for this chapter is to help the new generation of researchers bridge the gap between neuroscience and engineering, to use neuroengineering principles to create novel instruments and implement applications for studying neural activity in the brain.

7.2 From *Passive* to *Active* Multielectrode Arrays for Neural Interfacing

Implantable microwire arrays developed in the 1950s can be considered the precursors of modern integrated multielectrode arrays and current implantable probes for brain recordings [38]. These microwire arrays were mostly handmade and their performance was limited by the reproducibility of controlling the exposed area of the electrode tip and the interelectrode separation. In the 1970s, these issues motivated the application of microfabrication technologies, including photolithographic processes and thin-film technologies, to realize microelectrode arrays on silicon

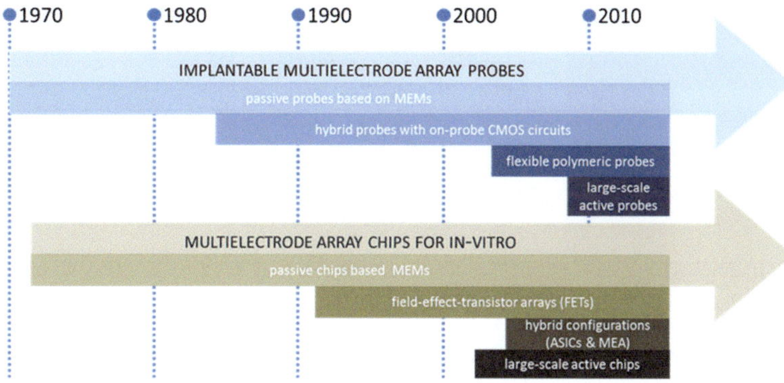

Fig. 7.3 Overview of the main progresses in the development of multielectrode array (MEA) technology applied in vivo and in vitro, from *passive* to *active* devices

substrates with well-controlled positions and exposed electrode-tip areas (Fig. 7.3). Such devices are here referred to as *passive electrode arrays* since each microelectrode is individually wired on the array by means of a "passive" metal lead and connected to a separate electronic circuit for signal conditioning, multiplexing, and analogue-to-digital conversion (Fig. 7.2d).

As reviewed in the next paragraphs, this passive approach was used to realize integrated chips for in vitro studies and brain-implantable probes. The development of microelectronic circuits adapted to these passive electrode-array devices started very early and was mainly oriented toward shortening the electrode-to-amplifier distances to lower noise and to reduce the whole system size for chronic behavioral studies with implantable probes [39]. However, with developments in low-power microelectronics in the beginning of the twenty-first century, dedicated CMOS circuits started to be intensively developed [40]. This gave rise to a wide range of *hybrid electrode-array* systems, consisting of dedicated CMOS front-end circuits for signal amplification [41] connected to passive MEAs [42] or integrated on-probe for signal conditioning and multiplexing of a few tens of electrodes (Fig. 7.2e).

More recently, *active electrode arrays* consisting of monolithic CMOS chips were proposed in order to manage the electrode connectivity on the array and to achieve recordings from a much higher number of closely integrated microelectrodes (Fig. 7.2f). These devices are realized by microelectronic processes of standard integrated circuits (ICs) and have thus also a high potential for commercialization due to their lower production costs when compared with *passive* devices realized with microelectromechanical (MEM) technologies [43]. The Active Pixel Sensor (APS) multielectrode array [35] that will be described in detail in the next section is one example of CMOS architectures that have been proposed for the integration of scalable high-density electrode arrays.

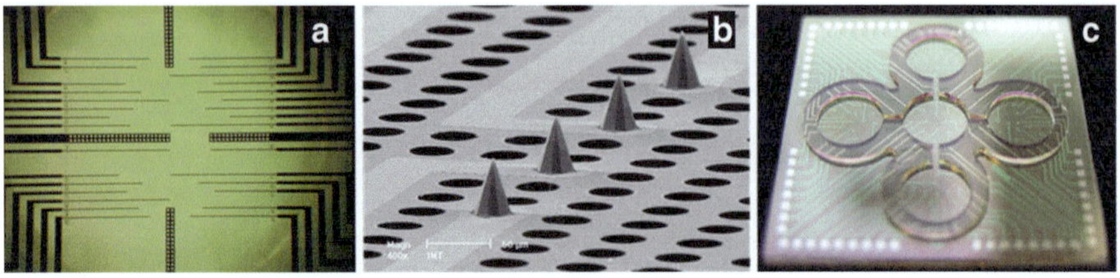

Fig. 7.4 Examples of passive multielectrode arrays (MEAs) for in vitro. (**a**) MEA with 32 planar electrodes and 4 on-chip pseudo-references [62]. (**b**) View of a porous Si-MEA with tip-shaped microelectrodes for electrophysiology on organotypic brain slices. (**c**) MEA with EPON SU-8 clustering structures for investigating interconnected neuronal populations [63]. (**a**) [62] © (1997) IEEE; (**c**) reproduced with permission from [63], © (2006) Elsevier

7.2.1 Multielectrode Arrays for In Vitro Neural Interfacing

Multielectrode arrays (MEAs) for in vitro neuroscience have become a widely used neurotechnology that has a promising potential for biomedical research and diagnostics [44]. Research on MEAs led to improvements in the recording and stimulation performances of chip-integrated microelectrodes, which in turn facilitated research on cultured neuronal networks [45–49] and ex vivo brain tissue [50–52]. The use of in vitro MEA chips for neuropharmacological and neurotoxicological screenings was also proposed [53–55] and is nowadays a promising approach for electrical readouts applied to human-derived cell culture preparations [56].

As it has been reviewed by J. Pine [57], one of the pioneers in MEAs, in vitro electrode arrays were independently demonstrated by C. Thomas [58], G. Gross [59], and J. Pine [60]. Briefly, C. Thomas and his colleagues performed the first experiment in 1972 and recorded peak-to-peak signals of 20–2,000 μV in amplitude from dissociated chick myocytes, using an array of 2×15 planar gold electrodes (100 μm pitch) that were plated with platinum black to lower their impedance [61]. In 1977, G. Gross proposed the use of multielectrode arrays of 36 gold electrodes (10 μm in diameter, 100 or 200 μm pitch) that were structured by laser etching.

Gross et al. used these arrays to record single-action potentials (up to 3 $mV_{p\text{-}p}$) from snail ganglion cells [59]. Three years later, J. Pine reported the first recording from rat dissociated neurons cultured on-chip [60]. Pine and colleagues' multielectrode array featured 16 platinized microelectrodes (250 μm pitch) insulated with silicon dioxide and with a square electrode opening of 10 μm.

The devices implemented in this early work do not differ much from more recent passive devices and from chips that are currently commercially available. These recent MEAs are the fruit of research that was mainly focused on optimizing the electrode and recording performance [64, 65] by applying different electrode materials (e.g., Au, Pt and black Pt, IrOx, TiN) [66–69] and by studying the interfacing properties of planar or three-dimensional electrode morphologies [62, 70] (Fig. 7.4). More recent approaches include the use of nanomaterials such as carbon nanotubes (CNTs) to enhance the recording and stimulation performance of microelectrodes

[16, 17, 71, 72]. The layout of the arrays has also been varied to fit different experimental models, from cultured networks to brain tissues. Finally, by taking advantage of developments in microfluidics and biosensors, additional on-chip functionality has been added to enhance capabilities for in vitro recordings from cultured neurons [73] and brain slices [74] and to organize the topology of cultured neuronal networks [63, 75–78]. However, the number of electrodes and their separation has not substantially changed: passive electrode arrays typically provide ~60–256 microelectrodes, with pitches of about 100 μm.

7.2.2 *Implantable Multielectrode Array Probes for In Vivo Neural Interfacing*

Implantable probes differ from in vitro probes mainly because of the three-dimensional shaping of the electrode substrate (Fig. 7.2a, b), which is necessary to optimize the geometry of the tip and the mechanical properties of the shaft to allow the probes to be implanted. In 1970, the first silicon implantable electrode-array probe was realized by K.D. Wise and his collaborators [79]. Since then, many other investigators and research companies have exploited advances in silicon processing to design and fabricate various silicon-based electrode arrays [12, 80].

Currently there are two predominant silicon electrode arrays used in neuroscience. The first comes from the original work of K. D. Wise and uses an "in-plane" technological approach to realize planar silicon combs with inline multiple shafts that are successively stacked one over the other during assembly to realize bidimensional arrays of shafts [80–82]. For microstructuring of the combs and of the electrodes on the shafts, this approach takes advantage from the resolution of photolithographic processes. In this way, each shaft can provide multiple microelectrodes that are individually wired to a contact pad with a metal line. The various microfabrication processes of the silicon substrate proposed in the literature included boron doping used as wet etch stop [39, 83], combined dry and wet etching of silicon [84], or only dry etching procedures [85, 86], including the use of silicon-on-insulator (SOI) substrates [87–89]. These processes can produce probes with multiple shafts, with length ranging from a few hundred micrometers up to several millimeters and with thicknesses in the range of 10–200 μm. Based on this approach, Wise and Najafi have developed a variety of probes with different layouts, including single-shaft, multi-shaft, and 3-D-stacked layouts [90, 91]. An advantage of this type of probes is their compatibility with on-probe integration of microelectronic circuits [80, 92–94], an approach that has been used to produce hybrid MEA probes since the 1980s.

The second silicon probe in use for more than 20 years is the so-called Utah electrode array developed by Normann and co-workers [7, 95, 96]. This probe differs from the previous one in the fabrication process, and it also provides only single electrodes on the apex of each shaft. Utah electrode arrays are realized with the "out-of-plane" fabrication process, which consists of structuring the array directly from a single block of silicon by using etching, doping, and heat treatments to realize a three-dimensional array of needlelike electrodes [97].

More recently, an alternative approach to realize implantable silicon probes was developed within the "NeuroProbes" European project [98]. This approach has been developed to realize multifunctional probes by the modular assembling of combs or individual shafts on a slim platform. Each shaft can provide up to 188 or 752 electrodes, with a pitch of 40 μm and a multiplexing active circuit that allows a subset of recording electrodes to be selected to optimize the experimental yield by discarding electrodes that are not effectively recording neural activity [99, 100]. It is also possible to include additional functionalities such as biosensors for in situ measurements or microfluidic injectors [101, 102]. The combination of electrical recordings with the simultaneous delivery of chemical substances is extremely desirable for in vivo studies. Specific drug administration can be used to stimulate, to inhibit, or to selectively lesion brain regions of interest. To relate these modulations with specific target areas, it is important that these injections are well localized and can deliver precise and small volumes (often <1 μl). Typically, microprobes and injectors are implanted independent of one another [103] or are glued directly together to fix their relative position. However, these handmade devices limit accuracy and reproducibility of the position of the fluidic outlet with respect to the microelectrode array. Therefore, fluid delivery channels were directly integrated into micromachined implantable needles made with silicon processing [104, 105] and more recently by using photo-structurable polymers such as the EPON SU-8 [106, 107] or polyimide [108].

While silicon technology has been widely used, the recent development of flexible polymeric technologies provides an alternative substrate [109–113]. Flexible materials reduce the mechanical mismatch between stiff microprobes and soft brain tissue that can cause injury and inflammation at the recording site, leading to signal and tissue loss in chronic and semi-chronic experiments [114]. Approaches for flexible probes have included the use of biopolymer coatings [115] or nanoparticle-embedded coatings [116].

7.2.3 Active Multielectrode Arrays for Large-Scale Neural Interfaces

Through the 1990s, there were only a few pioneering studies on developing microelectronic circuits for electrode arrays [40]. These circuits were integrated on implantable probes, to provide signal conditioning close to the microelectrodes and to reduce the number of output lines with multiplexing circuits or wireless data transmission [39, 117, 118]. R&D on CMOS circuits for neuroscience has expanded rapidly over the last decade. Due to achievements in low-power microelectronic integration [119, 120], novel CMOS electrode arrays are allowing chronic recordings from an increasingly larger number of single neurons.

The technological constraint of individually wiring each microelectrode to an amplifier is a major factor limiting the number of recoding electrodes in passive electrode-array configurations. Scaling up the number of electrodes faces the issues of routing the electrode interconnections and of managing the acquisition. As a first step to overcome this limitation, *hybrid* architectures consisting of CMOS-based application specific integrated circuits (ASICs) connected either on-chip or off-chip to passive electrode arrays were initially proposed [42, 121–124]. For example, Field et al. used this approach to simultaneously record from 512 closely spaced microelectrodes to investigate the functional connectivity of retinal ganglion cells (RGCs) in ex vivo retina at the resolution of individual neurons [125]. However, since this hybrid approach still relies on passive wiring of the electrodes, it is not a practical solution to drastically increase the number and density of microelectrodes over large recording areas (several square millimeters).

In order to solve this issue of electrode connectivity, *active* MEA chips were proposed [31]. In this case, on-chip digital microelectronic circuits and active matrices are integrated to address and multiplex a large number of microelectrodes. So far, these monolithic CMOS circuits have been validated in vitro and represent the most advanced large-scale arrays with respect to the number of simultaneously recording electrodes.

There have been three major developments using active MEA methodology, allowing integrated devices with up to 32,000 electrodes. These devices are referred to as CMOS-MEAs since they consist of monolithic microelectronic chips with active circuits for wiring the electrodes. However, there are substantial differences in their circuit architectures and in the type of electrode transducers (i.e., metal electrode or field-effect transistors), with consequent differences in performances with respect to the number of simultaneous recording electrodes, spatial resolution (i.e., interelectrode separation and electrode sizes), and signal-to-noise performance. Some of systems were designed specifically for continuously recording for long periods of time (hours, days), while others were designed for managing electrode arrays with very small pitches, down to cellular dimensions (<10 μm).

To our knowledge, the first report of high-resolution CMOS-MEAs was in 2001 by Berdondini and collaborators, to enable simultaneous recordings from 4,096 electrodes [129] (Fig. 7.5). In 2011, this platform became the first commercially available active MEA system. The electrode connectivity is managed by adopting an architecture based on the Active Pixel Sensor (APS) array concept that was originally applied for high-speed light-imaging sensors [36, 37]. For electrophysiological recordings, the original in-pixel circuitry for light sensing was entirely redesigned to instead provide sensitivity to small extracellular charge variations resulting from cellular activity. The active area is constituted by an array of "neuroelectronic pixels," each integrating a microelectrode and a circuit for the first-stage amplifier and low-pass filtering. The maximum currently available recording area is ~25 mm^2, with a density of electrodes up to 520 electrodes/mm^2 (square electrode of 21 μm × 21 μm in width and an electrode pitch of 42 μm). After three stages of amplification (in-pixel amplifier, column amplifier, and output amplifier) that

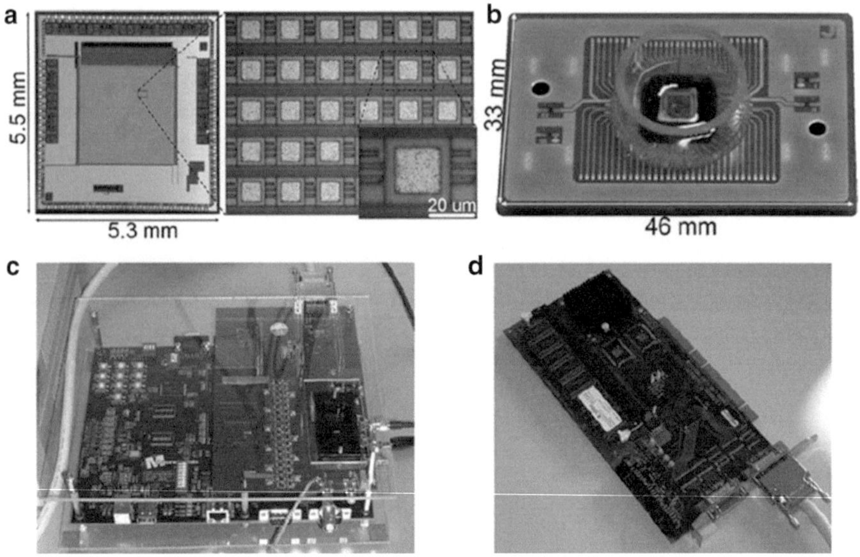

Fig. 7.5 First prototype of the Active Pixel Sensor MEA platform (APS-MEA). (**a**) Picture of the CMOS dye and close up on the electrode area. (**b**) Early version of the chip packaging on-PCB. (**c**) View of the acquisition board based on a commercial FPGA development kit. (**d**) View of a commercially available frame grabber used for data acquisition on a host computer. Reproduced with permission from [126] © (2008) IEEE

provide a total programmable gain of 52–76 dB, the readout is implemented by addressing and multiplexing the 4,096 electrode signals on 16 output lines. The major effort of this development was to enable whole-array recordings at a sufficiently high sampling rate to resolve single-action potentials (i.e., $\gtrsim$6 kHz). Thus a complete acquisition platform and software application were developed. An early version able to record from the entire array at 7.8 kHz/electrode was presented in 2008 [126]. Interestingly, because each electrode can be individually and randomly addressed, a selected *region of interest* (ROI) can be recorded at higher sampling rates. Additionally, while the first devices were only for recording, the new generation of chips now also allows on-chip electrical stimulation. This chip features 16 stimulating electrodes interlaced within an array of 64 by 64 recording sites (electrode density of ~156 electrodes/mm^2, 81 μm pitch, ~25 mm^2 active area).

A second major development in the field of monolithic CMOS-MEAs was carried out by the group of A. Hierlemann, with first results published in 2003 [127]. The device included an array of electrodes integrated with analogue and digital circuits realized in a standard CMOS technology. The first prototype comprised 16 electrodes (250 μm in pitch) that were connected to a voltage follower to lower the output impedance, an amplifier with band-pass filtering (gain of 40 dB, bandwidth of 100 Hz–50 kHz), and a switching circuit for configuring the electrodes in stimulation or recording modes. By 2007 [130, 131], this basic circuit had been scaled up to a CMOS chip with 11,016 microelectrodes (electrode size of 7 μm diameter and

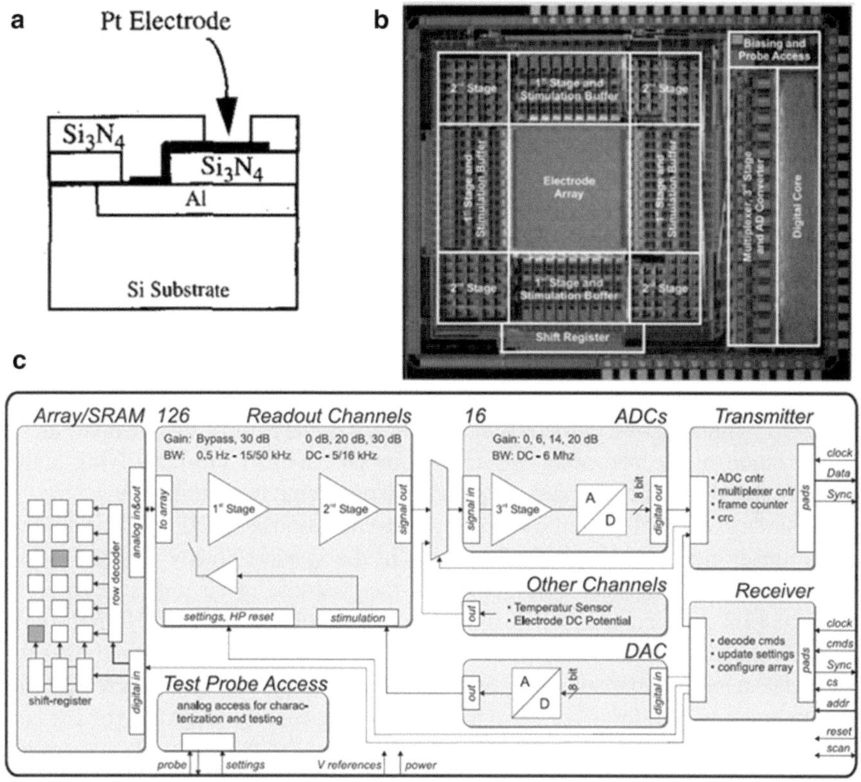

Fig. 7.6 High-resolution CMOS-MEA readout from 128 electrodes selected among 11,016. (**a**) Schematic description of the Pt electrode post-processing [127]. (**b**) View of the CMOS chip and location of the on-chip circuits [128]. (**c**) Block diagram of the on-chip components [128]. Reproduced with permission from (**a**) [127] © (2003) IEEE; (**b**) and (**c**) [128] © (2007) IEEE

electrode pitch of 18 μm), with on-chip circuits to select a subset of recording electrodes. The total recording area was of 2.0 × 1.75 mm^2 and with ~3,200 electrodes/mm^2. The complete schematic of the on-chip circuit is presented in Fig. 7.6c. As shown, electrophysiological signals are amplified and filtered through three stages (total gain of 0–80 dB; high-pass filter, 0.3–100 Hz; low-pass filter, 3.5–14 kHz), multiplexed to 16 lines and sampled at 20 kHz before on-chip analogue-to-digital conversion (ADC). The key feature of this CMOS design is to provide a very dense electrode array, for example, to enable recordings with subcellular spatial resolution for tracking signal propagations in the same neuron and among cells. To achieve this, the circuit architecture and performance of these chips are radically different compared to other APS-MEAs. This CMOS design does not integrate signal conditioning circuits underneath each electrode, and the chip is operated with a switching matrix circuitry that allows to address only up to 126 channels arbitrarily selected among the 11,016 metal electrodes. The amplification and filtering stages for the

126 channels are implemented around the area of the electrode array (Fig. 7.6b). In this way, large and noise-optimized circuits can be integrated. Second, the circuits for analogue-to-digital conversion (recording) and for digital-to-analogue conversion (electrical stimulation) are integrated on-chip (8-bit resolution). This CMOS circuit architecture thus provides a higher spatial resolution (smaller electrode pitch) compared with the APS-MEA platform, but it can only record from a portion of the whole array (more recently up to 256 electrodes). In this sense, the architecture is not scalable in the number of active recording electrodes. Increasing the number of simultaneously recorded electrodes implies a significant increase in the chip size to integrate the supplementary amplifiers for each additional output line. To circumvent this limited readout from the whole array, the device can be sequentially reset to reconfigure the switching matrix circuitry and to record from different subsets of electrodes, but, due to a reconfiguration delay, this approach does not allow to simultaneously record extracellular signals from the whole array. Optimization of the recording sequences, for instance, by pre-identifying active regions on the biological model, might help to overcome these limitations.

A third major development is based on field-effect transistor (FET) arrays instead of the metal electrode arrays implemented in the two previously reported active MEA systems [132]. This development is the outcome of an intensive research effort on FETs carried out by P. Fromhertz and his collaborators [133]. In the case of FETs, the extracellular electrical activity of the neuronal membrane modifies the open-gate region of the transistor and modulates the source–drain current [134]. The main motivation for using these transducers is to ensure the absence of any current flowing between the electronic and ionic sides of the electrode–electrolyte interface (i.e., direct faradic contact). Additionally, the integration of arrays of FETs takes advantage from the scalability of transistors and can be coupled with nanostructures [22]. Under these premises, in 2003 the Max Planck Institute for Biochemistry in Martinsried, in collaboration with Infineon Technologies AG (Belgium), presented the first implementation of a large-scale FET array (active area of ~1 mm^2, electrode pitch of 7.8 μm) with on-chip circuits for signal conditioning and multiplexing [132]. This chip was able to record from the entire array at a frame rate of 2 kHz, and similar to the APS-MEA, the sampling rate can be increased by reducing the number of recording electrodes from selected regions of interest (ROIs). A schematic cross section and the equivalent circuit of the oxide–semiconductor field-effect transistor (OSFET) used in this work are shown in Fig. 7.7a, while the on-chip circuit schematics and a micrograph of the CMOS chip are shown in Fig. 7.7b, c. Each of the 128 rows of the array is connected to a specific amplifier and the readout is performed through 16 output lines by multiplexing 8 electrodes on each output. In order to operate the FET electrodes, a calibration of the sensing transistors is performed typically every 100 frames on each column by applying a reference AC signal in the bath. An improved generation of this high-density FET electrode-array chip with higher sampling frequency (6 kHz) was presented in 2011 [135]. It was able to sequentially record from four regions of the whole active area with up to 4,096 electrodes at a sampling frequency of 24 kHz. The presented platform can record up to 600 ms consecutively, with interruptions of

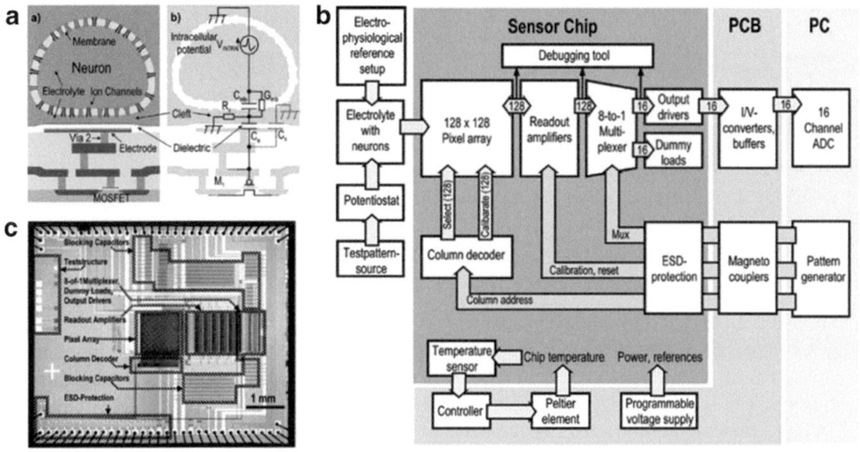

Fig. 7.7 High-resolution FET-MEA. (**a**) Schematic cross section and equivalent circuit of an oxide–semiconductor field-effect transistor (OSFET) used for neural interfacing. (**b**) Block diagram of the on-chip circuit. Each of the 128 rows of the electrode array is connected to a dedicated amplifier and the readout is performed through 16 multiplexers. (**c**) Micrograph of the FET-MEA CMOS chip. Reproduced with permission from [132] © (2003) IEEE

several seconds among each recording phase due to limitations in continuous streaming as well as in the need for a regular calibration of the FET transducers. Also in 2011, the same authors reported a redesigned chip providing 256×128 bidirectional FET recording/stimulation sites (total active area of $2.28 \times 1.14\ \text{mm}^2$) with a complete on-chip digital interface [136]. The entire array can be sampled at a rate of 2.4 kHz or can be operated at higher sampling rates by selecting a subset of recording sites.

7.3 The Active Pixel Sensor Multielectrode Array (APS-MEA) Technology

The Active Pixel Sensor multielectrode array platform (APS-MEA) was introduced in 2001 and has enabled simultaneous recordings in vitro from 4,096 microelectrodes integrated on a large-area array of several square millimeters, with electrode separations of a few tens of microns. This active electrode array is implemented using a highly scalable circuit architecture. The concept of the APS-MEA chip and the 4,096 electrode-array platform are shown in Fig. 7.8. The platform is composed of disposable chips, a real-time hardware unit, and a software application for recording, visualizing, and analyzing the acquired signals. In the next paragraphs, we will first describe in detail the hardware and software components and then review some of the experimental results and approaches used for the analysis of neural activity recorded at high resolution.

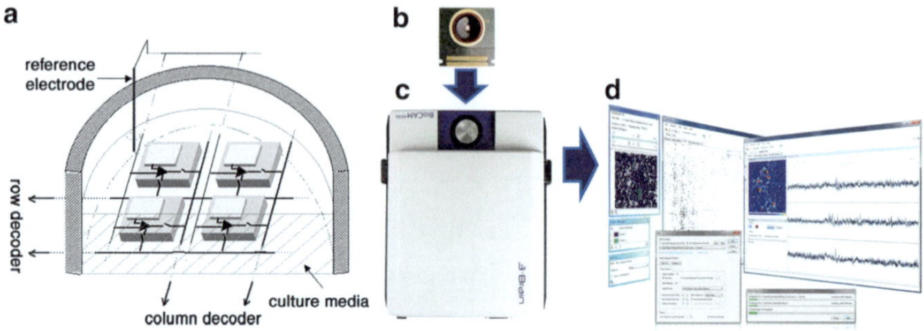

Fig. 7.8 Concept of the APS-MEA chip and overview of the current platform for simultaneous recordings from 4,096 microelectrodes. (**a**) Schematic representation of the APS-MEA circuit with in-pixel electrodes, amplifiers, and active circuits for the readout. (**b**) View of a CMOS APS-MEA chip mounted on its printed circuit board. (**c**) Real-time hardware for control and data acquisition. (**d**) Screenshots of the BrainWave software application used for managing the experiments. (**a**) Reproduced with permission from [137] © (2008) Elsevier

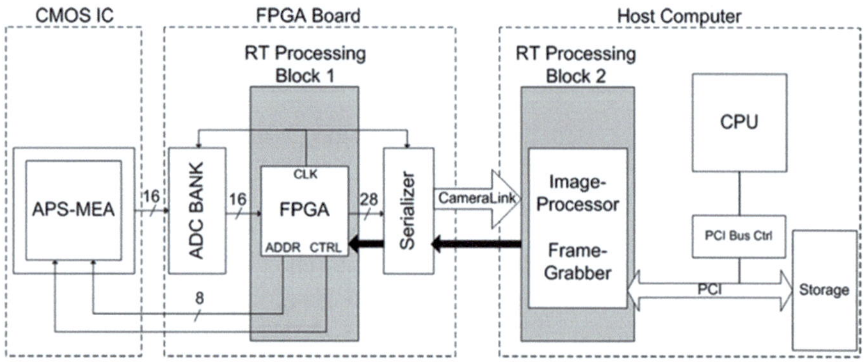

Fig. 7.9 Schematics of the APS-MEA hardware architecture. The platform is composed by a monolithic CMOS IC with 4,096 electrodes; a real-time FPGA board and a host computer equipped with a frame grabber and configured for online storage and visualization of the acquired data. Reproduced with permission from [126] © (2008) IEEE

7.3.1 The APS-MEA System Architecture and Its Hardware Components

The hardware architecture of the APS-MEA is organized in three main components (Fig. 7.9): (1) the CMOS integrated circuit (IC) providing 4,096 electrodes with a readout from 16 multiplexed analogue channels (256 electrodes each); (2) a field-programmable gate array (FPGA) board comprising a module of 16 analogue-to-digital converters (ADCs), a processor unit for operation and addressing of the CMOS chip as well as for the real-time filtering of the signals, and a serializer for high-speed communication through a camera-link standard protocol (data rate of

~60 MB/s); and (3) a host computer equipped with a high-speed frame grabber and with hard drives configured for fast data storage. The hardware is operated from the host computer, which communicates with the FPGA through serial communication (UART). The software running on the host computer implements tools for the online visualization of the data during recording and tools, for data mining, analysis, and management of the acquired datasets, such as exporting records of neuronal spiking activity to common file formats).

The first component of the platform is the *APS-MEA chip*, which is a solid-state CMOS integrated circuit (IC) fabricated in a standard five-metal-layer technology (0.5 μm in the early generations [137] and 0.35 μm successively [126]). It provides an array of 64×64 (4,096) squared electrodes of 21×21 μm^2 with pitches of 42 μm. The metal electrodes were initially made of silicon–aluminum alloy (layer of the CMOS process), and more recently they have been post-processed with noble metals (e.g., Au or Pt by electrodeposition) to improve the quality of the recordings [136]. A ~2 μm thick insulating layer of silicon oxide and silicon nitride protects the underlying CMOS circuitry. The microelectrodes are opened through this insulating layer and have thus a recessed morphology, as it is shown in the SEM picture of Fig. 7.10. The on-chip amplifier is composed of three stages. The first stage is integrated in each pixel underneath each microelectrode and provides a gain of 40 dB. The short distance between the electrode and the amplifier input guarantees a high signal-to-noise ratio since the signal is conditioned close to the source and is thus less sensitive to electronic perturbations introduced by the digital circuitry or by parasitic capacitive noise induced by long wiring. However, the design of this in-pixel circuit is constrained by the tight pixel area of 42×42 μm^2, which imposed a trade-off between noise performances and circuit complexity, transistor size, and the use of resistive or capacitive components for in-pixel filtering [126].

To limit the effects of the DC offset arising from the dynamic electrochemical equilibrium of the electrode–electrolyte interface, which could saturate the amplifiers, an in-pixel calibration circuit was included. Since integrating an AC coupling is not feasible in such a small pixel area, a DC-coupled architecture was chosen and an auto-zeroing circuit is used to manage the DC offset. This auto-zeroing circuit is based on a sampled feedback (SWAZ1 in Fig. 7.10), and it constitutes the in-pixel calibration circuit. For operation, a calibration sequence (SWAZ1 and SWAZ2 closed) lasting 4 μs is used for resetting the input to a defined potential that is equal to the electrolyte potential and which sets the feedback to the appropriate DC working point. At the end of the calibration sequence, the feedback path is opened and the sampling capacitor CAZ holds the corresponding value. The in-pixel circuits have to be regularly recalibrated due to drifts of the DC offsets introduced (mainly) by leakage currents in the capacitor and photoelectrically generated charges introduced by light (i.e., under standard operation, a typical calibration period of 1–2 s is sufficient). The first stage of the amplification circuit is completed by a column amplifier with programmable gain and by an output amplifier integrated on the sides of the electrode-array area. These circuits allow total gains of 52, 64, 70, or 76 dB, respectively, with the gain adjustable to signals from different samples. Additional on-chip circuits were integrated for controlling the nominal bias currents of the

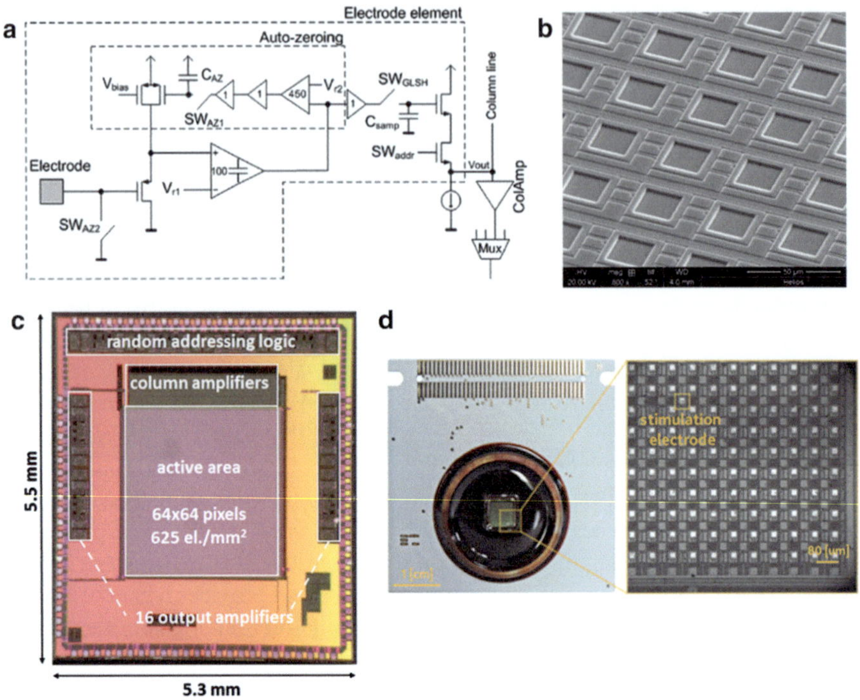

Fig. 7.10 View of the APS-MEA CMOS chip. (**a**) Schematics of the signal amplification chain (providing a first stage under each electrode) and of the electrode readout. (**b**) SEM micrograph of an area of the electrode array. Electrodes are recessed into the insulation layer. (**c**) Micrograph of the CMOS chip. (**d**) Most recent generation of APS-MEA providing 4,096 recording electrodes and 16 electrodes for electrical stimulation. (**a**) Reproduced with permission from [126] © (2008) IEEE

amplifiers, for digitally addressing each pixel with random addressing logic, and for multiplexing the electrode signals on 16 analogue output lines (256 electrodes each). The random addressing logic is sufficiently fast to enable acquisition at about 8 kHz/electrode for full frame recordings or up to 125 kHz/electrode for recordings from only 64 selected electrodes. While the first-generation APS-MEAs were designed for recording only, in 2012 we presented a new chip integrating electrodes for electrical stimulation (Fig. 7.10). Here, the recording and stimulation circuitry have been separated to avoid noise and artifacts induced by large amplitude stimulation signals on the recording circuits. The chip was redesigned to accommodate 16 stimulating electrodes positioned within the recording array to define a 4×4 grid with a pitch of 1,225 μm. To do so, the pitch of the recording electrodes was increased to 81 μm; this generation of chips thus features an electrode density of ~156 electrodes/mm² over a larger active area of ~25 mm².

The second hardware component is the real-time FPGA board that controls the APS-MEA chip during operation and is interfaced by a host computer. On the FPGA board, the 16 multiplexed analogue outputs of the chip are digitally converted with 12 bit resolution by a bank of analogue-to-digital converters (ADCs) operating at 3 MSamples/s. This allows sustained data streams of ~2 MSamples/s generated during full frame recordings. The FPGA processor (in this case a Cyclone from Altera) is in charge of controlling the timing of the ADCs and sends the calibration signals to the chip for the in-pixel amplifiers. Additionally, it implements a real-time high-pass digital filter (infinite impulse response) for the 4,096 electrodes that is used to remove the DC drift from the recordings. A digital implementation of this filter was chosen in order to enable programming of high-pass filtering between 0.1 and 10 Hz depending on the experimental needs [126, 127]. After filtering, signals are further multiplexed and sent to a serializer for transmission to the host computer with a camera-link standard protocol.

The receiver is a two-channel PCI frame grabber (Leonardo from Arvoo), operating at a transmission speed of 128 MB/s per channel. This board is installed on the host computer, which is the third hardware component of the system. Since the entire platform is able to record continuously the raw data from all the 4,096 electrodes, the stream cannot be stored entirely on the RAM of the host. Indeed, 10 min of recording generates a dataset of ~36 GB. Therefore, the software and hardware of the host computer have to allow writing the recorded data directly to hard drives. For this purpose, the host pc implements a RAID 0 hard-disk configuration (striping configuration) for fast parallel writing, guaranteeing an average speed of at least 128 MB/s. The maximal recording duration of raw data is only limited by the size of the available hard drives and can thus extend up to a few hours with currently available components. While data storage is continually improved by the availability on the market of larger and faster data storage solution, handling of these large datasets can be difficult. Therefore, real-time or online data compression solutions have been investigated [127] and are under development. Recently, the commercial version of this platform (BioCAM from 3Brain GmbH, Switzerland) has been upgraded with an online spike-detection algorithm that drastically reduces the size of the datasets, up to 1,000 times.

7.3.2 Overview of the Software Architecture

A software application was developed using the Visual Studio Environment, to record, analyze, and visualize online the data generated by the high-resolution active electrode arrays.

This environment was selected because it integrates different languages such as C, C++, and C#, enabling developers to adopt high-performance low-level languages for specific tasks (e.g., C and C++ to wrap the frame-grabber COM libraries or object-oriented C# to implement graphical user interfaces, etc.). In particular, the software architecture (Fig. 7.11) was designed to provide critical functionality related to large datasets, namely, (1) data handling for fast data streams (~62 MB/s), (2) online or off-line event detection, (3) graphical tools to visualize the spatiotemporal distribution of the activity, and (4) data mining to extract the most important

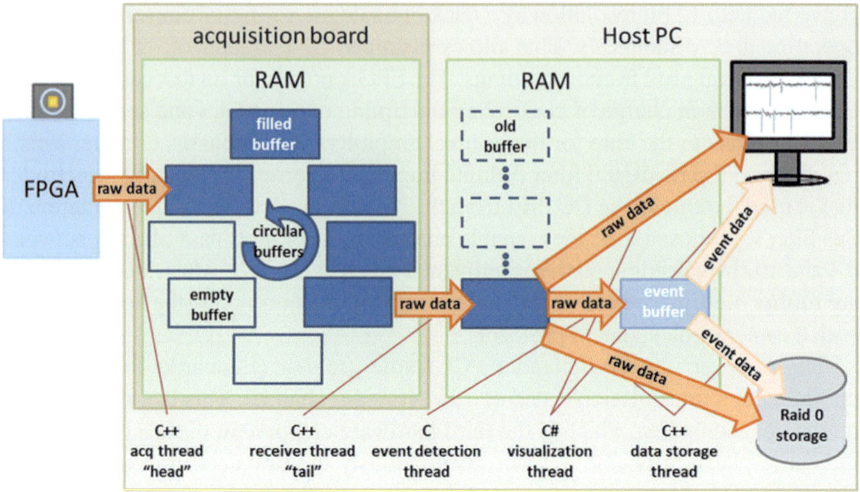

Fig. 7.11 Overview of the APS-MEA software architecture. This solution was developed to manage the continuous visualization, event detection, and storage of large datasets acquired from 4,096 simultaneously recording electrodes (~62 MB/s)

information from the acquired data. To manage these issues during continuous acquisitions, the recorded data are at first buffered on the 512 MB RAM of the frame grabber and then transferred to the memory of the host computer. By parallel multithreading, a chain of buffers is filled consecutively with an acquiring thread ("head"), while a receiver thread ("tail") is in charge of moving the filled buffers on the RAM of the host. This custom configuration, entirely developed in C++, maintains a balance between data acquisition and data handling. Once a buffer is available on the RAM of the host, a series of C# threads process the data for successive visualization, event detection (if required), and storage on the hard drive. It should be noted that this processing stream with event (spike) detection requires a host computer with a minimum of eight independent cores of 3 GHz and at least 4 GB of available RAM to process all 4,096 channels.

7.3.3 Visualization of Neural Activity Event Detection

The visualization of all signals acquired from several thousand electrodes would be difficult using conventional representations based on single-electrode traces available for lower-density multielectrode arrays (e.g., 60–120 microelectrodes). Instead, we have found that "image" and "movie" representations of the recorded extracellular potentials provide a useful way to exploit the high spatiotemporal resolution of active MEA. Signals measured by the $n \times m$ electrode array are visualized as sequences of images of $n \times m$ pixels. As a first pass the extracellular signal recorded by each electrode (pixel) can be directly encoded in false colors (Fig. 7.12).

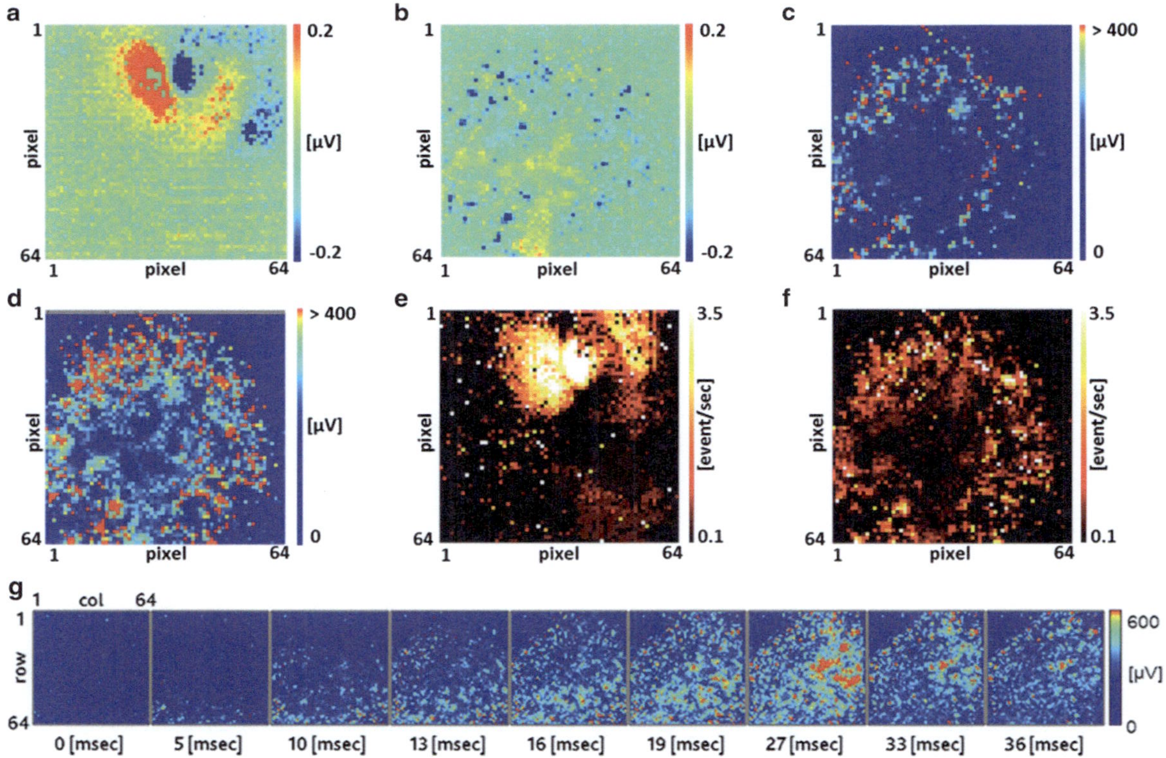

Fig. 7.12 Examples of visualization of 4,096 extracellular signals recorded by APS-MEAs as 64×64 color-coded images. (**a**) Instantaneous extracellular potential spatial distribution during an LFP propagation recorded in cortico-hippocampal brain slices (i.e., hyperpolarized region in *dark blue*, depolarized region in *bright red*) and (**b**) during spiking neural activity recorded in cultured hippocampal neurons. (**c**) Color-coded maximal signal variation representing spiking activity recorded in cultured hippocampal neurons (same as (**b**) over a time interval of 10 ms and of (**d**) 100 ms). (**e**) Representation of the spatial distribution of detected LFP events in a cortico-hippocampal brain slice and (**f**) of detected spike events in a cultured neural network. (**g**) Image sequence representing the time course of spiking activity during a network burst recorded in a cultured neural network and lasting ~40 ms

Movies of the recorded spiking activity as well as time-lapse images that represent the spatiotemporal pattern of activity can be generated to visualize the propagation of neuronal activity. This simple encoding is useful for slow and large amplitude signals of hundreds of microvolts, such as discriminating between hyperpolarization (dark blue) and depolarization (bright red) phases in the low-frequency potentials (LFPs) recorded from cortico-hippocampal brain slices (Fig. 7.12a). However, displays of raw extracellular potential are insufficient to capture neuronal action potentials, which typically last only 1–2 ms. A more effective and simple approach consists in color coding the maximal signal variation over a time interval. This is illustrated in Fig. 7.12c, d using time intervals of 10 and 100 ms, respectively. For the analysis of the recorded signals, the first step typically consists of detecting the timing of each neuronal signal event, such as spikes, spike bursts, multiunit spiking activity, or field potentials. Depending on the sparseness of these events, the detection can contribute to a drastic reduction of the dataset size. Different event detection algorithms have been developed for passive MEA and can be adapted for active electrode arrays. For the APS-MEA platform, a precise and robust algorithm has been adopted from previous work and provides a precise temporal identification of the peaks of the events with low false-positive detections [138]. After event detection, the results can be again displayed as images of 64×64 pixels with color codes that represent the activity parameter calculated for each single electrode. This can help in the data mining, in order to select relevant electrodes to be further analyzed. For instance, the visualization of the average firing rate can be used to select the most valuable electrodes, either by manual selection (Fig. 7.12e) or by thresholding of the firing rate, for example, by considering only electrodes recording signals within a selected spike rate interval (Fig. 7.12f).

7.3.4 Fundamentals of Spike Train Analysis

The spike-detection algorithm is an event detector for action potentials. Its output is the so-called spike train, which represents the time stamps of each spike for each electrode. Importantly, for extracellular recordings, each electrode samples the surrounding electrical field and will thus likely detect signals from more than one neuron. Therefore, before spike detection, a spike-sorting algorithm (SSA, such as those provided by the Offline Sorter from Plexon Inc. or as presented for CMOS-MEAs in [127, 139]) can be applied to the recorded signals to assign the clustered spike shapes to the different neural units [140, 141]. In principle, the spike sorting would constitute a fundamental step of the data-analysis process, but a careful understanding of the intrinsic limitations of these algorithms should be taken into account. First, most of the SSAs rely on the constancy of the action potential waveform to assign the spikes to a specific cluster. However, the shape and amplitude of an action potential recorded with extracellular electrodes can change along with time (e.g., during a burst of spikes), and the SSA may overestimate the number of units. Second, the SSAs may yield different sorted "neurons" when analyzing the same dataset and thus an objective choice of the properly sorted "neurons" might be hard to achieve. Rigorous approaches for the validation of the sorting algorithms have been recently introduced using synthetic ground truths generated by artificial neural networks [142] and are nowadays used to validate new spike-sorting techniques [143] that might better identify single neuronal "units" by taking advantage of multiple nearby electrodes recording the same neuronal signal.

After spike sorting, some insights on the network activity can be obtained by computing first-order statistics from the spike time stamps. Typical quantities are the firing rate (the number of spikes counted in a time window), inter-spike intervals, bursting rate, burst duration, and intra-burst frequency (number of spikes in a burst divided by the burst duration). These quantities can be computed for individual units or averaged across many units in the network. The network-average parameters have been used extensively to characterize neural development in cultures [144], to assess the effects of drugs (e.g., [53]), and to examine network plasticity [145]. High-density MEAs obviously allow for more reliable estimation of first-order parameters than low-density MEAs [146].

Further insights on network dynamics can be described by computing parameters based on the joint activities of two or more electrodes. These second-order statistical metrics are based on concepts derived from cross-correlation analysis or information theory to quantify statistical dependencies (or functional connection) among electrodes. The cross-correlation function is a measure of the number of coincident spikes found at different time lags. From its definition, the cross-correlation peak relays information about both the strength and the directionality of the functional connection between two neurons. The significance of the cross-correlation peaks can be assessed against null models obtained by shuffling the original spike trains in time [147]. Based on this analysis, a functional connectivity map representing the significant links among electrode sites can be built and further analyzed to assess the topological properties of a network [148]. Similar functional connectivity analyses have been extensively applied to fMRI recordings to interpret the interactions between different brain regions during specific tasks, but the information gathered by fMRI data has a typical limited spatial (~1 mm^3) and temporal (~seconds) resolution and is thus less informative of the underlying processing at the cellular scale. There is an increasing interest in understanding brain function in terms of its elemental properties, at the level of neural circuits [149]. Multielectrode array devices such as the APS-MEA accessing thousands of single neurons thus have high potential to also reveal functional connectivity maps at the cellular scale in vivo [139].

Finally, the high-dimensional datasets generated by high-resolution MEAs are motivating the development of novel data-analysis techniques adapted to extract information from these multielectrode arrays [150]. Although high-dimensional datasets are maximally informative, a simplified description of the datasets is normally required. In fact, high-dimensional spaces pose a set of issues for statistical and data analysis collectively known as the "curse of dimensionality." Therefore, dimensional reduction techniques that retain the most informative views on the data, such as principal component analysis, have been used. Multiunit activity has also been successfully described using Ising models [151], in which network activities

Table 7.1 Specifications of the current generations of APS-MEA chips integrating recording and stimulation microelectrodes

	APS-MEA (recording)	APS-MEA (recording and stimulation)
Number of recording electrodes	64 × 64 (4,096)	64 × 64 (4,096)
Recording electrode size	21 × 21 μm²	21 × 21 μm²
Recording electrode pitch	42 μm	81 μm
Active area	7.1 mm²	26.9 mm²
Number of stimulating electrodes	None	4 × 4 (16)
Stimulating electrode size	None	21 × 21 μm²
Stimulating electrode pitch	None	1225 μm
Supply voltage	3.3 V	3.3 V
Overall chip power consumption	132 mW	132 mW
ADC resolution (off-chip)	12 bit	12 bit
Input-referred noise (static)	11 μVrm	11 μVrm
Input-referred noise (dynamic)	21.2 μVrm	21.2 μVrm
Average peak-to-peak noise	63.6 μV	63.6 μV
Programmable total gain (on-chip)	52, 64, 70, or 76 dB	52, 64, 70, or 76 dB
Full frame rate (4,096 electrodes)	8 kHz	8 kHz
Frame rate on 64 electrodes	125 kHz	125 kHz
Data throughput	490 Mbit/s	490 Mbit/s

are fit considering only pairwise interactions in the model or by using maximum entropy models that generalizes the same idea to higher orders (>2) of interaction.

The development of descriptors of neuronal activity can be driven by the availability of high-resolution experimental data, particularly in the area of reducing the size of high-dimensional datasets. For example, simplified descriptors might suffice to describe network bursts spreading through an entire network of cultured neurons. An example is the *center activity trajectory* (CAT) algorithm [152] that reduces each time frame of the network activity to the center of mass of the activity. The CAT algorithm has been applied to APS-MEA recordings [153] and was used to identify the trajectory of each propagating event. Successively, the trajectories were classified to determine the major patterns of propagating network bursts.

7.3.5 Sensing and Actuating Performances of Current Active Pixel Sensor MEAs

The specifications achieved with current APS-MEA chips are summarized in Table 7.1. The main challenge in achieving an adequate input-referred noise for recording action potentials was the small pixel area available for the integration of the first-stage amplifier circuit. The noise measured in phosphate buffer solution was 11 μV_{rms} in static operation (single-electrode output) and of ~21 μV_{rms} during full array readout.

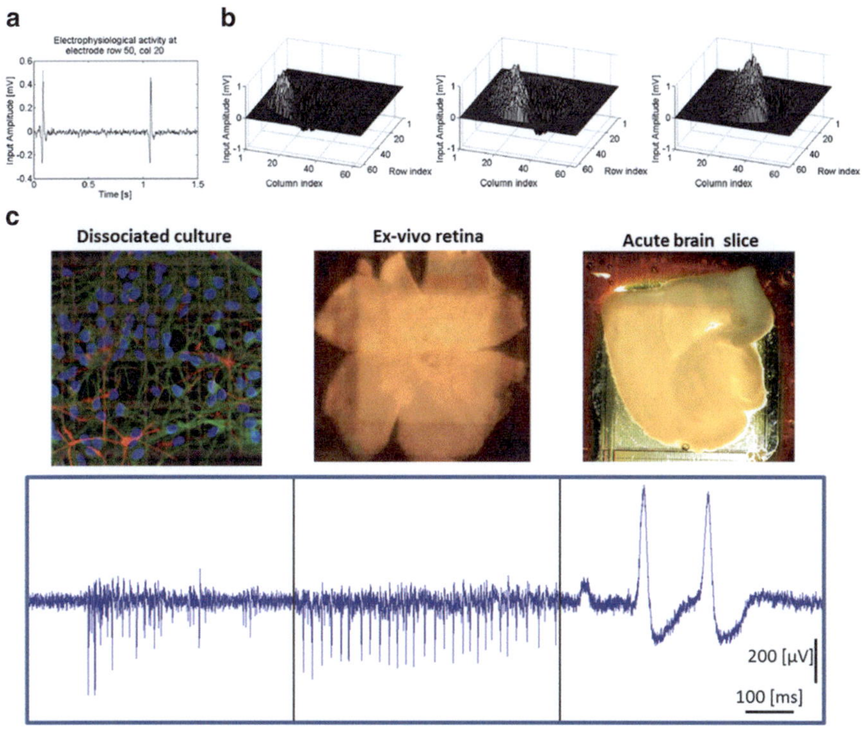

Fig. 7.13 Overview of electrophysiological signals recorded with APS-MEA on different in vitro models. (**a**) Action potentials recorded from cardiac cell cultures [126]. (**b**) Three-dimensional reconstruction of a propagating cardiac wave [126]. (**c**) Examples of extracellular electrophysiological signals recorded on cultured primary neurons, ex vivo retina, and cortico-hippocampal brain slices. Pictures of the experimental model on APS-MEA chips are shown on the *top*. Picture in (**a**) and (**b**) reproduced with permission from [126] © (2008) IEEE

The resulting peak-to-peak noise of ~63 μV is compatible with measuring small amplitude spike signals of ~100 μV. In addition to noise, an additional challenge to resolve action potentials and their spatiotemporal propagations was to implement a system able to simultaneously record with a sampling rate of several kHz from 4,096 electrodes. To illustrate the performance of the APS-MES, in the next paragraphs we will describe electrophysiological signals recorded in vitro, on cultured cardiomyocytes and neural networks, brain slices, and ex vivo retina preparations. The first experiments for validating the APS-MEA were performed on dissociated cardiomyocytes. After a few days in culture, cardiomyocyte preparations typically show rhythmic (~1 s period) large amplitude extracellular signals (0.5–1 mV) that propagate as waves spreading across the culture [126] (Fig. 7.13). For neural activity recordings, the APS-MEAs were used on dissociated cultures from rat primary hippocampal neurons, ex vivo retina tissue from mouse, and acute brain slices of cortico-hippocampal circuit of mouse. Some examples of typical signals recorded from these experimental models are shown in Fig. 7.13c.

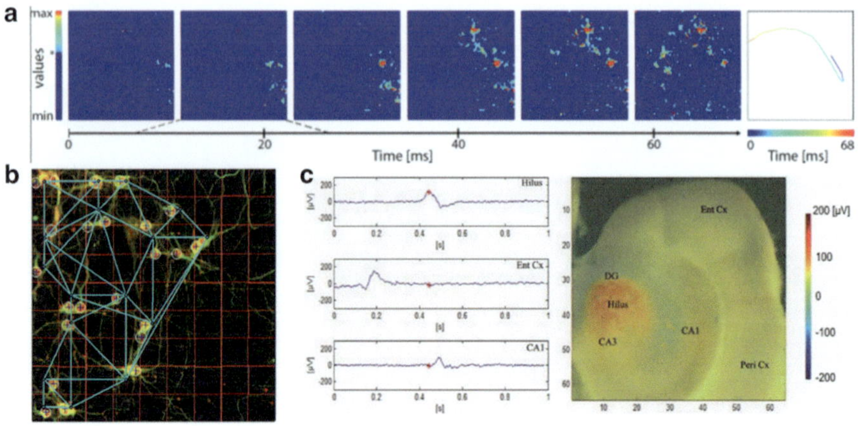

Fig. 7.14 Some examples of studies taking advantage from the spatiotemporal performances of APS-MEAs. (**a**) Studying burst activity propagations in cultured networks based on CATs and classification. (**b**) Investigating functional connectivity estimations by APS-MEA recordings and optical microscopy. (**c**) Superimposed anatomical picture and recorded activity used for studying fast LFP propagations among multiple brain regions in brain slices

The average signal amplitude of spikes typically ranges between 180 μV_{p-p} for dissociated neural cultures up to 319 μV_{p-p} for recordings form the ganglion cell layer of explanted retina. Recordings of low-frequency field potentials (LFPs) in acute brain slice show a higher average signal amplitude of ~440 μV_{p-p}. Interestingly, even though the noise of the on-chip amplifier is not as low as for low-density MEA systems, the neuron–electrode coupling and the short distance between the electrodes and the in-pixel amplifiers allow for an excellent signal-to-noise ratio.

A key feature of these large-scale electrode-array devices is that they can resolve propagating activity within networks with high spatiotemporal resolution. With an electrode pitch of 21 μm between thousands of recording electrodes, the spatial resolution is high enough to literally image the distribution of extracellular signals and to finely track signal propagation. Some illustrative examples from neuronal cell cultures and on brain slices are shown in Fig. 7.14. As shown, APS-MEAs allow tracking of network bursts to investigate functional connectivity in combination with optical microscopy or to characterize LFP propagations in brain slices. In particular, the APS-MEA chip can uniquely track fast field-potential propagations across and between hippocampal and cortical circuits, over a large field of view (up to 5.1×5.1 mm^2). This is useful for finely characterizing spontaneous or drug-induced propagations [154], for example, activity propagating from regions of the cortex to the DG, hilus, CA3, and CA1 areas of the hippocampus, and to estimate the speed of propagation (e.g., 17 mm/s CA3 to CA1 and 166 mm/s DG to hilus).

On-chip electrical stimulation has recently been tested on dissociated hippocampal cultures. Biphasic electrical stimuli of 600 μs and 300 μA_{p-p} were applied to evoke electrical responses, and the evoked spatiotemporal propagation of neuronal activity can be finely described (Fig. 7.15). Interestingly, the stimulation artifact is only

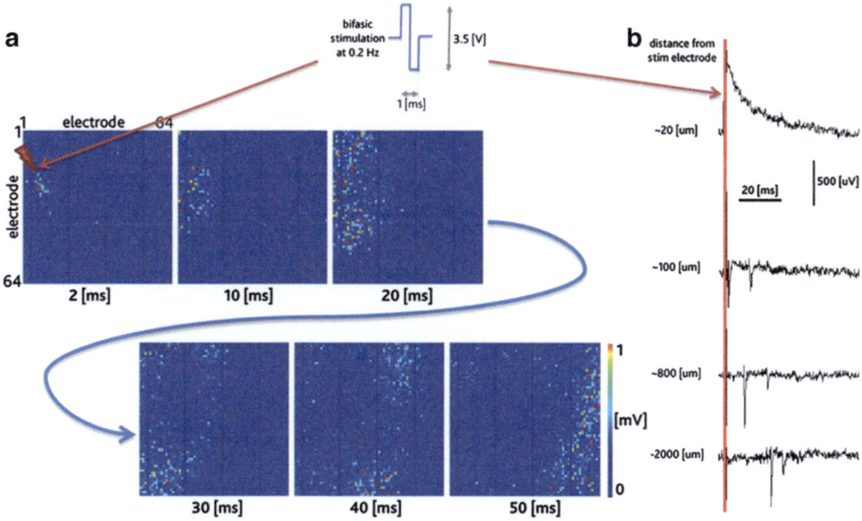

Fig. 7.15 On-chip electrical stimulation with APS-MEA. (**a**) Example of evoked response using on-chip stimulation electrodes. (**b**) The stimulation artifact is confined to electrodes close to the stimulating electrodes (<100 μm)

visible on the recording electrodes within ~20 μm of the stimulation electrodes, and the evoked responses can be recorded without artifacts with a latency after stimulation lower than 3 ms (e.g., see electrode at ~100 μm distance).

7.4 Conclusions

Neurotechnologies need to rapidly evolve to provide single-unit recordings from a much larger number of neurons than today possible. In this respect, here we have discussed active multielectrode arrays (MEAs) as an emerging methodology that uses CMOS microelectronic circuits to implement electrode-based devices for electrophysiological recording from thousands of microelectrodes. Instead of individually wiring each electrode with passive metal lines as in previous generations of passive and hybrid MEAs, this novel MEA integrates microelectronic circuits to actively connect and read out each electrode within dense and large arrays. The Active Pixel Sensor multielectrode array (APS-MEA) discussed in detail in this chapter is one of the current implementations of active MEAs and provides recordings at milliseconds resolutions from 4,096 closely spaced electrodes. Designed currently for in vitro applications, we have used this platform to record from acute brain tissues and neuronal cultures, literally imaging propagation of neuronal activity.

The knowledge acquired thus far in the design of these circuits could allow in the coming years the development of large-scale recording capabilities for in vivo studies, with minimally invasive and highly scalable devices [18] and with an overall power dissipation smaller than the estimated safe maximum of 80 mW/cm^2 for brain implantation [155]. Such a development would ideally provide flexible and biomimetic properties to reduce damage to brain tissue. In this respect, novel interfacing modalities and materials emerging from nanotechnologies and flexible biodegradable materials could be combined with the functionalities provided by advanced CMOS microelectronic circuits. This will require intensive multidisciplinary R&D converging both engineering and neuroscience, including the development and validation of devices and computational methods adapted to large array recordings.

References

1. Averbeck, B.B., and Lee, D.: Coding and transmission of information by neural ensembles. Trends in Neurosciences. **27**(4), 225–230 (2004)
2. Schwartz, A.B., Cortical neural prosthetics. Annual Review of Neuroscience. **27**, 487–507 (2004)
3. Hatsopoulos, N.G., and Donoghue, J.P.: The science of neural interface systems. Annual Review of Neuroscience. **32**, 249–266 (2009)
4. Stieglitz, T., et al.: Brain-computer interfaces: an overview of the hardware to record neural signals from the cortex. In: Verhaagen, J., et al. (eds) Neurotherapy: Progress in Restorative Neuroscience and Neurology, pp. 297–315. Elsiever, Amsterdam (2009)
5. Csicsvari, J., et al.: Mechanisms of gamma oscillations in the hippocampus of the behaving rat. Neuron **37**(2), 311–322 (2003)
6. Normann, R.A., et al.: A neural interface for a cortical vision prosthesis. Vision Research, **39**(15), 2577–2587 (1999)
7. Maynard, E.M., Nordhausen, C.T., and Normann, R.A.: The Utah intracortical electrode array: a recording structure for potential brain-computer interfaces. Electroencephalography and Clinical Neurophysiology. **102**(3), 228–239 (2003)
8. Normann, R.A.: Technology insight: future neuroprosthetic therapies for disorders of the nervous system. Nature Clinical Practice Neurology. **3**(8), 444–452 (2007)
9. Donoghue, J.P.: Connecting cortex to machines: recent advances in brain interfaces. Nature Neuroscience. **5,** 1085–1088 (2002)
10. Wise, K.D.: Silicon microsystems for neuroscience and neural prostheses. IEEE Engineering in Medicine and Biology Magazine. **24**(5), 22–29 (2005)
11. Pearce, T.M. and Williams, J.C.: Microtechnology: meet neurobiology. Lab on a Chip **7**(1), 30–40 (2007)
12. HajjHassan, M., Chodavarapu, V., and Musallam, S.: NeuroMEMS: Neural Probe Microtechnologies. Sensors **8**(10), 6704–6726 (2008)
13. Park, J.W., et al.: Advances in microfluidics-based experimental methods for neuroscience research. Lab on a Chip **13**(4), 509–521 (2013)
14. Alivisatos, A.P., et al.: Nanotools for neuroscience and brain activity mapping. Acs Nano **7**(3), 1850–1866 (2013)
15. Tian, B., et al.: Three-dimensional, flexible nanoscale field-effect transistors as localized bioprobes. Science **329**(5993), 830–834 (2010)
16. Keefer, E.W., et al.: Carbon nanotube coating improves neuronal recordings. Nature Nanotechnology. **3**(7), 434–439 (2008)
17. Lovat, V., et al.: Carbon nanotube substrates boost neuronal electrical signaling. Nano Letters. **5**(6), 1107–1110 (2005)

18. Du, J., et al.: Multiplexed, high density electrophysiology with nanofabricated neural probes. Plos One. **6**(10), e26204 (2011)
19. Spira, M.E., and Hai, A.: Multi-electrode array technologies for neuroscience and cardiology. Nature Nanotechnology. **8**(2), 83–94 (2013)
20. Hai, A., Shappir, J., and Spira, M.E.: In-cell recordings by extracellular microelectrodes. Nature Methods. **7**(3), 200–202 (2010)
21. Martiradonna, L., et al.: Beam induced deposition of 3D electrodes to improve coupling to cells. Microelectronic Engineering **97**, 365–368 (2012)
22. Xie, C., et al.: Intracellular recording of action potentials by nanopillar electroporation. Nature Nanotechnology. **7**(3), 185–190 (2012)
23. Robinson, J.T., et al.: Vertical nanowire electrode arrays as a scalable platform for intracellular interfacing to neuronal circuits. Nature Nanotechnology. **7**(3), 180–184 (2012)
24. Buzsaki, G.: Large-scale recording of neuronal ensembles. Nature Neuroscience. **7**(5), 446–451 (2004)
25. Stevenson, I.H. and Kording, K.P.: How advances in neural recording affect data analysis. Nature Neuroscience. **14**(2), 139–142 (2011)
26. Kerr, J.N.D. and Denk, W.: Imaging in vivo: watching the brain in action. Nature Reviews Neuroscience. **9**(3), 195–205 (2008)
27. Maschio, M.D., et al.: Two-photon calcium imaging in the intact brain. Advances in Experimental Medicine and Biology. **740,** 83–102 (2012)
28. Csicsvari, J., et al.: Massively parallel recording of unit and local field potentials with silicon-based electrodes. Journal of Neurophysiology. **90**(2), 1314–1323 (2003)
29. Viventi, J., et al.: Flexible, foldable, actively multiplexed, high-density electrode array for mapping brain activity in vivo. Nature Neuroscience. **14**(12), 1599–1605 (2011)
30. BeMent, S.L., et al.: Solid-state electrodes for multichannel multiplexed intracortical neuronal recording. IEEE Transactions on Bio-Medical Engineering. **33**(2), 230–241 (1986)
31. Hierlemann, A., et al.: Growing cells atop microelectronic chips: interfacing electrogenic cells in vitro with CMOS-based microelectrode arrays. Proceedings of the IEEE. **99**(2), 252–284 (2011)
32. Franke, F., et al.: High-density microelectrode array recordings and real-time spike sorting for closed-loop experiments: an emerging technology to study neural plasticity. Frontiers in Neural Circuits. **6**, 105 (2012). doi: 10.3389/fncir.2012.00105
33. Eldawlatly, S., Jin, R., and Oweiss, K.G.: Identifying functional connectivity in large-scale neural ensemble recordings: a multiscale data mining approach. Neural Computation. **21**(2), 450–477 (2009)
34. Wise, K.D., Integrated sensors, MEMS, and microsystems: reflections on a fantastic voyage. Sensors and Actuators A-Physical. **136**(1), 39–50 (2007)
35. Berdondini, L., et al.: Active pixel sensor array for high spatio-temporal resolution electro-physiological recordings from single cell to large scale neuronal networks. Lab on a Chip **9**(18), 2644–2651 (2009)
36. Fossum, E.R.: CMOS image sensors: electronic camera-on-a-chip. IEEE Transactions on Electron Devices **44**(10), 1689–1698 (1997)
37. Mendis, S.K., et al.: CMOS active pixel image sensors for highly integrated imaging systems. IEEE Journal of Solid-State Circuits **32**(2), 187–197 (1997)
38. Cheung, K.C.: Implantable microscale neural interfaces. Biomedical Microdevices **9**(6) 923–938 (2007)
39. Najafi, K. and Wise, K.D.: An implantable multielectrode array with on-chip signal processing. Journal of Solid-State Circuits. **sc-21**(6), 1035–1044 (1986)
40. Jochum, T., Denison, T., and Wolf, P.: Integrated circuit amplifiers for multi-electrode intracortical recording. Journal of Neural Engineering. **6**(1), 012001 (2009)
41. Denison, T., Molnar, G., and Harrison, R.: Integrated amplifier architectures for efficient coupling to the nervous system. In: Steyaert, M., Roermund, A.M.V., and Casier, H. (eds) Analog Circuit Design, pp. 167–191. Springer, The Netherlands (2009)

42. Dabrowski, W., Grybos, P., and Litke, A.M.: A low noise multichannel integrated circuit for recording neuronal signals using microelectrode arrays. Biosensors & Bioelectronics **19**(7), 749–761 (2004)
43. Graham, A.H.D., et al.: Commercialisation of CMOS integrated circuit technology in multi-electrode arrays for neuroscience and cell-based biosensors. Sensors **11**(5), 4943–4971 (2011)
44. Jones, I.L., et al.: The potential of microelectrode arrays and microelectronics for biomedical research and diagnostics. Analytical and Bioanalytical Chemistry **399**(7), 2313–2329 (2011)
45. Chiappalone, M., et al.: Dissociated cortical networks show spontaneously correlated activity patterns during in vitro development. Brain Research **1093**, 41–53 (2006).
46. Morin, F.O., Takamura, Y., and Tamiya, E.: Investigating neuronal activity with planar microelectrode arrays: achievements and new perspectives. Journal of Bioscience and Bioengineering **100**(2), 131–143 (2005)
47. Potter, S.M.: Closing the loop between neurons and neurotechnology. Frontiers in Neuroscience **4**, (2010). doi: 10.3389/fnins.2010.00015
48. van Pelt, J., et al.: Long-term characterization of firing dynamics of spontaneous bursts in cultured neural networks. IEEE Transactions on Biomedical Engineering **51**(11), 2051–2062 (2004)
49. Wagenaar, D.A., Pine, J., and Potter, S.M.: An extremely rich repertoire of bursting patterns during the development of cortical cultures. BMC Neuroscience **7**, 11 (2006). doi:10.1186/1471-2202-7-11
50. Rutten, W., et al.: Neuroelectronic interfacing with cultured multielectrode arrays toward a cultured probe. Proceedings of the IEEE **89**(7), 1013–1029 (2001)
51. Egert, U., et al.: A novel organotypic long-term culture of the rat hippocampus on substrate-integrated multielectrode arrays. Brain Research Protocols **2**(4), 229–242 (1998)
52. Meister, M., et al.: Synchronous bursts of action-potentials in ganglion-cells of the developing mammalian retina. Science **252**(5008), 939–943 (1991)
53. Frega, M., et al.: Cortical cultures coupled to Micro-Electrode Arrays: a novel approach to perform in vitro excitotoxicity testing. Neurotoxicology and Teratology **34**(1), 116–127 (2012)
54. Johnstone, A.F.M., et al.: Microelectrode arrays: a physiologically based neurotoxicity testing platform for the 21st century. Neurotoxicology **31**(4), 331–350 (2010)
55. Stett, A., et al.: Biological application of microelectrode arrays in drug discovery and basic research. Analytical and Bioanalytical Chemistry **377**(3), 486–495 (2003)
56. Mandenius, C.-F., et al.: Cardiotoxicity testing using pluripotent stem cell-derived human cardiomyocytes and state-of-the-art bioanalytics: a review. Journal of Applied Toxicology **31**(3), 191–205 (2011)
57. Pine, J.: A history of MEA development. In: Taketani, M., and Baudry, M. (eds) Advances in Network Electrophysiology, pp. 3–23. Springer, New York, USA (2006)
58. Thomas, C.A., Jr., et al.: A miniature microelectrode array to monitor the bioelectric activity of cultured cells. Experimental Cell Research **74**(1), 61–66 (1972)
59. Gross, G.W., et al.: A new fixed-array multi-microelectrode system designed for long-term monitoring of extracellular single unit neuronal activity in vitro. Neuroscience Letters **6**(2–3), 101–105 (1977)
60. Pine, J.: Recording action potentials from cultured neurons with extracellular microcircuit electrodes. Journal of Neuroscience Methods. **2**(1), 19–31 (1980)
61. Robinson, D.A.: The electrical properties of metal microelectrodes. Proceedings of the IEEE **56**(6), 1065–1071 (1968)
62. Thiebaud, P., et al.: Microelectrode arrays for electrophysiological monitoring of hippocampal organotypic slice cultures. IEEE Transactions on Biomedical Engineering **44**(11), 1159–1163 (1997)
63. Berdondini, L., et al.: A microelectrode array (MEA) integrated with clustering structures for investigating in vitro neurodynamics in confined interconnected sub-populations of neurons. Sensors and Actuators B-Chemical **114**(1), 530–541 (2006)

64. Rutten, W.L.C.: Selective electrical interfaces with the nervous system. Annual Review of Biomedical Engineering **4**, 407–452 (2002)
65. Martinoia, S., et al.: A general-purpose system for long-term recording from a microelectrode array. Journal of Neuroscience Methods **48**(1–2), 115–121 (1993)
66. Norlin, A., Pan, J., and Leygraf, C.: Investigation of interfacial capacitance of Pt, Ti and TiN coated electrodes by electrochemical impedance spectroscopy. Biomolecular Engineering **19**(2–6), 67–71 (2002)
67. Cogan, S.F.: Neural stimulation and recording electrodes. Annual Review of Biomedical Engineering. **10**, 275–309 (2008)
68. Navarro, X., et al.: A critical review of interfaces with the peripheral nervous system for the control of neuroprostheses and hybrid bionic systems. Journal of the Peripheral Nervous System **10**(3), 229–258 (2005)
69. Polikov, V.S., Tresco, P.A., and Reichert, W.M.: Response of brain tissue to chronically implanted neural electrodes. Journal of Neuroscience Methods **148**(1), 1–18 (2005)
70. Heuschkel, M.O., et al.: A three-dimensional multi-electrode array for multi-site stimulation and recording in acute brain slices. Journal of Neuroscience Methods. **114**(2), 135–148 (2002)
71. Kotov, N.A., et al.: Nanomaterials for neural interfaces. Advanced Materials **21**(40), 3970–4004 (2009)
72. Bareket-Keren, L. and Hanein, Y.: Carbon nanotube-based multielectrode arrays for neuronal interfacing: progress and prospects. Frontiers in Neural Circuits **6**, 122 (2013). doi: 10.3389/fncir.2012.00122.
73. Soe, A.K., Nahavandi, S., and Khoshmanesh, K.: Neuroscience goes on a chip. Biosensors & Bioelectronics **35**(1), 1–13 (2012)
74. Huang, Y., Williams, J.C., and Johnson, S.M.: Brain slice on a chip: opportunities and challenges of applying microfluidic technology to intact tissues. Lab on a Chip **12**(12), 2103–2117 (2012)
75. Claverol-Tinture, E., et al.: Multielectrode arrays with elastomeric microstructured overlays for extracellular recordings from patterned neurons. Journal of Neural Engineering **2**(2), L1–7 (2005)
76. Wang, L., et al.: Biophysics of microchannel-enabled neuron-electrode interfaces. Journal of Neural Engineering **9**(2), 026010 (2012)
77. Suzuki, I., et al.: Stepwise pattern modification of neuronal network in photo-thermally-etched agarose architecture on multi-electrode array chip for individual-cell-based electrophysiological measurement. Lab on a Chip **5**(3), 241–247 (2005)
78. E. Marconi, A. Maccione, T. Nieus, P. L. Valente, M. Messa, P. Baldelli, L. Berdondini and F. Benfenati, "Investigating emergent functional properties of spontaneously active neuronal networks with controlled topology", PLoS One, 2012;7(4):e34648, 2012
79. Wise, K.D., Angell, J.B., and Starr, A.: An integrated-circuit approach to extracellular microelectrodes. IEEE Transactions on Bio-Medical Engineering **17**(3), 238–247 (1970)
80. Wise, K.D., et al.: Microelectrodes, microelectronics, and implantable neural microsystems. Proceedings of the IEEE **96**(7), 1184–1202 (2008)
81. Hoogerwerf, A.C. and Wise, K.D., A 3-dimensional microelectrode array for chronic neural recording. IEEE Transactions on Biomedical Engineering **41**(12), 1136–1146 (1994)
82. Wise, K.D., et al.: Wireless implantable microsystems: high-density electronic interfaces to the nervous system. Proceedings of the IEEE **92**(1), 76–97 (2004)
83. Leong, K.H., et al.: Multichannel microelectrode probes machined in silicon. Biosensors & Bioelectronics **5**(4), 303–310 (1990)
84. Kewley, D.T., et al.: Plasma-etched neural probes. Sensors and Actuators A-Physical **58**(1), 27–35 (1997)
85. Errachid, A., et al.: New technology for multi-sensor silicon needles for biomedical applications. Sensors and Actuators B-Chemical **78**(1–3), 279–284 (2001)
86. Wassum, K.M., et al.: Silicon wafer-based platinum microelectrode array biosensor for near real-time measurement of glutamate in vivo. Sensors **8**(8), 5023–5036 (2008)
87. Cheung, K.C., et al., Implantable multichannel electrode array based on SOI technology. Journal of Microelectromechanical Systems **12**(2), 179–184 (2003)

88. Ensell, G., et al.: Silicon-based microelectrodes for neurophysiology, micromachined from silicon-on-insulator wafers. Medical & Biological Engineering & Computing **38**(2), 175–179 (2000)
89. Norlin, P., et al.: A 32-site neural recording probe fabricated by DRIE of SOI substrates. Journal of Micromechanics and Microengineering **12**(4), 414–419 (2002)
90. Merriam, S.M.E., et al.: A three-dimensional 64-site folded electrode array using planar fabrication. Journal of Microelectromechanical Systems **20**(3), 594-600 (2011)
91. Wise, K.D. and Najafi, K.: Microfabrication techniques for integrated sensors and microsystems. Science **254**(5036), 1335–1342 (1991)
92. Ji, J., Najafi, K., and Wise, K.D.: A low-noise demultiplexing system for active multichannel microelectrode arrays. IEEE Transactions on Biomedical Engineering **38**(1), 75–81 (1991)
93. Sodagar, A.M., Wise, K.D., and Najafi, K.: A wireless implantable microsystem for multichannel neural recording. IEEE Transactions on Microwave Theory and Techniques **57**(10), 2565–2573 (2009)
94. Tanghe, S.J. and Wise, K.D., A 16-channel CMOS neural stimulating array. IEEE Journal of Solid-State Circuits **27**(12), 1819–1825 (1992)
95. Campbell, P.K., et al.: A silicon-based, 3-dimensional interface-manufacturing processes for an intracortical electrode array. IEEE Transactions on Biomedical Engineering **38**(8), 758–768 (1991)
96. Rousche, P.J. and Normann R.A.: Chronic recording capability of the Utah Intracortical Electrode Array in cat sensory cortex. Journal of Neuroscience Methods **82**(1), 1–15 (1998)
97. Nordhausen, C.T., Rousche, P.J., and Normann, R.A.: Optimizing recording capabilities of the Utah-intracortical-electrode-array. Brain Research, **637**(1-2), 27-36 (1994)
98. Neves, H.P., et al.: The NeuroProbes project: a concept for electronic depth control. Paper presented at the 30th Annual International IEEE EMBS Conference, British Columbia, Vancouver, Canada, 20–24 August, pp. 1857–1857 (2008)
99. Ruther, P., et al.: Recent progress in neural probes using silicon MEMS technology. IEEE Transactions on Electrical and Electronic Engineering **5**(5), 505–515 (2010)
100. Seidl, K., et al.: CMOS-based high-density silicon microprobe arrays for electronic depth control in intracortical neural recording. Journal of Microelectromechanical Systems **20**(6), 1439–1448 (2011)
101. Aarts, A.A., et al.: A 3D slim-base probe array for in vivo recorded neuron activity. Paper presented at the 30th Annual International IEEE EMBS Conference, British Columbia, Vancouver, Canada, 20–24 August, pp. 5798–5801 (2008)
102. Herwik, S., et al.: Fabrication technology for silicon-based microprobe arrays used in acute and sub-chronic neural recording. Journal of Micromechanics and Microengineering **19**(7), 074008 (2009)
103. Oldenziel, W.H., et al.: In vivo monitoring of extracellular glutamate in the brain with a microsensor. Brain Research **1118**, 34–42 (2006)
104. Chen, J.K., et al.: A multichannel neural probe for selective chemical delivery at the cellular level. IEEE Transactions on Biomedical Engineering **44**(8), 760–769 (1997)
105. Spieth, S., et al.: A floating 3D silicon microprobe array for neural drug delivery compatible with electrical recording. Journal of Micromechanics and Microengineering **21**(12), 125001 (2011)
106. Frey, O., et al.: Biosensor microprobes with integrated microfluidic channels for bi-directional neurochemical interaction. Journal of Neural Engineering **8**(6), 066001 (2011)
107. Psoma, S.D., et al.: A novel enzyme entrapment in SU-8 microfabricated films for glucose micro-biosensors. Biosensors & Bioelectronics **26**(4), 1582–1587 (2010)
108. Rousche, P.J., et al.: Flexible polyimide-based intracortical electrode arrays with bioactive capability. IEEE Transactions on Biomedical Engineering **48**(3), 361–371 (2001)
109. Chen, C.-H., et al.: A three-dimensional flexible microprobe array for neural recording assembled through electrostatic actuation. Lab on a Chip **11**(9), 1647–1655 (2011)
110. Chen, Y.-Y., et al.: Design and fabrication of a polyimide-based microelectrode array: application in neural recording and repeatable electrolytic lesion in rat brain. Journal of Neuroscience Methods **182**(1), 6–16 (2009)

111. Cheung, K.C., et al.: Flexible polyimide microelectrode array for in vivo recordings and current source density analysis. Biosensors & Bioelectronics **22**(8), 1783–1790 (2007)
112. Hassler, C., Boretius, T., and Stieglitz, T.: Polymers for neural implants. Journal of Polymer Science Part B-Polymer Physics **49**(1), 18–33 (2011)
113. Lai, H.-Y., et al.: Design, simulation and experimental validation of a novel flexible neural probe for deep brain stimulation and multichannel recording. Journal of Neural Engineering **9**(3), 036001 (2012)
114. Mercanzini, A., et al.: Demonstration of cortical recording using novel flexible polymer neural probes. Sensors and Actuators A-Physical **143**(1), 90–96 (2008)
115. Chow, W.W.Y., et al.: Bio-polymer coatings on neural probe surfaces: Influence of the initial sample composition. Applied Surface Science **258**(20), 7864–7871 (2012)
116. Mercanzini, A., et al.: Controlled release nanoparticle-embedded coatings reduce the tissue reaction to neuroprostheses. Journal of Controlled Release **145**(3), 196–202 (2010)
117. Perlin, G.E. and Wise K.D.: An ultra compact integrated front end for wireless neural recording microsystems. Journal of Microelectromechanical Systems **19**(6), 1409–1421 (2010)
118. Sodagar, A.M., et al.: An implantable 64-channel wireless microsystem for single-unit neural recording. IEEE Journal of Solid-State Circuits **44**(9), 2591–2604 (2009)
119. Harrison, R.R.: The design of integrated circuits to observe brain activity. Proceedings of the IEEE **96**(7), 1203–1216 (2008)
120. Harrison, R.R., et al.: A low-power integrated circuit for a wireless 100-electrode neural recording system. IEEE Journal of Solid-State Circuits **42**(1), 123–133 (2007)
121. Blum, R.A., et al.: An integrated system for simultaneous, multichannel neuronal stimulation and recording. IEEE Transactions on Circuits and Systems I-Regular Papers **54**(12), 2608–2618 (2007)
122. Grybos, P., et al.: 64 channel neural recording amplifier with tunable bandwidth in 180 nm CMOS technology. Metrology and Measurement Systems **18**(4), 631–643 (2011)
123. Bottino, E., et al., Low-noise low-power CMOS preamplifier for multisite extracellular neuronal recordings. Microelectronics Journal **40**(12), 1779–1787 (2009)
124. Gunning, D.E., et al.: High spatial resolution probes for neurobiology applications. Nuclear Instruments & Methods in Physics Research Section A-Accelerators Spectrometers Detectors and Associated Equipment **604**(1–2), 104–107 (2009)
125. Field, G.D., et al.: Functional connectivity in the retina at the resolution of photoreceptors. Nature **467**(7316), 673–677 (2010)
126. Imfeld, K., et al.: Large-scale, high-resolution data acquisition system for extracellular recording of electrophysiological activity. IEEE Transactions on Biomedical Engineering **55**(8), 2064–2073 (2008)
127. Franks, W., et al.: CMOS monolithic microelectrode array for stimulation and recording of natural neural networks. IEEE Transducers Dig. Tech. Papers **2**, 963–966 (2003)
128. Frey, U., et al.: Cell recordings with a CMOS high-density microelectrode array. Paper presented at the 29th Annual International IEEE EMBS Conference, Lyon, Canada, 22–26 August, pp. 167–170 (2007)
129. Berdondini, L., et al.: High-density microelectrode arrays for electrophysiological activity imaging of neuronal networks. Paper presented at the 8th IEEE International Conference on Electronics, Circuits and Systems, Malta (2001)
130. Hafizovic, S., et al.: A CMOS-based microelectrode array for interaction with neuronal cultures. Journal of Neuroscience Methods **164**(1), 93–106 (2007)
131. Heer, F., et al.: Single-chip microelectronic system to interface with living cells. Biosensors & Bioelectronics **22**(11), 2546–2553 (2007)
132. Eversmann, B., et al.: A 128x128 CMOS biosensor array for extracellular recording of neural activity. IEEE Journal of Solid-State Circuits **38**(12), 2306–2317 (2003)
133. Fromherz, P., et al.: A neuron-silicon junction - a retzius cell of the leech on an insulated-gate field-effect transistor. Science **252**(5010), 1290–1293 (1991)
134. Fromherz, P.: Electrical interfacing of nerve cells and semiconductor chips. Chemphyschem **3**(3), 276–284 (2002)

135. Lambacher, A., et al.: Identifying firing mammalian neurons in networks with high-resolution multi-transistor array (MTA). Applied Physics A **102**(1), 1–11 (2011)
136. Eversmann, B., et al: A neural tissue interfacing chip for in-vitro applications with 32k recording/stimulation channels on an active area of 2.6 mm^2. ESSCIRC **41**, 211–214 (2011)
137. Berdondini, L., et al.: High-density electrode array for imaging in vitro electrophysiological activity. Biosensors & Bioelectronics **21**(1), 167–174 (2005)
138. Maccione, A., et al.: A novel algorithm for precise identification of spikes in extracellularly recorded neuronal signals. Journal of Neuroscience Methods **177**(1), 241–249 (2009)
139. Maccione, A., et al.: Multiscale functional connectivity estimation on low-density neuronal cultures recorded by high-density CMOS Micro Electrode Arrays. Journal of Neuroscience Methods **207**(2), 161–171 (2012)
140. Lewicki, M.S.: A review of methods for spike sorting: the detection and classification of neural. Network: Computation in Neural Systems **9**(4), 53–78 (1998)
141. Delescluse, M. and Pouzat, C.: Efficient spike-sorting of multi-state neurons using inter-spike intervals information. Journal of Neuroscience Methods **150**(1), 16–29 (2006)
142. Quiroga, R.Q., Nadasdy, Z., and Ben-Shaul, Y.: Unsupervised spike detection and sorting with wavelets and superparamagnetic clustering. Neural Computation **16**(8), 1661–1687 (2004)
143. Takekawa, T., Isomura, Y., and Fukai, T.: Spike sorting of heterogeneous neuron types by multimodality-weighted PCA and explicit robust variational Bayes. Frontiers in Neuroinformatics **6**, 5 (2012). doi: 10.3389/fninf.2012.00005
144. Bologna, L.L., et al.: Low-frequency stimulation enhances burst activity in cortical cultures during development. Neuroscience **165**(3), 692–704 (2010)
145. Chiappalone, M., Massobrio, P., and Martinoia, S.: Network plasticity in cortical assemblies. European Journal of Neuroscience **28**(1), 221–237 (2008)
146. Maccione, A., et al.: Experimental investigation on spontaneously active hippocampal cultures recorded by means of high-density MEAs: analysis of the spatial resolution effects. Frontiers in Neuroengineering **3**, 4 (2010). doi: 10.3389/fneng.2010.00004
147. Grün, S., and Rotter, S. (eds): Analysis of parallel spike trains. Springer Series in Computational Neuroscience, vol. 7, p. 444. Springer, Berlin (2010)
148. Bullmore, E. and O. Sporns, Complex brain networks: graph theoretical analysis of structural and functional systems. Nature Reviews Neuroscience **10**(3), 186–198 (2009)
149. D'Angelo, E.: Toward the connectomic era. Functional Neurology **27**(2), 77 (2012)
150. Churchland, M.M., et al.: Techniques for extracting single-trial activity patterns from large-scale neural recordings. Current Opinion in Neurobiology **17**(5), 609–618 (2007)
151. Nirenberg, S.H. and Victor, J.D.: Analyzing the activity of large populations of neurons: how tractable is the problem? Current Opinion in Neurobiology **17**(4), 397–400 (2007)
152. Zenas, C.C., Douglas, J.B., and Steve, M.P.: Region-specific network plasticity in simulated and living cortical networks: comparison of the center of activity trajectory (CAT) with other statistics. Journal of Neural Engineering **4**(3), 294 (2007)
153. Gandolfo, M., et al.: Tracking burst patterns in hippocampal cultures with high-density CMOS-MEAs. Journal of Neural Engineering, **7**(5), 056001 (2010). doi: 10.1088/1741-2560/7/5/056001
154. Ferrea, E., et al.: Large-scale, high-resolution electrophysiological imaging of field potentials in brain slices with microelectronic multielectrode arrays. Frontiers in Neural Circuits **6**, 80 (2012). doi: 10.3389/fncir.2012.00080
155. Kim, S., et al.: Thermal impact of an active 3-D microelectrode array implanted in the brain. IEEE Transactions on Neural Systems and Rehabilitation Engineering **15**(4), 493–501 (2007)

Chapter 8
Multielectrode and Multitransistor Arrays for In Vivo Recording

Stefano Vassanelli

8.1 Introduction

In recent years the experimental toolkit at disposal of neuroscientists to investigate electrophysiologically the brain "in vivo" down to the level of neuronal microcircuits circuits and to elucidate their fundamental mechanisms for mapping and processing information has grown rapidly and even beyond expectations [1]. Driven by the compelling need of recording large numbers of neurons within the cortex and deeper structures, in a minimally invasive manner and over long time periods [2–4], the development of implantable brain probes based on microelectromechanical systems (MEMS) with arrays of microelectrodes has experienced a significant boost, leading to substantial optimization of pioneering approaches conceived in the 1970s [5] as well as to the development of novel technologies. *Multielectrode arrays* (*MEAs*) and *multitransistor arrays* (*MTAs*) integrated in silicon microchips constitute two major representatives from this class of brain implantable probes. Originally developed as "in vitro" prototypes for recording dissociated neurons or brain slices and other excitable cells [5–7], MEA and MTA reflect two different philosophies for transducing a neuronal electrical signal to a semiconductor chip, that is, either through a metal microelectrode or by means of an *electrolyte–oxide–semiconductor field-effect transistor* (*EOSFET*), a modified version of the metal–oxide–semiconductor field-effect transistor (MOSFET) that is widely used in integrated circuits [8] (Fig. 8.1).

MEA and, more recently, MTA are at the basis of new generations of brain implantable probes, whose common and ultimate goal is the recording of extracellular signals generated by neurons in the brain tissue, in the form of either spike correlates or local field potentials (LFP), at multiple sites and at high spatial

S. Vassanelli (✉)
University of Padova, Italy
e-mail: stefano.vassanelli@unipd.it

M. De Vittorio et al. (eds.), *Nanotechnology and Neuroscience: Nano-electronic, Photonic and Mechanical Neuronal Interfacing*, DOI 10.1007/978-1-4899-8038-0_8,

Fig. 8.1 MTA chip for "in vitro" recording. Neurons from the rat hippocampus are cultured on the surface of a silicon microchip covered by a thin layer of silicon oxide. An MTA is integrated in the chip. This linear array of oxide-insulated field-effect transistors is used to record neuronal extracellular signals. Small dark squares are oxide-insulated gates of the individual transistors, which are integrated in the bulk silicon beneath the oxide, and work as voltage sensors. Extracellular voltages generated by a neuron growing on the transistor gate are transduced through the oxide into a modulation of the source–drain current of the transistor and amplified for recording. Modified from [9], © The Author(s)

resolution. Similar to single electrode and tetrode technologies [10, 11], most state-of-the-art chip-based neuronal probes generally share as a common feature a needlelike design for limiting tissue damage, minimizing interference with the physiological function of neuronal circuits "in vivo," and favoring long-lasting recording in chronic implants [4, 12]. From the application point of view, MEA- and MTA-based approaches improve more conventional single electrode or tetrode recording as they provide a large number of electrodes spaced by small distances, each one gathering independently information on electrical activity from neighboring brain tissue. As such, they provide exhaustive information on the spatial distribution of extracellular signals, taking advantage of the large scale and high density of their microelectrode recording arrays. Nowadays, advanced probes can feature even hundreds of recording sites that are spaced by distances in the micrometer's range, and next generations of MEA and MTA implants may be expected that are performing a *functional* "*electrical*" *imaging* of brain regions at high spatiotemporal resolution. To this endeavor, a joint contribution of electronic engineering, material science, and neurophysiology has proven to be essential to face the severe challenges related to reliable detection of low-amplitude voltages and currents generated by neurons in the extracellular environment (typically in the range of microvolts and nanoamperes, respectively) and to long-term biocompatibility of the brain implants. A further hurdle is the need to guarantee high measurement accuracy. As reliable recording of extracellular neuronal signals in terms of shape and amplitude is fundamental to extract correct and exhaustive electrophysiological information on neuronal activity both when considering LFPs and extracellular spikes [13–16], neuronal probes and their transducers must be conceived to minimally interfere with the signal itself.

As a complement to recording, arrays of microelectrodes are also employed for *microstimulation*, that is, for the spatially distributed extracellular stimulation of neuronal networks [17, 18]. In principle, both MEA and *electrolyte–oxide–semiconductor capacitors* (*EOSCs*) arranged in *multicapacitor arrays* (*MCA*) can be used for the stimulation purpose. Particularly when integrated at large scale and high density, they represent a tool for probing neuronal circuits in experimental neuroscience and with potentially tremendous impact when moving to application fields such as brain–machine interfacing, neuroprosthetics, and treatment of neurological disorders. Although optical stimulation methods have gained momentum for "in vivo" studies with the advent of optogenetics and its ability of targeting specific neurons, electrical stimulation remains a cardinal approach and with closer application to clinics. Compared to recording, and despite success of peripheral implants such as the cochlear [19, 20] and, more recently, first retinal implants [21], microstimulation of brain neuronal networks "in vivo" by chip-based microelectrode arrays remains a largely unexploited approach. Reasons may be ascribed in some measure to the need of high-voltage or high-current delivery to the tissue (up to volts or hundreds of microamperes) in order to excite neurons from the extracellular side and the difficulties to develop microelectrodes capable to reliably sustain such stimulation over time without corrosion or tissue damage. It is a matter of fact that, in view of the demand from neuroscientists for "near-physiological" stimulation methods "in vivo" [18], current probes based on microelectrodes appear mostly unsatisfactory, lacking of neuronal selectivity and being far away from mimicking synaptic inputs. Despite that, stimulation through microelectrodes has large room for improvement, for example, by attaining higher spatiotemporal resolution and by imitating physiological inputs through a finely controlled opening of voltage-gated ion channels [22].

In this chapter we present the state of the art and comment on perspectives of chip-based microelectrode arrays with respect both to recording and stimulation "in vivo." Two different "philosophies" of electrical brain–chip interfacing, by metal electrodes and electrolyte–oxide–semiconductor transducers (i.e., EOSFET and EOSC), are presented with a description of fundamental technical aspects and examples of their use in neuroscience.

The core difference between the two interfaces lies deeply in their electrochemical nature: as a first rough approximation, in the case of MEA, electrons work as charge carriers between metal electrode and electrolyte generating *faradaic* currents. In electrolyte–oxide–semiconductor transducers, electrical interfacing between semiconductor and electrolyte is mediated by capacitive *nonfaradaic* currents through a thin oxide layer. Thus, in this case, electrons in the semiconductor and ions in the electrolyte play separately the role of charge carriers. It is highly advisable to understand and to keep in mind the fundamentally different nature and behavior of the two interfaces when designing and performing electrophysiological experiments. Therefore, we start by providing a brief introduction on the electrochemistry of faradaic and nonfaradaic electrodes. For more information on the subject, the interested reader can refer to [23–29].

8.2 Electrode–Electrolyte Interfaces

As mentioned above, amplitude and shape of extracellular potentials convey important information on the dynamics of neuronal signaling. This holds for local field potentials (LFPs) and for spikes, which makes high recording accuracy a fundamental goal for modern electrophysiology with implantable neuronal probes. On the other hand, when aiming for near-physiological neuronal microstimulation, achieving a high-quality spatiotemporal control of extracellular potentials and currents generated in the brain tissue by implanted microelectrodes is of primary importance. Thus, a key factor to attain high recording accuracy and, especially, near-physiological microstimulation is a well-characterized, stable, and controllable electrode–electrolyte interface. The design and use of microelectrode arrays in implantable probes must therefore take into account this fundamental need.

8.2.1 The Metal Electrode–Electrolyte Interface

Once a metal electrode is inserted in the brain tissue, an interface is formed between the solid phase of the electrode and the liquid phase of the electrolyte in the extracellular environment. In the case of the metal electrode, as in the electronic circuitry within the probe, charge is carried by electrons, while in the extracellular solution, charge is carried by ions, mainly sodium, potassium, and chloride. During recording and stimulation of neuronal activity, current flows between electrode and electrolyte. Two types of processes occur at the interface. In the first one, electrons are transferred across the metal–solution interface and cause oxidation or reduction of chemical species to occur in the electrolyte. Since such reactions are governed by Faraday's law (i.e., the amount of chemical reaction caused by the flow of current is proportional to the amount of electricity passed), they are called *faradaic* processes. Electrodes at which faradaic processes occur are sometimes called *charge-transfer electrodes*. In reality, a given metal–solution interface will usually show a range of potentials where no reactions occur, because they are thermodynamically or kinetically unfavorable. Instead, processes such as charges adsorption and desorption at the

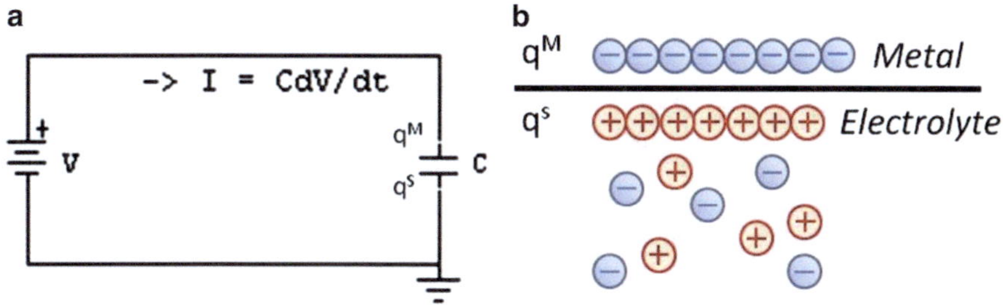

Fig. 8.2 The IPE as a capacitor. (**a**) The IPE is represented by a capacitor (C) as equivalent electrical element. The capacitor is charged by a battery and a displacement current (I) proportional to the capacitance and to the first derivative of the voltage flows at the interface, without the occurrence of faradaic reactions. At a given voltage, equal charges with opposite sign ($q^M = -q^S$) build up on the capacitor plates. (**b**) Schematic representation of the metal–electrolyte interface with charge accumulation at the metal, q^M, and electrolyte surface, q^s

electrode surface can still happen, changing the structure of the electrode–electrolyte interface. Although charge does not cross the interface, capacitive currents can flow (at least transiently) when the potential electrode area or solution composition changes. These processes are called *nonfaradaic* processes, as they do not involve faradaic oxidation/reduction reactions associated to current flow. Both faradaic and nonfaradaic processes can take place when changing the potential of the electrolyte (as during recording) or of the electrode (as in the case of stimulation), and both must be taken into account when designing probes, conceiving experiments, and interpreting data. At first, we discuss the simple ideal case of a metal electrode–electrolyte interface where only nonfaradaic processes occur: the *ideal polarized electrode*. We should emphasize that, in real experimental conditions, we are not dealing with one single interface but rather with at least two electrodes (one of which is commonly chosen with constant interface properties such as the silver–silver chloride and used as a reference) and two interfaces separated by an electrolytic phase. Such a system is called an *electrochemical cell.*

8.2.2 The Ideal Polarizable Electrode

A metal electrode at which no charge transfer can occur across the metal–solution interface, regardless of the potential imposed by an outside source of voltage, is called an *ideal polarized* (or *ideal polarizable*) *electrode* (*IPE*). While no real electrode can behave as an IPE over the whole potential range available in a solution, some electrode–solution systems can approach ideal polarizability over limited potential ranges. Thus, the only faradaic current that flows in this region is due to charge-transfer reactions of trace impurities (e.g., metal ions, oxygen, and organic species), and this current is quite small in clean systems. Since charge cannot cross the IPE interface when the potential across it is changed, only displacement currents occur and the behavior of the electrode–solution interface resembles that of a capacitor (Fig. 8.2a).

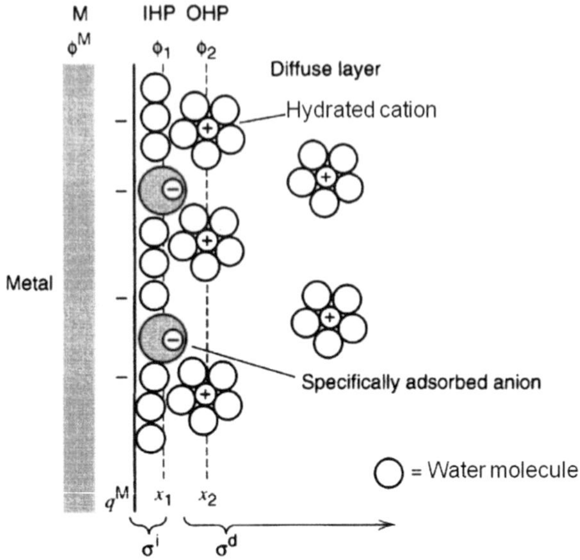

Fig. 8.3 Proposed model of the double-layer region under conditions where anions are specifically adsorbed. Reproduced with permission from [23], © (2001) Wiley

At a given potential, there will exist a charge on the metal electrode, q^M, and a charge in the solution, q^s (Fig. 8.2b). Whether the charge on the metal is negative or positive with respect to the solution depends on the potential across the interface and the composition of the solution. At all times, however, $q^M = -q^s$. The charge on the metal, q^M, represents an excess or deficiency of electrons and resides in a very thin layer (<0.1 Å) on the metal surface. The charge in solution, q^s, is made up of an excess of either cations or anions in the vicinity of the electrode surface. The charges q^M and q^s are often divided by the electrode area and expressed as *charge densities*, such as $\sigma^M = q^M/A$, usually given in μC/cm², where A is the electrode area. In reality, a whole ensemble of charged species and oriented dipoles exists at the metal–solution interface forming the so-called electrical double layer, although its structure only very loosely resembles two charged layers as discussed below. At a given potential, the electrode–solution interface is characterized by a double-layer capacitance, C_d, typically in the range of 10–40 μF/cm². Noteworthy, and unlike real capacitors whose capacitances are independent of the voltage across them, C_d is therefore a function of potential.

8.2.3 The Electrical Double Layer

Several layers are thought to form the electrical double layer at the solution side of the electrode–electrolyte interface. In electrophysiological experiments, the layer closest to the electrode, the *inner layer*, contains water molecules and other species (ions or molecules) that are said to be *specifically adsorbed* (Fig. 8.3). This inner layer is also called the *compact*, *Helmholtz*, or *Stern layer*. The locus of the electrical centers of the specifically adsorbed ions is called the *inner Helmholtz plane* (IHP), which is at a distance x_1. The total charge density from specifically adsorbed ions in this inner layer is σ^i (μC/cm^2). Solvated ions can approach the metal only to a distance x_2; the locus of centers of these nearest solvated ions is called the *outer Helmholtz plane* (OHP). The interaction of the solvated ions with the charged metal involves only long-range electrostatic forces, so that their interaction is essentially independent of the chemical properties of the ions. These ions are said to be *nonspecifically adsorbed*. Because of thermal agitation in the solution, the nonspecifically adsorbed ions are distributed in a three-dimensional region called the *diffuse layer*, which extends from the OHP into the bulk of the solution. The excess charge density in the diffuse layer is σ^d; hence, the total excess charge density on the solution side of the double layer, σ^s, is given by (8.1):

$$\sigma^S = \sigma^i + \sigma^d = -\sigma^M \tag{8.1}$$

The thickness of the diffuse layer depends on the total ionic concentration in the solution; for concentrations greater than 10^{-2} M, the thickness is less than ~100 Å. As charges distribution and the associated potential profile across IHP, OHP, and diffuse layer account for the double-layer capacitance, any perturbation causing a change in the double-layer structure will affect the C_d value. This is the case, for example, of voltages applied to the metal for electrophysiological stimulation or changes in the potential of the electrolyte during recording. Similarly, the double-layer structure can affect the rate of faradaic processes at the interface by influencing the accessibility to the electrode of electroactive species in solution.

8.2.4 Faradaic Processes and the Ideal Nonpolarizable Electrode

In electrochemistry, two different systems may account for faradaic currents: *galvanic* or *electrolytic* cells. A *galvanic cell* is one in which reactions occur spontaneously at the electrodes when they are connected externally by a conductor. These cells are often employed in converting chemical energy into electrical energy as in the case of commercial batteries, but they are generally unwanted in electrophysiological measurements as they generate potential offsets and spurious currents. An *electrolytic cell*, instead, is one in which reactions are caused by the imposition of an external voltage greater than the so-called open-circuit potential[1] of the cell, while no reaction occurs if such voltage is not applied. The presence of galvanic and electrolytic cells in electrophysiological measurements is mostly influenced by the

[1] The *open-circuit potential* (or *zero-current potential* or *rest potential*) is the potential measured when a high impedance voltmeter is placed across the cell. This potential is established at equilibrium when a pair of redox forms linked by a given half-reaction (i.e., a redox couple) is present at each electrode [23].

choice of reference electrodes, a choice that is important for both MEA and MTA recordings [24]. As far as phenomena occurring at the working electrode (i.e., the measuring or stimulating electrode) are concerned, treatment can be simplified by concentrating our attention on only one-half of the cells where *electrolysis* is the general term that include chemical changes accompanying faradaic reactions.

Let us consider more closely the nature of current and potential in an electrochemical cell, assuming a "perfect" reference electrode and considering only the half-reaction occurring at the working electrode. Being E_o the open-circuit potential, when a voltage equal to E_o is applied to the working electrode by an external power ($E_{appl}=E_0$), the current at the electrode–electrolyte interface is zero. When $E_{appl}>E_0$, the cell behaves as an electrolytic cell and current caused by redox reactions flows. Since the current flowing, i, is a faradaic current, it is proportional to the electrolysis rate (8.2–8.4):

$$i = {dQ}/{dt} \tag{8.2}$$

$$Q/{nF} = N \tag{8.3}$$

$$R(mol/s) = {dN}/{dt} = i/{nF} \tag{8.4}$$

where Q is the mobilized charge, n the number of electrons consumed in the electrode reaction, F the Faraday constant, N the number of moles electrolyzed, and R the reaction rate. Contrary to *homogeneous reaction* that is occurring everywhere within the medium at a uniform rate, an electrode process is a *heterogeneous reaction*, which is occurring only at the electrode–electrolyte interface. Here, the current is governed by the rates of processes such as:

1. Mass transfer (e.g., of an oxidized species from the bulk solution to the electrode surface)
2. Electron transfer at the electrode surface
3. Chemical reactions preceding or following the electron transfer
4. Other surface reactions, such as adsorption, desorption, or crystallization (electrodeposition)

Independently from the complexity of the reaction occurring at the electrode, if a cell has a defined open-circuit potential (E_{eq}), this potential is an important reference point of the system. The departure of the electrode potential (or cell potential) from the equilibrium value upon passage of faradaic current is termed *polarization*. The extent of polarization is measured by the *overpotential*, η (8.5), which is interrelated with the current density at the electrode.

$$\eta = E - E_{eq} \tag{8.5}$$

Information on the behavior of the electrode–electrolyte interface can be gained by determining current as a function of potential by obtaining i–E curves.

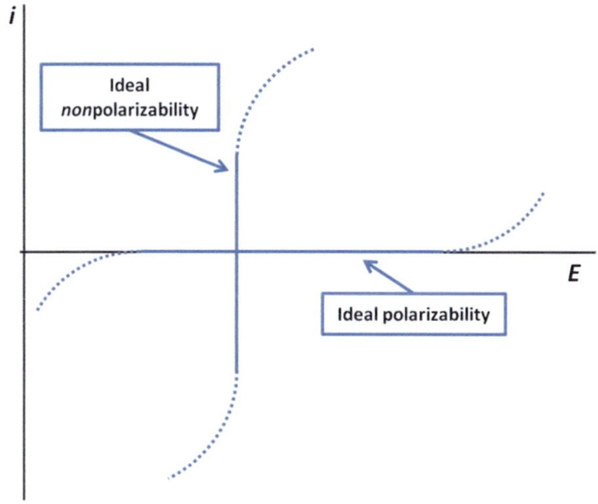

Fig. 8.4 Schematic examples of steady-state *i–E* curves for ideally polarizable and nonpolarizable electrodes. The polarizable electrode changes its potential without the occurrence of faradaic reactions (i.e., $i = 0$). Thus, the behavior of the electrode is capacitive. In the ideal nonpolarizable electrode, instead, faradaic current passes but the electrode remains at a zero potential. In this ideal case, all resistive components (electrode and electrolyte) are neglected. Modified with permission from [23], © (2001) Wiley

Current–potential curves, particularly those obtained under steady-state conditions, are sometimes called *polarization curves*. We have seen that an ideal polarizable electrode shows a capacitive-like behavior, with storage of charge and change in potential upon the passage of current. Thus, ideal polarizability is characterized by a horizontal region of an *i–E* curve. Instead, an *ideal nonpolarizable electrode* is an electrode whose potential does not change upon passage of current, that is, an electrode of fixed potential. Therefore, nonpolarizability is characterized by a vertical region on an *i–E* curve (Fig. 8.4).

With respect to these ideal cases, electrode–electrolyte interfaces in real experimental electrophysiological conditions behave quite differently, commonly showing intermediate characteristics between polarizable and nonpolarizable electrodes and nonlinear responses to voltages. Detailed studies by impedance spectroscopy show that metal electrode–electrolyte interfaces can be modeled by equivalent electrical circuits with capacitive elements representing the double-layer capacitance, C_d, and impedance elements, Z_f, to simulate faradaic components of the interface (Fig. 8.5) [23, 25–27]. However, in contrast to electronic elements, the components of the interface impedance usually strongly deviate from ideal behavior, which suggested breaking down the generic faradaic impedance in subcomponents, including a "Warburg" capacitance and impedance [23]. Nevertheless, the dependency of these elements from amplitude and frequency of driving voltages and from the specific combination of material and geometry of the electrode makes modeling of metal microelectrode–electrolyte interfaces quite cumbersome. As a rule of thumb, in the case of metal electrodes commonly used in neuronal probes, for small driving

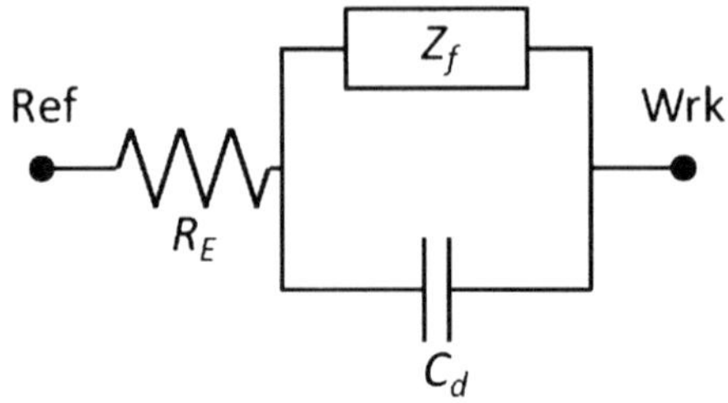

Fig. 8.5 Equivalent circuit of metal electrode–electrolyte interface with R_E, the solution and reference electrode resistance; C_d, the double-layer capacitance; and Z_f, the faradaic impedance. *Ref* and *Wrk* indicate the reference and the working (recording/stimulating) electrode, respectively. Note that the faradaic impedance represents the effect of the heterogeneous electron-transfer process between metal and electrolyte

voltages at the working electrode, faradaic processes are negligible and current mainly flows through the capacitive branch of the circuit (i.e., Z_f tends to infinity). This situation is typical for recording, since neuronal extracellular signals, wherever spikes or LFPs, typically reach only a few mV amplitude at best. For substantial potential changes, as during stimulation, when driving voltages in the volts range are applied, the faradaic current may become relevant, and current starts to flow in the impedance branch of the circuit [23, 25]. Bearing these basic concepts in mind, we can now examine some general features of MEA-based implantable probes, with respect to recording and stimulation.

8.3 Recording and Stimulation with MEA Probes

"In vivo" experiments produced undoubtful evidence that MEA-based implants are a very promising tool to investigate brain microcircuits [4, 10, 28] and, moving to clinical areas, to aid paralyzed and other neurological patients through prosthetic devices [3, 18, 29–31]. MEA probe application is not limited to neuronal recording. Electrodes capable of both recording and stimulation have been also developed aimed for two-way brain machine interfacing [18] or adaptive deep brain stimulation (DBS) [32]. Despite the increasing variety of available MEA probes, all of them obey some general principles of operation and face common hurdles that we can identify. Neuronal activity is recorded by implanted MEA microelectrodes in the form of "spikes" (i.e., the usual action potential-related extracellular signals also known as single units) or LFPs. Although the exact origin and mechanisms of spatial spread in the brain tissue of these signals are not precisely known [1, 13, 33–35], in the perspective of dissecting neuronal network activity, it would be desirable that extracellular potentials are mapped by "punctiform" electrodes, i.e., having dimensions that are negligible with respect to signal sources, which are the neurons and their current generating compartments [16]. In state-of-the-art probes, this challenge remains partially unmet. Nowadays, microelectrodes have dimensions that are comparable or at best slightly smaller than neuronal bodies and with a surface area that is in the range between a hundred and thousand micrometers squared. An additional challenge comes from low signal amplitude. Recording single units and LFPs with a decent signal-to-noise ratio (SNR), and to do this chronically, is technologically demanding. The amplitude of action potentials recorded in the CNS by extracellular electrodes is typically in the order of 100 μV or smaller, except exceptional cases when they can be a few hundred microvolts. On the other hand, if LFPs are frequently found whose maximum amplitude lies in the millivolt range, smaller amplitudes are common. Furthermore, the information content of the whole LFP shape is very relevant [1, 33], which implies the need of recording also the low-amplitude signal components. Thus, low-noise MEAs and readout electronics are mandatory for "in vivo" recording. In general, part of the background noise in single-unit recording is in reality "neuronal noise" from the multitude of background action potentials. Similarly, at lower frequencies, brain oscillations and background neuronal ensembles activity affect SNR for LFP recording. However, metal microelectrode impedance does contribute to noise. In particular, high electrode impedance will increase noise and, in combination with stray capacitances between the electrode and the recording amplifier, will reduce the electrodes' high-frequency response by RC filtering. Thus, in practice, recording metal microelectrodes are typically characterized by their impedance at 1 kHz, which is quite variable, ranging in vivo from approximately 50 kΩ to 1 MΩ. Despite this general principle, studies suggest a less certain correlation between electrode impedance (at 1 kHz) and recording quality. In the attempt of optimizing size and performance, microelectrodes have been developed based on a wide range of materials, including stainless steel, gold, tungsten, platinum, platinum–iridium alloys, iridium oxide, titanium nitride, and poly(ethylenedioxythiophene) (PEDOT). An exact knowledge of the behavior of the different electrode–electrolyte interfaces is obviously crucial for optimal operation. As far as recording is concerned, extracellular spikes [14], but to some extend also LFP signals [33], are associated to tiny potentials and small currents at the recording electrode. Thus, no high-rate electrochemical challenges should be expected: faradaic reactions are negligible, and the metal microelectrode behaves more closely to an ideal polarizable double-layer capacitor (see Sect. 8.2). When electrodes are used for stimulation, large voltages have to be applied and faradaic reactions may come into play, depending also on the quality of the interface [21, 23, 24]. When addressing long-term experiments, consistent neural recordings with chronically implanted microelectrodes have been difficult and for reasons that may have little to do with the properties of the electrodes themselves. Tissue reaction with gliosis and formation of a tissue scar interposing between probe and neurons is probably the major obstacle [37, 38], and it is still unclear how materials, dimensions, and geometry of silicon probes affect the glia response [37]. Overall, as micromotion of the probe relative to the brain can cause tissue damage in chronic implants, tethering to the skull should be made through flexible structures. In any case, having a small neuron-to-electrode distance (i.e., below about 100 μm) is critical for recording single units [4], as the field strength rapidly decreases moving away from the spike source [16]. Perhaps mainly for this reason, maintaining good and stable recording for more than 1–2 years has proven to be a challenge both in monkey [39] and in first clinical trials [30], which has also led to the adoption of cortical ensemble or multiunit recording methods to provide more stable control signals for prosthetic devices [40, 41]. Gliosis also hampers stimulation efficiency, but additional issues have to be considered when metal microelectrodes are to be employed for eliciting neuronal activity. Due to the large driving voltages, selection of adequate electrode materials is crucial to avoid as much as possible irreversible redox reactions, with production of chemical species potentially toxic for the brain tissue. On the other hand, reactions that could lead to electrode corrosion must be minimized. Thus, in principle, capacitive charge injection is more desirable than the faradaic one, although high-capacitive current delivery is only possible with capacitor electrodes that are porous or employ high dielectric constant coatings. Among faradaic electrodes that can be used for both recording and stimulation in neuronal probes, platinum (Pt) and iridium (Ir) oxide electrodes are the most common. Pt (and PtIr alloys) can inject charge both by faradaic and capacitive double-layer charging, with the faradaic component usually dominating over the capacitive one. However, the voltage stimulation window is limited from −0.6 to 0.8 V (Ag|AgCl) at pH 7 if reduction or oxidation of water has to be avoided, which constraints current density limits. Nevertheless, cochlear implants (among the most successful and experimented implants in patients) still use Pt or PtIr microelectrodes for stimulation, even though relatively large electrode areas are required to keep current densities safe [36]. The need for microelectrodes with higher charge-injection capacities led to development of thin-film coating of hydrated Ir oxide, obtained through Ir activation [42]. The film, once deposited on the metal surface of an electrode, greatly increases charge-injection capacity through a fast and reversible faradaic reaction between Ir^{3+} and Ir^{4+} states of the oxide [43]. Activated iridium oxide microelectrodes were also developed for implantable silicon probes taking advantage of thin-film micromachining technology and even tested in clinical studies requiring intracortical stimulation [44]. Yet at present, iridium oxide microelectrodes are considered among the best when neuronal stimulation is concerned.

Titanium nitride (TiN) and tantalum/tantalum oxide (Ta/Ta_2O_5) electrodes are the mostly known solutions in the world of capacitive stimulation. TiN, a chemically stable and biocompatible metallic conductor, offers large charge-injection capacity through the electrical double layer, particularly because highly porous surfaces can be obtained by sputter deposition that enhance electrode capacitance. Although limited by series resistance of the pores, current injection capacity is remarkable, but inferior to best-performing faradaic electrodes such as Ir oxide. On the other hand, TiN electrodes provide a wider voltage operating window (−0.9 to 0.9 V). Coating a metal surface with a thin dielectric layer is another strategy to enhance charge-injection capability. Dielectric coating, particularly when using high dielectric constant materials, increases specific capacitance and, therefore, the current density that can be supplied for a given driving voltage. This is the case of Ta/Ta_2O_5 electrodes, whose use, however, has never been extensive [45].

Over the years, and mostly recently, a host of different technologies, designs, and electrode materials has been proposed to overcome obstacles and improve performance of implantable silicon probes. These efforts are described in detail in specific literature, and the interested reader can refer to [5, 28, 46–49] for extensive reviews on the subject. In the future, emergent materials and innovative nanostructures, such

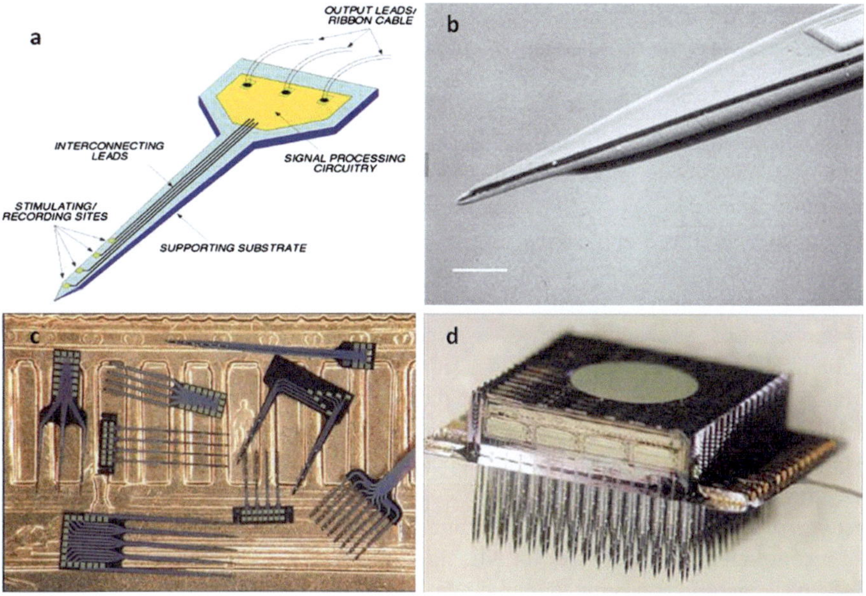

Fig. 8.6 The *Michigan* probes. (**a**) Schematic drawing of the basic shank structure of the Michigan probe. The silicon probe has a needlelike shape to favor implantation. While the needle region of the probe is vertically inserted in the tissue, the larger non-recording part contains signal processing circuitry and metallic pads for output lead connections. Integrated in the shank and close to the tip of the chip are recording (stimulating) sites, represented, in this case, by four planar metal microelectrodes arranged in a linear array. (**b**) Scanning electron microscopy picture of the tip of a Michigan probe manufactured by a boron etch-stop process (scale bar: 10 μm). (**c**) Several probe designs, including multi-shank structures forming 2D recording arrays. (**d**) Multi-shank structure with 1,024 sites for 3D recording. Reproduced with permission from (**a**) and (**c**) [49], © (2004) IEEE; (**b**) and (**d**) [62], © (2008) IEEE

as silicon nanowires [50], carbon nanotubes [51–54], conductive polymers as poly(ethylenedioxythiophene) (PEDOT), or other polymeric compounds [55, 56], may find their way for application on neuronal implants "in vivo." Nevertheless, and despite the wealth of innovations, two basic architectures have inspired nearly all implantable MEAs that are today available to the neuroscience community: the "Michigan" and the "Utah" array. Thus, we will proceed by discussing their fundamental concepts and technological features.

8.3.1 The "Michigan" Probe

One of the first silicon-based implantable multielectrode arrays was made by Wise, Starr, and Angell [57–59]. These structures have evolved into several subsequent devices whose basic architecture is commonly referred to as "Michigan" probe [60]. In this probe, several planar microelectrodes are patterned on one of the two faces of a shank structure, providing a set of spatially arranged recording sites (Fig. 8.6).

In brief, the probe structure consists of a selectively etched (micromachined) silicon substrate having needlelike (shank) geometry. Conducting leads are insulated above and below by inorganic dielectrics, while recording (and in some cases also stimulating) microelectrodes with approximately planar surface are formed by an area of exposed metal. Starting in the 1970s, the manufacturing of the Michigan probe relied on single-sided processing of silicon wafers that were becoming available in the electronic industry [49, 61], including technologies such as diffused boron etch-stops, reactive ion etching (RIE), and silicon-on-insulator (SOI) wafer technology. Since then, most of the present approaches to probe manufacturing use silicon substrates [49], relying on expertise and equipment developed for industrial microelectronic circuits and achieving an incomparable degree of miniaturization and reliability of yield. Thermally grown silicon dioxide and low-pressure chemical-vapor-deposited silicon nitride are thin dielectrics typically adopted for insulation, while conductor leads can be formed using metals, metal silicides, or polysilicon. In spite of its relatively high resistance, polysilicon is adequate for most recording electrodes, while stimulating electrodes require leads with lower resistance to reduce stimulation voltage amplitudes and RC filtering. A more detailed description of technologies and processes for Michigan probe manufacturing can be found in [49, 62], including advanced implementations such as 2D and 3D multi-shank structures (Fig. 8.6) and CMOS signal processing circuitry [63] for signal processing and wireless data acquisition. Single-shank and 2D multi-shank structures inspired by this work are commercially available, and they have been reported to work quite successfully in acute and chronic implantation [28].

In the perspective of offering a tool for microstimulation, the same architecture can be used. Typical current thresholds for stimulation with one single microelectrode are about 10–30 μA, but currents of 100 μA or more are very common. Facing the need of such a relatively high current, a potential drawback is interchannel crosstalk. In the Michigan probe, the problem is virtually negligible as the conducting silicon substrate below the leads and the electrolyte above them both act as ground planes shunting parasitic currents. For what concerns stray capacities, instead, the small area of the lithographically defined leads minimizes shunt capacitance so that voltage filtering and attenuation are negligible as well. Recording sites can be gold or platinum, but anodically formed iridium oxide [64–66] is increasingly used as it produces significantly lower recording impedances than other materials. Employing Ir oxide microelectrodes was essential for stimulating sites, as it enabled more than 20 times the charge delivery to tissue than platinum or gold at the same voltage [15, 42].

When designing implantable neuronal probes, choosing proper geometry and dimensions to minimize brain damage during penetration is a central issue. In the Michigan probe series, the tendency was to keep the shank width and thickness as small as possible following the basic principle that the smaller is the body implanted, the smaller is the resulting tissue damage. Furthermore, theoretical analysis of extracellular potential fields has shown that the presence of a wide probe can distort the extracellular current flow around a neuron [60] and this distortion can be minimized by scaling the probe width on the order of the soma radius. On the other hand,

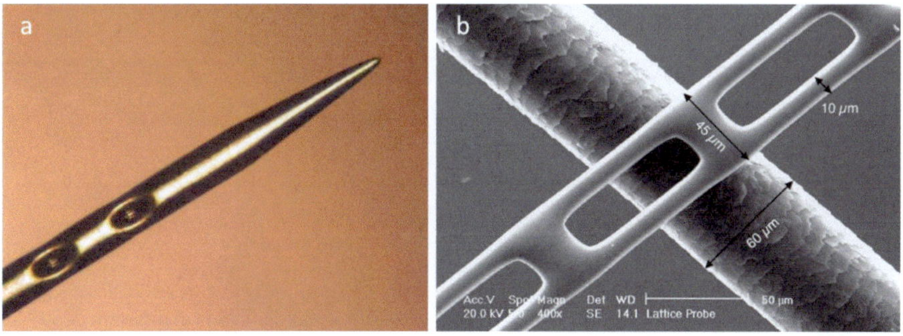

Fig. 8.7 Double-sided and lattice Michigan structures. (**a**) Backside view of an 18-μm-wide shank providing some double-looking sites (top and back). (**b**) Multisite lattice recording probe on a human hair. Modified with permission from [49], © (2004) IEEE

neurosurgery requires that the probe maintains some strength and stiffness to avoid rupture or bending. When inserting the probe in the brain for recording, an incision is typically made into the tough dura mater, leaving the thinner arachnoid and pia meninges intact before probe insertion. The strength characteristics of the Michigan probe have been measured experimentally [67] and show that a cross-sectional area 15 μm thick × 30 μm wide can sustain arachnoid and pia layers penetration with no breakage.

Thanks to these technological advances, implantable 3D structures are now available that can contain up to 1,024 recording sites (Fig. 8.6) and that are poised to revolutionize the way we investigate and understand the functional connectivity of cortical microcircuits. By omitting the boron diffusion in selected areas of the silicon substrate, combinations of top- and back-looking sites can be arranged on the same shank (Fig. 8.7), further enriching the toolbox available for acute recordings [49]. Long-term recording remains a major challenge due to the foreign body response with gliosis. Whereas stimulation can perform well even for years, by applying stimuli that are sufficiently large to bypass the gliotic barrier, long-term single-unit recording remains challenging. A variety of new approaches are being explored in the attempt of better coping with the glia response, including surface functionalization with organic compounds or manufacturing of latticelike shanks allowing the growth of neuronal processes through the probe (Fig. 8.7).

8.3.2 The "Utah" Probe

The Utah Electrode Array (UEA) (Fig. 8.8) is the most known implantable silicon MEA probe having a 3D structure and the first one that has been implanted chronically in the cortex of paralyzed patients [30]. While the Michigan electrode array has been built to take advantage of the planar photolithographic manufacturing techniques used in the semiconductor industry, the UAE was designed "from

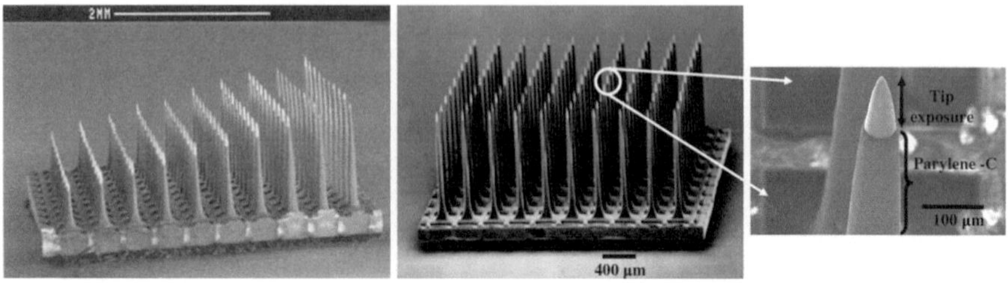

Fig. 8.8 The *Utah* Electrode Array. (*Left*) This three-dimensional array features 100 silicon-based microelectrodes. (*Middle*) The slanted array has been used to stimulate fibers at different depths in the cochlear nerve [73]. (*Right*) Enlarged view of the array showing platinum-coated tips. *Left* reproduced with permission from [98] © (2001) American Physiological Society. *Middle right* reproduced with permission from [99] © (2009) IOP Publishing. All rights reserved

scratch" to meet the need of a neural interface for multisite recording within a horizontal plane in the cerebral cortex. As such, new manufacturing techniques had to be developed in order to build this device, and they are described in detail in [68]. It consists of a silicon-based, 3D structure of a 10×10 array of tapered silicon electrodes with a base width of 80 μm, a length up to 1,500 μm, and an electrode-to-electrode horizontal distance of 400 μm. Insulation between electrodes is provided by glass in the base plane of the array. The starting material is a thick *n*-type wafer, through which p^+ trails are created by thermomigration and then chemically etched and sawed to expose the p^+ trails as thin needles. The tip of each needle electrode is coated with metals: gold, platinum, or iridium [68]. Particularly, activated iridium oxide is used when also stimulation is required [69]. A backside electrical contact to each electrode is formed by sputter deposition and patterned by photolithography. The UEA is encapsulated by a biocompatible polymer, Parylene-C, which is deposited by chemical-vapor deposition (CVD). The UEA is connected to an external data acquisition system by thin Pt–Ir wires that are soldered to the bond pads of each of 100 electrodes on one side and a connector system on the other side [70]. These arrays have been used in the cat auditory and visual cortex [71], cochlear nerve, and dorsal root ganglion. Recent work has proven it can be used for chronic recording in the cortex [30, 39]. It is now ascertained that the array can support single-unit recording for 1–5 years, depending on the subject implanted and despite suffering from signal loss from several electrodes. A graded variation of the Utah Electrode Array, in which the microelectrode length increased from 0.5 to 1.5 mm to permit focal excitation of fibers at different depths, was implanted in the cochlear [72] and sciatic peripheral nerve [73].

UEA developers have also engineered tools that enable the probe to be implanted in cortical tissues by limiting tissue damage. Even though the individual needle-shaped electrodes are extremely sharp, early attempts at implanting large numbers of them into the visual cortex only deformed the cortical surface and resulted in incomplete implantation. Further, the compression of the cortical surface produced by slow mechanical insertion can injure blood vessels, causing intracranial hemorrhage and cortical edema. Because the brain is a viscoelastic material, it behaves in a much more rigid fashion if electrodes are inserted at a very high velocity. A surgical instrument that appears to circumvent the abovementioned problems has been developed based upon this concept and pneumatically inserts the probe in about 200 μs [74].

8.4 Recording and Stimulation with MTA Probes

The idea of using a transistor instead of a metal electrode as the sensing element for recording biological electrochemical signals dates back to the 1970s, when ion-sensitive field-effect transistors (ISFETs) were first introduced [75]. The concept relies on the modification of a standard MOSFET [8] where the gate metal is replaced by an insulating oxide for the actual contact with the electrolyte of the biological environment, thus leading to an electrolyte–oxide–semiconductor field-effect transistor (i.e., EOSFET) structure [6]. While recording with ISFETs was centered on measuring changes of ion concentration in solution (e.g., H^+ and Na^+) [76], Fromherz and coworkers provided the first experimental evidence that the EOSFET can measure extracellular potential transients generated by spiking neurons [6, 77]. Large Retzius cells from *Hirudo medicinalis* (leech) placed on a silicon chip with an integrated planar multi-EOSFET array (i.e., MTA) were used for this groundbreaking work. In subsequent experiments, the approach was further optimized to record dissociated mammalian neurons [4, 77, 78] and brain slices [3, 79, 80, 82]. Aiming for a two-way semiconductor–neuron interface, an EOSFET counterpart was developed for stimulation: the electrolyte–oxide–semiconductor capacitor (i.e., EOSC). The device was sharing with the EOSFET the concept of establishing a capacitive (i.e., nonfaradaic) electrical coupling between the semiconductor silicon and the electrolyte, enabling the stimulation of neurons through displacement currents in neuronal cultures and brain slices [20, 81, 83]. Only recently, and taking the cue from implantable MEA probes, EOSFET/EOSC arrays were exploited for "in vivo" recording [84], opening novel opportunities for the manufacturing of neuronal probes for high-resolution and large-scale recording of neuronal networks "in vivo."

Before presenting the latest MTA[2] developments and applications, we provide some hints on EOSFET and EOSC structure and introduce their working mechanism. Further details on theoretical and technical aspects of these devices are available in [6].

[2] The term MTA, literally meant to indicate transistor arrays, in this context is used in a more general sense to indicate arrays of EOSFET and EOSC elements.

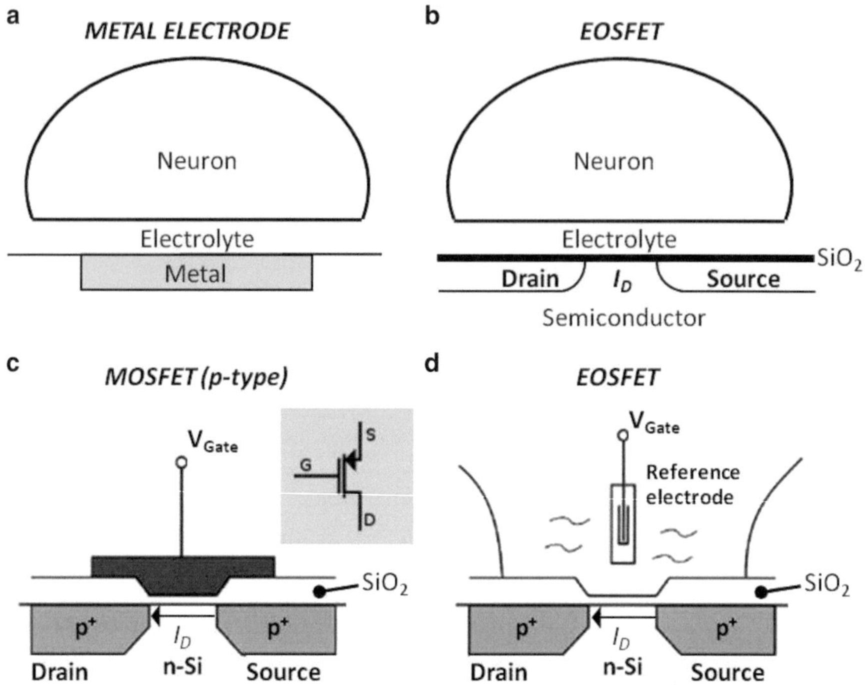

Fig. 8.9 Metal electrode and EOSFET recording. (**a**) A planar metal electrode of an MEA is depicted beneath a neuron forming a metal–electrolyte interface. (**b**) In the EOSFET, the metal electrode is replaced by a field-effect transistor integrated in a semiconductor, which is insulated from the electrolyte by a layer of dielectric material (e.g., silicon oxide—SiO_2). Extracellular potentials in the electrolyte are "sensed" as they modulate the source–drain current (I_D) of the transistor, similarly to the "field-effect" gate modulation in a conventional MOSFET. (**c**) Scheme of a pMOSFET realized in n-type bulk silicon (n-Si). A p-type conductive channel beneath the SiO_2 is modulated by the gate potential through a "field effect," which affects the source–drain current, I_d. The symbol of a pMOSFET as used in electronic circuits is shown in the *inset*. (**d**) In the EOSFET, the metal gate is replaced by the electrolyte, grounded at the periphery through a reference electrode. Local changes of electrical potential at the gate due to extracellular neuronal signals modulate the source–drain current as in a common MOSFET. Note that, similarly to the MOSFET, also in the EOSFET we do not assist to electrons exchange between electrolyte and semiconductor, and the current at the interface is nonfaradaic

8.4.1 *EOSFET and EOSC: Structure and Working Principle*

As anticipated, the EOSFET and ISFET structures are superimposable. In the EOSFET, and differently from a metal electrode, electrolyte and transistor are interfaced through a dielectric (Fig. 8.9). Ions in the electrolyte and electrons (or positive holes) in the semiconductor give rise to displacement currents without reduction or oxidation of chemical species in solution and are, therefore, intrinsically nonfaradaic.

Biological signals generating potential changes in the electrolyte are transduced into a modulation of a source–drain current in the semiconductor device. Thus, from the operational point of view, the EOSFET closely resembles a standard MOSFET where the metal gate is replaced by the conductive electrolyte (Fig. 8.9). First, we briefly examine the structure and operating principle of a standard MOSFET. The MOSFET is commonly used in electronics for amplifying or switching electronic signals. Although the MOSFET has four terminals with source (S), gate (G), drain (D), and body (B), the body (or bulk substrate) of the MOSFET often is connected to the source, making it a three-terminal device. Because these two terminals are normally connected to each other (short-circuited) internally, only three terminals appear in electrical diagrams. In enhancement mode, a voltage drop across the oxide induces a conducting channel between the source and drain contacts via a "field effect." The term "enhancement mode" refers to the increase of conductivity observable when increasing the oxide field, which adds carriers to the channel. The channel can contain electrons (nMOSFET) or holes (pMOSFET), opposite in type to the substrate. So the pMOSFET is made with a p-type substrate. Under these conditions, in a pMOSFET the source–drain current I_{D} is controlled by the voltage V_{DS} between drain and source and the voltage V_{GS} between metal gate and source. Particularly, above a threshold $|V_{GS}|>|V_T|$ of strong inversion and below pinch-off ($|V_{GS}|<|V_{DS}+V_T|$), the pMOSFET works in "linear mode," thus resembling a resistor. Provided that gate voltage perturbations are small (as it is certainly the case for extracellular neuronal signals), the drain current, $\mathrm{I_D}$, varies linearly with $\mathrm{V_G}$ for a given $\mathrm{V_D}$ and $\mathrm{V_S}$ (i.e., a so-called transistor "working point") [6]. In the EOSFET, the electrolyte replaces the metal gate, and it is joined to an external metallic contact by a reference electrode (usually Ag/AgCl) that transforms ionic into electronic current. The source–drain current is controlled by the voltage $V_{ES}=V_E-V_S$ applied to the electrolyte as in a common pMOSFET with $V_{GS}\rightarrow V_{ES}$. Here, the voltage threshold V_{T} is determined by the work function of silicon, the redox potential of Ag/AgCl, the contact potential of the Ag/AgCl electrode, and the electrical double layer at the interface electrolyte/silicon dioxide [24]. Any local change of the potential in the electrolyte (either extracellular spike or field potential) occurring in the proximity of the EOSFET during neuronal activity modulates the I_D. Noteworthy, such potential change occurs far beyond the electrical double layer which has a thickness of 1 nm in 100 mM NaCl. Thus, we are dealing with a genuine modulation of the gate voltage. It is now clear how local voltage recording by an *EOSFET* has to be distinguished from the application of an *EOSFET* as an *ISFET*. Their, indeed, chemical interactions of ions in the electrical double layer are responsible for modulation of the threshold voltage. In practice, the characteristics $I_{\mathrm{D}}(V_{DS}, V_{ES})$ of an *EOSFET* are measured in a calibration experiment by variation of the bath potential without cell. The transconductance $\left(\partial I_D / \partial V_{ES}\right)_{V_{DS}}$ is determined at a working point defined by the potentials V_E, V_D and V_S. Once we assume that the transconductance of the calibration experiment is valid for the local recording of a potential in the electrolyte, we can obtain from the experimental ΔI_D the extracellular potential, V_{Local}, according to (8.6).

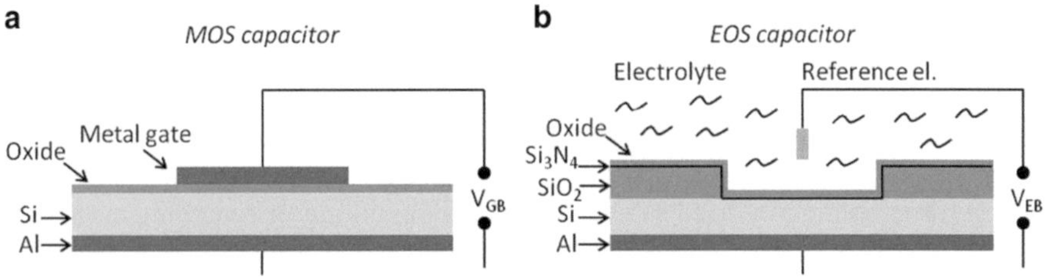

Fig. 8.10 The EOS capacitor. (*Left*) Cross section of an ideal MOS capacitor. The semiconductor (Si) is contacted by an Al pad, while the dielectric (oxide) through a metal gate. The capacitor is charged by a potential difference applied between gate and body (V_{GB}). (*Right*) EOS capacitor structure as in [85]. The device resembles a MOS capacitor with a substitution of the metal gate by conductive electrolyte. A driving potential between the body Al contact and the reference electrode (V_{EB}) generates a nonfaradaic ionic current and a local potential change in the electrolyte, particularly in the proximity of the dielectric. When a neuron is close enough to the EOSC, carefully tailored voltage transients can be used to activate voltage-gated ion channels through membrane depolarization [22, 83]. Modified with permission from [85], © (2006) AIP Publishing LLC

$$\Delta I_D = \left(\partial I_D / \partial V_{ES} \right)_{V_{DS}} V_{Local} \tag{8.6}$$

Thus, in a common recording configuration, the terminals of the p-type EOSFET are held at a positive voltage with respect to the electrolyte bathing the neurons with $V_{ES} = V_E - V_S < 0$, $V_{DS} > V_{ES}$ and with bulk silicon on source potential $V_B = V_S$. The electrolyte is grounded at the periphery through the reference electrode, while locally the potential generated extracellularly by neuronal activity, V_{Local}, is sensed by the transistor. The positive bias voltages at the working point prevent cathodic corrosion of the chip and an invasion of sodium ions into the transistors. The thickness of the gate oxide is in the order of a few tens of nanometers.

Whereas the EOSFET is used for recording, a similar device, the EOS capacitor (EOSC), can be used for injecting current and stimulating neurons. The EOSC emulates a conventional MOS capacitor [8] where the metal is replaced by the electrolyte. With respect to the EOSFET, as a transistor is not required for stimulation, the EOSC features a simpler structure. It consists of a highly doped conductive semiconductor insulated from the electrolyte by a dielectric, typically SiO_2 or a material with higher dielectric constant to enhance capacitance (e.g., TiO_2, HfO_2) [85]. Application of a voltage transient to the semiconductor causes a capacitive nonfaradaic current to flow at the semiconductor–electrolyte interface and the generation of an extracellular voltage for neuronal stimulation. As shown in cell cultures "in vitro," careful design of voltage transients allows for a tuning of activation of voltage-gated ion channels, thus leading to controlled neuronal firing or intracellular Ca^{2+} signals [22, 83, 86]. EOSC stimulation has been successfully tested on MTA coupled to brain slices [81], while only very recently this stimulation approach has been applied "in vivo" (see Sect. 8.4.2) (Fig. 8.10).

MTA developments aiming for large-scale and high-resolution recording have taken advantage of complementary metal–oxide–semiconductor (CMOS) technology to integrate up to about 16,000 transistors in a 1 mm^2 array [82, 87, 88] and,

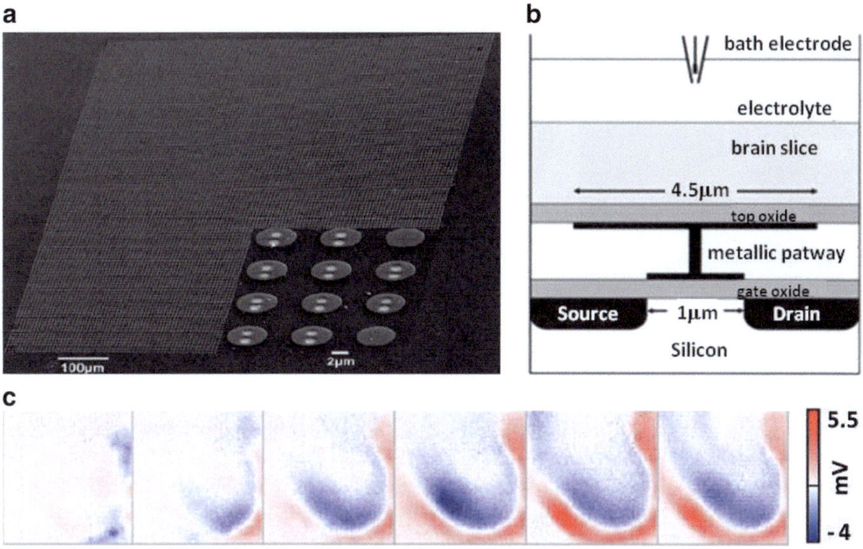

Fig. 8.11 MTA electrical imaging. (**a**) 128 × 128 MTA with a total surface of 1 mm^2 (transistor pitch: 7.8 μm). Each transistor has a circular top gate of 4.5 μm diameter (visible in the *inset* blowup at the *bottom right corner* of the picture). The whole surface is covered by a layer of TiO_2. (**b**) Schematic cross section of a sensor transistor with electrolyte–oxide–metal–oxide–semiconductor (EOMOS) configuration (not to scale). A brain slice is attached to the top layer of titanium dioxide, which is connected by a metallic pathway to the gate oxide of a field-effect transistor with source and drain. (**c**) MTA electrical imaging of extracellular field potentials in a slice of rat hippocampus attached to the chip surface. Six representative frames at 0.5 ms intervals are represented from left to right after stimulation of the CA3 region (stratum pyramidale). The color code of the potential recorded by each pixel is reported on the *right*. [82] The American Physiological Society

more recently, reaching the remarkable density of 32,000 transistors in a 2.6 mm^2 area [89]. In these prototypes, the original EOS structure has evolved, with the introduction of an additional metal layer, to an electrolyte–metal–oxide–semiconductor (EMOS) configuration. Thus, although the original oxide–electrolyte interface is conserved for near-ideal polarizability and nonfaradaic recording and stimulation, these devices are based on dielectric coating of metal and not semiconductor silicon. This MTA technology and similar high-resolution large-scale MEA approaches [90, 91] are poised to revolutionize the way neuronal networks are investigated, by imaging, from the extracellular side, their electrical activity (Fig. 8.11). The first steps towards application of this technology for "in vivo" recording are summarized in the next section.

8.4.2 MTA Recording and Stimulation "In Vivo"

Only recently, and with a time lag of nearly 40 years with respect to metal microelectrodes, EOS elements (transistors and capacitors) have appeared in the first prototypes of implantable neuronal probes for "in vivo" recording.

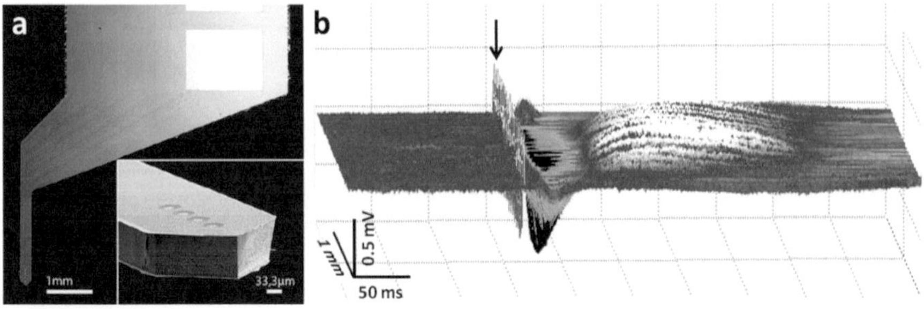

Fig. 8.12 MTA implantable probe. (**a**) First MTA prototype with integrated four EOSFETs. The chip features a needle, with the transistors at the tip, that is suitable for implantation and recording in the brain tissue. The remaining and wider part of the chip contains metallic bond pads (*white rectangles*) for contacting the external electronics. The whole chip is surrounded by a thin (13 nm) insulating layer of TiO_2 for a better chemical stability and biocompatibility of the implant. *Inset*: blowup of the needle tip with the four gates of the transistors shining at the surface. Modified with permission from [92], © (2011) Springer. (**b**) Depth profile of extracellular field potentials measured in the rat barrel cortex by a new-generation high-resolution (10 μm) linear MTA (50 sweeps average). Upon whisker stimulation (*black arrow*), extracellular potential waves (LFPs) are generated which vary across the cortex (from the surface to the deepest layers) in amplitude and shape (Vassanelli et al., unpublished results)

Basing on the technology developed for "in vitro" applications and described in the previous section, a needle-shaped chip with a small EOSFET array was engineered and tested, in a first instance, with slices from the rat hippocampus [92] (Fig. 8.12). The implantable MTA was manufactured from silicon-on-insulator (SOI) wafers following a process inspired by previous "in vitro" devices [93] and consisted of two functionally distinct areas: a needle (2 mm long) with an array of four transistors (gate area 10 μm × 10 μm, pitch 80 μm) and a contact plate with the bond pads for connection to an external amplifier and acquisition system. After manufacturing, the needles were post-processed by atomic layer deposition (ALD) of a 13-nm-thick TiO_2 insulting layer covering the whole needle surface to improve chemical stability and biocompatibility [81, 85]. A proof of principle demonstration of the EOSFET capability to record neuronal activity "in vivo" was provided by recording LFP responses upon whisker stimulation in the somatosensory barrel cortex of the rat [84]. Taking the cue from these first results, we have further developed, within the *CyberRat* project (http://neurochiplab.cyberrat.eu/CyberRat/CyberRat.html), a series of MTA implantable probes for high-resolution large-scale recording (Fig. 8.12), which represent a step towards the engineering of devices for the electrical imaging of neuronal networks in the brain.

Implantable MTAs featuring EOS capacitors have been also tested first "in vitro" and then "in vivo" with promising outcomes. In chronic implants, EOSCs were allowing a reliable stimulation of the rat brain cortex for at least 3 months (Cohen and Vassanelli, unpublished results). At present, EOSFET/EOSC MTAs for bidirectional interfacing are available and under testing in our laboratory. As a matter of fact, however, further development is required to achieve a reliable and efficient two-way interfacing with neuronal networks "in vivo." Recording is still underperforming with respect to MEAs as it is suffering from a higher background noise whose nature is currently under investigation. Stimulation through the EOSC semiconductor–oxide interface is still limited in terms of maximum current injection with respect to activated Ir oxide or porous capacitive electrodes. Further improvements are therefore required, e.g., through surface nanostructuring or by use of novel high-κ materials. In perspective, one advantage of the EOS approach is the relatively simple and nonfaradaic electrochemical nature of the electrolyte–oxide interface [24, 75]. EOS transistors and capacitors, behaving closely to ideal polarizable electrodes, may allow for accurate monitoring and control of extracellular currents and potentials in the brain tissue, which will be particularly advantageous for near-physiological neuronal stimulation and two-way interfacing with neuronal networks "in vivo."

Finally, the use of MTA for very high-resolution electrocorticography (ECoG) has been recently explored in our laboratory by placing a living anaesthetized rat with the brain surface in contact with a large-scale MTA of 16,000 transistors [84]. Although it is currently still unclear how much information can be drawn from very closely spaced electrodes placed on the cortex and what are the mechanisms of spreading and spatial attenuation of recorded LFPs [1, 28], our recent attempts appear to show promising results with the capability of resolving field potential events that are separated by only a few tens of micrometers.

8.5 Conclusions and Outlook

Thanks to recent MEA and MTA developments, imaging neuronal networks "in vivo" by high-resolution recording of extracellular potentials is coming into reach. However, several relevant issues are still to be solved. Above all, the power management for large-scale and high-resolution implants will have to be optimized in order to avoid dissipation and overheating of the brain tissue. Furthermore, keeping dimensions of large MTAs within a window compatible with brain implantation will be a true engineering challenge. New avenues are being explored in the attempt to make probes even smaller, more biocompatible, and capable of recording for many years. Silicon nanowire FETs have been arranged in two- and three-dimensional microporous structures that may facilitate tissue integration [50, 94]. Moving away from silicon, other emerging materials such as carbon nanotubes or graphene [51–54, 81, 95] or conductive polymers [96, 97] promise to revolutionize the way multielectrode arrays are manufactured and to better support probe integration in the brain tissue. Despite these fascinating developments, we feel that MEA- and MTA-based implantable probes will represent the option of choice for "in vivo" recording in the near future and, perhaps, for many years to come. Taking the maximum advantage from processes and technologies developed for silicon in the microelectronic industry, many opportunities may arise to optimize their performance well beyond current limitations. The new probes will provide neuroscientists with novel tools for investigating neuronal networks "in vivo" and in near-physiological conditions, combining high-resolution brain imaging and behavioral studies and, in the long term, offering a gateway for interfacing the brain with artificial neuroelectronic circuits for rehabilitation of neurological patients.

References

1. Buzsáki, G., Anastassiou, C.A., and Koch, C.: The origin of extracellular fields and currents-EEG, ECoG, LFP and spikes. Nat. Rev. Neurosci. **13**(6), 407–420 (2012)
2. Nicolelis, M.A.L., and Lebedev, M.A.: Principles of neural ensemble physiology underlying the operation of brain-machine interfaces. Nat. Rev. Neurosci. **10**(7), 530–540 (2009)
3. Lebedev, M.A., and Nicolelis, M.A.L.: Toward a whole-body neuroprosthetic. Prog. Brain Res. **194**, 47–60 (2011)
4. Buzsáki, G.: Large-scale recording of neuronal ensembles. Nat. Neurosci. **7**(5), 446–451 (2004)
5. Pine, J.: A history of MEA development. In: Taketani, M., and Baudry, M. (eds.) Advances in Network Electrophysiology, pp. 3–23. Springer, New York, USA (2006)
6. Fromherz, P.: Neuroelectronic Interfacing: Semiconductor Chips with Ion Channels, Nerve Cells, and Brain. Wiley-VHC, Weinheim (2003)
7. Fromherz, P.: Three levels of neuroelectronic interfacing. Ann. N. Y. Acad. Sci. **1093**(1), 143–160 (2006)
8. Nicollian, E.H., and Brews, J.R.: MOS (Metal Oxide Semiconductor) Physics and Technology. Wiley and Sons, New York (2003)
9. Vassanelli, S., and Fromherz, P.: Transistor probes local potassium conductances in the adhesion region of cultured rat hippocampal neurons. J. Neurosci. **19**(16), 6767–6773 (1999)
10. Kelly, R.C., Smith, M.A., Samonds, J.M., Kohn, A., Bonds, A.B., Movshon, J.A., and Lee, T.S.: Comparison of recordings from microelectrode arrays and single electrodes in the visual cortex. J. Neurosci. **27**(2), 261–264 (2007)
11. Jog, M.S., Connolly, C.I., Kubota, Y., Iyengar, D.R., Garrido, L., Harlan, R., and Graybiel, A.M.: Tetrode technology: advances in implantable hardware, neuroimaging, and data analysis techniques. J. Neurosci. Methods **117**(2), 141–152 (2002)
12. Felderer, F., and Fromherz, P.: Transistor needle chip for recording in brain tissue. Appl. Phys. **104**(1), 1–6 (2011)
13. Fee, M.S., Mitra, P.P., and Kleinfeld, D.: Variability of extracellular spike waveforms of cortical neurons. J. Neurophysiol. **76**(6), 3823–3833 (1996)
14. Gold, C., et al.: On the origin of the extracellular action potential waveform: a modeling study. J. Neurophysiol. **95**(5), 3113–3128 (2006)
15. Diwakar, S., Lombardo, P., Solinas, S., Naldi, G., and D'Angelo, E.: Local field potential modeling predicts dense activation in cerebellar granule cells clusters under LTP and LTD control. PLoS One **6**(7), e21928 (2011)
16. Pettersen, K.H., Lindén, H., Dale, A.M., and Einevoll, G.T.: Extracellular spikes and current-source density. In: Brette, R., and Destexhe, A. (eds.) Handbook of Neural Activity Measurement. Cambridge University Press, Cambridge (2012)

17. Hanson, T., Fitzsimmons, N., and O'Doherty, J.E.: Technology for multielectrode microstimulation of brain tissue. In: Nicolelis, M.A.L. (ed.) Methods for Neural Ensemble Recordings, 2nd ed. CRC Press, Boca Raton, FL (2008)
18. O'Doherty, J.E., Lebedev, M.A., Ifft, P.J., Zhuang, K.Z., Shokur, S., Bleuler, H., and Nicolelis, M.A.L.: Active tactile exploration using a brain-machine-brain interface. Nature **479**(7372), 228–231 (2011)
19. Gaylor, J.M., Raman, G., Chung, M., Lee, J., M. Rao, J. Lau, and D. S. Poe, "Cochlear implantation in adults: a systematic review and meta-analysis," JAMA Otolaryngol.– Head Neck Surg., **139**(3), 265–272 (2013)
20. Wilson, B. S., and Dorman, M. F.: Cochlear implants: a remarkable past and a brilliant future. Hear. Res. **242**(1–2), 3–21 (2008)
21. Chader, G.J., Weiland, J., and Humayun, M.S.: Artificial vision: needs, functioning, and testing of a retinal electronic prosthesis. In: Verhaagen, J.M.H., et al. (eds.) Progress in Brain Research, vol. 175, pp. 317–332. Elsevier, Amsterdam (2009)
22. Schoen, I., and Fromherz, P.: Extracellular stimulation of mammalian neurons through repetitive activation of Na+channels by weak capacitive currents on a silicon chip. J. Neurophysiol. **100**(1), 346–357 (2008)
23. Bard, A.J., and Faulkner, L.R.: Electrochemical Methods: Fundamentals and Applications. Wiley, New York (2001)
24. Fromherz, P.: Threshold voltage of the EOSFET: reference electrode and oxide-electrolyte interface. Phys. Status Solidi **209**(6), 1157–1162 (2012)
25. Merrill, D.R., Bikson, M., and Jefferys, J.G.R.: Electrical stimulation of excitable tissue: design of efficacious and safe protocols. J. Neurosci. Methods **141**(2), 171–198 (2005)
26. Mayer, S., Geddes, L.A., Bourland, J.D., and Ogborn, L.: Faradic resistance of the electrode/electrolyte interface. Med. Biol. Eng. Comput. **30**(5), 538–542 (1992)
27. Wei, X.F., and Grill, W.M.: Impedance characteristics of deep brain stimulation electrodes in vitro and in vivo. J. Neural Eng. **6**(4), 046008 (2009)
28. Kipke, D.R., Shain, W., Buzsáki, G., Fetz, E., Henderson, J.M., Hetke, J.F., and Schalk, G.: Advanced neurotechnologies for chronic neural interfaces: new horizons and clinical opportunities. J. Neurosci. **28**(46), 11830–11838 (2008)
29. Carmena, J.M., Lebedev, M.A., Crist, R.E., O'Doherty, J.E., Santucci, D.M., Dimitrov, D.F., Patil, P.G., Henriquez, C.S., and Nicolelis, M.A.L.: Learning to control a brain-machine interface for reaching and grasping by primates. PLoS Biol. **1**(2), e2 (2003)
30. Hochberg, L.R., Serruya, M.D., Friehs, G.M., Mukand, J.A., Saleh, M., Caplan, A.H., Branner, A., Chen, D., Penn, R.D., and Donoghue, J.P.: Neuronal ensemble control of prosthetic devices by a human with tetraplegia. Nature **442**(7099), 164–171 (2006)
31. Lebedev, M.A., and Nicolelis, M.A.L.: Brain-machine interfaces: past, present and future. Trends Neurosci. **29**(9), 536–546 (2006)
32. McCreery, D., Lossinsky, A., Pikov, V., and Liu, X.: Microelectrode array for chronic deep-brain microstimulation and recording. IEEE Trans. Biomed. Eng. **53**(4), 726–737 (2006)
33. Lindén, H., Tetzlaff, T., Potjans, T.C., Pettersen, K.H., Grün, S., Diesmann, M., and Einevoll, G.T.: Modeling the spatial reach of the LFP. Neuron **72**(5), 859–872 (2011)
34. Kajikawa, Y., and Schroeder, C.E.: How local is the local field potential? Neuron **72**(5), 847–858 (2011)
35. Lindén, H., Pettersen, K.H., and Einevoll, G.T.: Intrinsic dendritic filtering gives low-pass power spectra of local field potentials. J. Comput. Neurosci. **29**(3), 423–444 (2010)
36. Cogan, S.F.: Neural stimulation and recording electrodes. Annu. Rev. Biomed. Eng. **10**(1), 275–309 (2008)
37. Moxon, K.A., Hallman, S., Sundarakrishnan, A., Wheatley, M., Nissanov, J., and Barbee, K.A.: Long-term recordings of multiple, single-neurons for clinical applications: the emerging role of the bioactive microelectrode. Materials **2**(4), 1762–1794 (2009)
38. Polikov, V.S., Tresco, P.A., and Reichert, W.M.: Response of brain tissue to chronically implanted neural electrodes. J. Neurosci. Methods **148**(1), 1–18 (2005)

39. Suner, S., Fellows, M.R., Vargas-Irwin, C., Nakata, G.K., and Donoghue, J.P.: Reliability of signals from a chronically implanted, silicon-based electrode array in non-human primate primary motor cortex. IEEE Trans. Neural Syst. Rehabil. Eng. Publ. IEEE Eng. Med. Biol. Soc. **13**(4), 524–541 (2005)
40. Hatsopoulos, N.: Decoding continuous and discrete motor behaviors using motor and premotor cortical ensembles. J. Neurophysiol. **92**(2), 1165–1174 (2004)
41. Supèr, H., and Roelfsema, P.R.: Chronic multiunit recordings in behaving animals: advantages and limitations. Prog. Brain Res. **147**, 263–282 (2005)
42. Robblee, L.S., Lefko, J.L. and Brummer, S.B.: Activated Ir: an electrode suitable for reversible charge injection in saline solution. J. Electrochem. Soc. **130**(3), 731–733 (1983)
43. Mozota, J., and Conway, B.E.: Surface and bulk processes at oxidized iridium electrodes—I. Monolayer stage and transition to reversible multilayer oxide film behaviour. Electrochimica Acta. **28**(1), 1–8 (1983)
44. Schmidt, E.M., Bak, M.J., Hambrecht, F.T., Kufta, C.V., O'Rourke, D.K., and Vallabhanath, P.: Feasibility of a visual prosthesis for the blind based on intracortical microstimulation of the visual cortex. Brain J. Neurol. **119**(2), 507–522, 1996
45. Rose, T.L., Kelliher, E.M., and Robblee, L.S.: Assessment of capacitor electrodes for intracortical neural stimulation. J. Neurosci. Methods **12**(3), 181–193 (1985)
46. Cheung, K.C.: Implantable microscale neural interfaces. Biomed. Microdevices **9**(6), 923–938, (2007)
47. Zhou, D.D., and Greenbaum, E.S.: Implantable neural prostheses 1: devices and applications. Springer, Berlin (2009)
48. HajjHassan, M., Chodavarapu, V., and Musallam, S.: NeuroMEMS: neural probe microtechnologies. Sensors **8**(10), 6704–6726 (2008)
49. Wise, K.D., Sodagar, A.M., Ying Yao, Gulari, M.N., Perlin, G.E., and Najafi, K.: Microelectrodes, microelectronics, and implantable neural microsystems. Proc. IEEE **96**(7), 1184–1202 (2008)
50. Tian, B., Liu, J., Dvir, T., Jin, L., Tsui, J.H., Qing, Q., Suo, Z., Langer, R., Kohane, D.S., and Lieber, C.M.: Macroporous nanowire nanoelectronic scaffolds for synthetic tissues. Nat. Mater. **11**(11), 986–994 (2012)
51. Keefer, E.W., Botterman, B.R., Romero, M.I., Rossi, A.F., and Gross, G.W.: Carbon nanotube coating improves neuronal recordings. Nat. Nanotechnol. **3**(7), 434–439 (2008)
52. Ballerini, L., Prato, M., Giugliano, M., Gambazzi, L., Mazzatenta, A., Businaro, L., Markram, H., and Campidelli, S.: Interfacing neurons with carbon nanotubes: electrical signal transfer and synaptic stimulation in cultured brain circuits. J. Neurosci. **27**(26), 6931–6936 (2007)
53. Ciofani, G., Raffa, V., Vittorio, O., Cuschieri, A., Pizzorusso, T., Costa, M., and Bardi, G.: In vitro and in vivo biocompatibility testing of functionalized carbon nanotubes. In: Balasubramanian, K., and Burghard, M. (eds.) Carbon Nanotubes, pp. 67–83. Humana Press, Totowa (2010)
54. Bareket-Keren, L., and Hanein, Y.: Carbon nanotube-based multi electrode arrays for neuronal interfacing: progress and prospects. Front. Neural Circuits **6**, 122 (2013)
55. Ghezzi, D., Antognazza, M.R., Dal Maschio, M., Lanzarini, E., Benfenati, F., and Lanzani, G.: A hybrid bioorganic interface for neuronal photoactivation. Nat. Commun. **2**, 166 (2011)
56. Benfenati, V., Toffanin, S., Bonetti, S., Turatti, G. Pistone, A., Chiappalone, M., Sagnella, A., Stefani, A., Generali, G., Ruani, G., Saguatti, D., Zamboni, R., and Muccini, M.: A transparent organic transistor structure for bidirectional stimulation and recording of primary neurons. Nat. Mater. **12**(7), 672–680 (2013)
57. Wise, K.D., Angell, J.B., and Starr, A.: An integrated-circuit approach to extracellular microelectrodes. IEEE Trans. Biomed. Eng. **BME-17**(3), 238–247 (1970)
58. Wise, K., and Angell, J.: A microprobe with integrated amplifiers for neurophysiology. In: Solid-State Circuits Conference. Digest of Technical Papers. 1971 IEEE International, vol. XIV, pp. 100–101 (1971)

59. Wise, K.D., and Angell, J.B.: A low-capacitance multielectrode probe for use in extracellular neurophysiology. IEEE Trans. Biomed. Eng. **BME-22**(3), 212–219 (1975)
60. Drake, K.L., Wise, K.D., Farraye, J., Anderson, D.J., and BeMent, S.L.: Performance of planar multisite microprobes in recording extracellular single-unit intracortical activity. IEEE Trans. Biomed. Eng. **35**(9), 719–732 (1988)
61. Najafi, K., Wise, K.D., and Mochizuki, T.: A high-yield IC-compatible multichannel recording array. IEEE Trans. Electron Devices **32**(7), 1206–1211 (1985)
62. Wise, K.D., Anderson, D.J., Hetke, J.F., Kipke, D.R., and Najafi, K.: Wireless implantable microsystems: high-density electronic interfaces to the nervous system. Proc. IEEE **92**(1), 76–97 (2004)
63. Bai, Q., and Wise, K.D.: Single-unit neural recording with active microelectrode arrays. IEEE Trans. Biomed. Eng. **48**(8), 911–920 (2001)
64. Beebe, X., and Rose, T.L.: Charge injection limits of activated iridium oxide electrodes with 0.2 ms pulses in bicarbonate buffered saline (neurological stimulation application). IEEE Trans. Biomed. Eng. **35**(6), 494–495 (1988)
65. Weiland, J.D., and Anderson, D.J.: Chronic neural stimulation with thin-film, iridium oxide electrodes. IEEE Trans. Biomed. Eng. **47**(7), 911–918 (2000)
66. Weiland, J.D., Anderson, D.J., and Humayun, M.S.: In vitro electrical properties for iridium oxide versus titanium nitride stimulating electrodes. IEEE Trans. Biomed. Eng. **49**(12), 1574–1579 (2002)
67. Najafi, K., Ji, J., and Wise, K.D.: Scaling limitations of silicon multichannel recording probes. IEEE Trans. Biomed. Eng. **37**(1), 1–11 (1990)
68. Jones, K.E., Campbell, P.K., and Normann, R.A.: A glass/silicon composite intracortical electrode array. Ann. Biomed. Eng. **20**(4), 423–437 (1992)
69. Yoo, J.-M., Negi, S., Tathireddy, P., Solzbacher, F., Song, J.-I., and Rieth, L.W.: Excimer laser deinsulation of Parylene-C on iridium for use in an activated iridium oxide film-coated Utah electrode array. J. Neurosci. Methods **215**(1), 78–87 (2013)
70. Rousche, P.J., and Normann, R.A.: Chronic recording capability of the Utah Intracortical Electrode Array in cat sensory cortex. J. Neurosci. Methods **82**(1), 1–15 (1998)
71. Rousche, P.J., and Normann, R.A.: Chronic intracortical microstimulation (ICMS) of cat sensory cortex using the Utah Intracortical Electrode Array. IEEE Trans. Rehabil. Eng. Publ. IEEE Eng. Med. Biol. Soc. **7**(1), 56–68 (1999)
72. Badi, A.N., Kertesz, T.R., Gurgel, R.K., Shelton, C., and Normann, R.A.: Development of a novel eighth-nerve intraneural auditory neuroprosthesis. The Laryngoscope **113**(5), 833–842 (2003)
73. Clark, G.A., Ledbetter, N.M., Warren, D.J., and Harrison, R.R.: Recording sensory and motor information from peripheral nerves with Utah Slanted Electrode Arrays. Paper presented at 2011 Annual International Conference of the IEEE Engineering in Medicine and Biology Society, EMBC, pp. 4641–4644, 30 August–3 September 2011
74. Rousche, P.J., and Normann, R.A.: A method for pneumatically inserting an array of penetrating electrodes into cortical tissue. Ann. Biomed. Eng. **20**(4), 413–422 (1992)
75. Bergveld, P.: Thirty years of ISFETOLOGY: what happened in the past 30 years and what may happen in the next 30 years. Sensors Actuators B Chem. **88**(1), 1–20 (2003)
76. Bergveld, P.: Development, operation, and application of the ion-sensitive field-effect transistor as a tool for electrophysiology. IEEE Trans. Biomed. Eng. **BME-19**(5), 342–351 (1972)
77. Fromherz, P., Offenhausser, A., Vetter, T., and Weis, J.: A neuron-silicon junction: a Retzius cell of the leech on an insulated-gate field-effect transistor. Science **252**(5010), 1290–1293 (1991)
78. Vassanelli, S., and Fromherz, P.: Transistor records of excitable neurons from rat brain. Appl. Phys. **66**(4), 459–463 (1998)

79. Voelker, M., and Fromherz, P.: Signal transmission from individual mammalian nerve cell to field-effect transistor. Small **1**(2), 206–210 (2005)
80. Besl, B., and Fromherz, P.: Transistor array with an organotypic brain slice: field potential records and synaptic currents. Eur. J. Neurosci. **15**(6), 999–1005 (2002)
81. Hutzler, M., and Fromherz, P.: Silicon chip with capacitors and transistors for interfacing organotypic brain slice of rat hippocampus. Eur. J. Neurosci. **19**(8), 2231–2238 (2004)
82. Hutzler, M.: High-resolution multitransistor array recording of electrical field potentials in cultured brain slices. J. Neurophysiol. **96**(3), 1638–1645 (2006)
83. Schoen, I., and Fromherz, P.: The mechanism of extracellular stimulation of nerve cells on an electrolyte-oxide-semiconductor capacitor. Biophys. J. **92**(3), 1096–1111 (2007)
84. Vassanelli, S., Mahmud, M., Girardi, S., and Maschietto, M.: On the way to large-scale and high-resolution brain-chip interfacing. Cogn. Comput. **4**(1), 71–81 (2012)
85. Wallrapp, F., and Fromherz, P.: TiO[sub 2] and HfO[sub 2] in electrolyte-oxide-silicon configuration for applications in bioelectronics. J. Appl. Phys. **99**(11), 114103 (2006)
86. Giacomello, M., Girardi, S., Scorzeto, M., Peruffo, A., Maschietto, M., Cozzi, B., and Vassanelli, S.: Stimulation of Ca2+ signals in neurons by electrically coupled electrolyte-oxide-semiconductor capacitors. J. Neurosci. Methods **198**(1), 1–7 (2011)
87. Eversmann, B., Jenkner, M., Hofmann, F., Paulus, C., Brederlow, R., Holzapfl, B., Fromherz, P., Merz, M., Brenner, M., Schreiter, M., Gabl, R., Plehnert, K., Steinhauser, M., Eckstein, G., Schmitt-Landsiedel, D., and Thewes, R.: A 128 x 128 cmos biosensor array for extracellular recording of neural activity. IEEE J. Solid-State Circuits, **38**(12), 2306–2317 (2003)
88. Lambacher, A., Vitzthum, V., Zeitler, R., Eickenscheidt, M., Eversmann, B., Thewes, R., and Fromherz, P.: Identifying firing mammalian neurons in networks with high-resolution multitransistor array (MTA). Appl. Phys. **102**(1), 1–11 (2011)
89. Eversmann, B., Lambacher, A., Gerling, T. Kunze, A., Fromherz, P., and Thewes, R.: A neural tissue interfacing chip for in-vitro applications with 32k recording / stimulation channels on an active area of 2.6 mm2. In ESSCIRC (ESSCIRC), 2011 Proceedings of the **41**, 211–214 (2011)
90. Ferrea, E., Maccione, A., Medrihan, L., Nieus, T., Ghezzi, D., Baldelli, P., Benfenati, F., and Berdondini, L.: Large-scale, high-resolution electrophysiological imaging of field potentials in brain slices with microelectronic multielectrode arrays. Front. Neural Circuits **6**, 80 (2012). doi: 10.3389/fncir.2012.00080
91. Frey, U., Sedivy, J., Heer, F., Pedron, R., Ballini, M., Mueller, J., Bakkum, D., Hafizovic, S., Faraci, F. D., Greve, F., Kirstein, K.-U., and Hierlemann, A.: Switch-matrix-based high-density microelectrode array in CMOS technology. IEEE J. Solid-State Circuits **45**(2), 467–482 (2010)
92. Felderer, F., and Fromherz, P.: Transistor needle chip for recording in brain tissue. Appl. Phys. **104**(1), 1–6 (2011)
93. Weis, R., and Fromherz, P.: Frequency dependent signal transfer in neuron transistors. Phys. Rev. E **55**(1), 877 (1997)
94. Liu, J., Xie, C., Dai, X., Jin, L., Zhou, W., and Lieber, C.M.: Multifunctional three-dimensional macroporous nanoelectronic networks for smart materials. Proc. Natl. Acad. Sci. **110**(17), 6694–6699 (2013)
95. Yang, W., Ratinac, K.R., Ringer, S.P., Thordarson, P., Gooding, J.J., and Braet, F., Carbon nanomaterials in biosensors: should you use nanotubes or graphene? Angew. Chem. Int. Ed. **49**(12), 2114–2138 (2010)
96. Seymour, J.P.: Advanced Polymer-based Microfabricated Neural Probes Using Biologically Driven Designs. Dissertation, University of Michigan (2009).
97. Kozai, T.D.Y., Langhals, N.B., Patel, P.R., Deng, X., Zhang, H., Smith, K. L., Lahann, J., Kotov, N.A., and Kipke, D.R.: Ultrasmall implantable composite microelectrodes with bioactive surfaces for chronic neural interfaces. Nat. Mater. **11**(12), 1065-1073 (2012)

98. Brenner, A., Stein R.B., and Normann, R.A.: Selective stimulation of cat sciatic nerve using an array of varying length microelectrodes. Journal of Neurophysiology **85**, 1585–1594 (2001)
99. Bhandari, R., Negi, S., Rieth, L., Normann, R.A., and Solzbacher, F.: A novel masking method for high aspect ratio penetrating microelectrode arrays. J. Micromech. Microeng. **19**, 035004 (8pp) (2009)

Chapter 9
Optogenetics

Allison Quach, Nicholas James, and Xue Han

9.1 Introduction

Optogenetics is the use of light and genetically encoded light-sensitive proteins to modulate the activity of specific cells. Here, we focus on the use of optogenetics in neuroscience, where it is used to rapidly activate or inhibit a specific population of neurons in the brain. The basic components required are (1) photosensitive molecules targeted to specific cells, (2) light delivery to the brain, and (3) electrophysiological or behavioral readouts for optogenetic manipulation of a system.

9.2 Microbial Opsins

Opsins were identified in bacteria and algae in the 1970s and have been studied extensively since then [1, 2], but were only recently tested in mammalian neurons for millisecond-scale control of action potentials [3]. Type I (microbial) opsins are seven-transmembrane domain proteins found in archaea, bacteria, and algae. The main classes of microbial opsins that directly convert light to ion transport are channelrhodopsins, halorhodopsins, and archaerhodopsins (Fig. 9.1). Other classes of rhodopsins recruit intracellular and secondary messenger systems that do not directly lead to ion flux.

For opsins to be photosensitive, they require an all-*trans* retinal cofactor bound to a lysine residue in the retinal binding pocket of the opsin [4]. This complex is referred to as rhodopsin. In the presence of light, all-*trans* retinal photoisomerizes to 13-*cis* retinal, resulting in conformational changes in the opsin that allow ion transport.

A. Quach • N. James • X. Han (✉)
Biomedical Engineering Department, Boston University, Boston, MA, USA
e-mail: xuehan@bu.edu

M. De Vittorio et al. (eds.), *Nanotechnology and Neuroscience: Nano-electronic, Photonic and Mechanical Neuronal Interfacing*, DOI 10.1007/978-1-4899-8038-0_9,

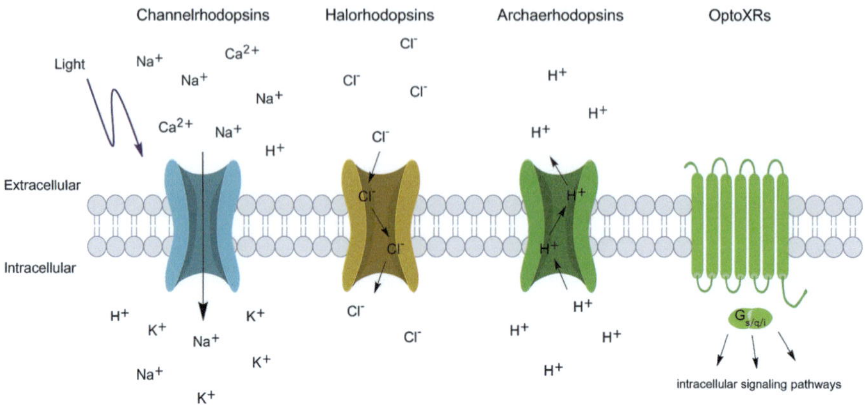

Fig. 9.1 Classes of optogenetic molecules. Channelrhodopsins are ion channels that passively transport cations such as Na^+, K^+, H^+, and Ca^{2+} down their electrochemical gradients to depolarize neurons. Halorhodopsins are chloride pumps that transport ions into the cell, resulting in hyperpolarization. Archaerhodopsins are a second class of inhibitory opsins that pump protons out of the cell. OptoXRs were developed from mammalian opsins and trigger intracellular G-protein cascades

Fortunately, it turns out that mammalian tissue contains sufficient all-trans retinal for microbial opsins to function, and the use of opsin-based optogenetic tools has been demonstrated in various mammalian brains, i.e., rodents and nonhuman primates. In some species, all-*trans* retinal needs to be supplemented for opsins to work, such as in the neural systems of *Caenorhabditis elegans* and zebrafish.

9.2.1 Excitation

Channelrhodopsin2 (ChR2) was identified from the green algae *Chlamydomonas reinhardtii* as a photoactivated ion channel [5] and was the first opsin used to demonstrate optogenetic control of mammalian neurons [3]. At low light intensities, ChR2 functions as a calcium channel [6]; however, at higher light intensities, the flux through ChR2 is predominately cations (including Na^+, K^+, H^+, and Ca^{2+}) flowing down the electrochemical gradient. One characteristics of ChR2 is that ion conduction is uncoupled from photodetection [7]. Rather, the channel closes based on channel kinetics, which makes ChR2 more efficient compared to opsins that are ion pumps.

Since the demonstration of ChR2 as a method of controlling neural activity, many variants have been developed. Mutations have increased photocurrent, varied kinetics, or shifted sensitivity spectrum [8–11].

9.2.2 Silencing

Halorhodopsins (HR) are archaebacterial chloride pumps that transport chloride ions into the cell [12]. The first used opsin in this family was Halo/NpHR, from

Natronomonas pharaonis [13–15], a yellow-light-activated opsin. However, it was poorly targeted to membranes in mammalian cells, and therefore, subsequent modifications with membrane targeting sequences were necessary to create variants (e.g., eNpHR, eNpHr3.0) that successfully inhibit mammalian neurons [16].

Unlike ChR2, Halo requires constant light to function as a pump. Halo also has slow recovery due to nonconducting intermediate protein states [17], but the use of blue light can accelerate recovery [14, 18]. Another consideration when using Halo is a potential secondary effect on GABA receptors due to the change in Cl^- concentration. It has been shown that the $GABA_A$ reversal potential can be significantly changed by activating Halo [19], increasing the excitability of modified neurons.

Archaerhodopsins (Arch) are a class of outward-moving proton pumps that are used to silence neural activity [18]. Archaerhodopsin-3, from *Halorubrum sodomense*, is a yellow-green light-activated proton pump that hyperpolarizes neurons. In contrast to Halo, Arch recovers within seconds after inactivation, allowing for finer-timescale control of photoinactivation. Subsequent variants of Arch have increased light sensitivity, allowing silencing of a larger area (ArchT; [20]), or shifted activation spectra (Mac; [18]), sensitive to green-blue light.

It has been shown that the flux of protons out of the cell does not change cellular pH [18]. This is thought to be due to intracellular buffering mechanisms; however, it is possible that the increase of protons extracellularly may have undesired effects on cells in a non-cell-type-specific manner.

9.2.3 Other

Type II (mammalian) opsins are also photosensitive seven-transmembrane domain proteins. However, they are G-protein-coupled receptors that function through second messengers, and light sensitivity is not directly coupled to ion channels. Instead, activation of type II opsins triggers intracellular G-proteins. These so-called "optoXRs" can be used to modulate intracellular signaling and investigate biochemical signaling pathways such as those downstream of alpha-1 and beta-2 adrenergic receptors [21] and the 5-HT1a receptor [22].

9.3 Opsin Targeting

Optogenetics depends on the expression of microbial opsins in a specific subset of cells. The main methods of gene delivery are viral injections and transgenic animal lines (Fig. 9.2). Opsins are targeted to a specific subset of cells using cell-type-specific promoters or anatomical projection targeting. Further specificity can be gained by spatially restricting viral or light delivery [23].

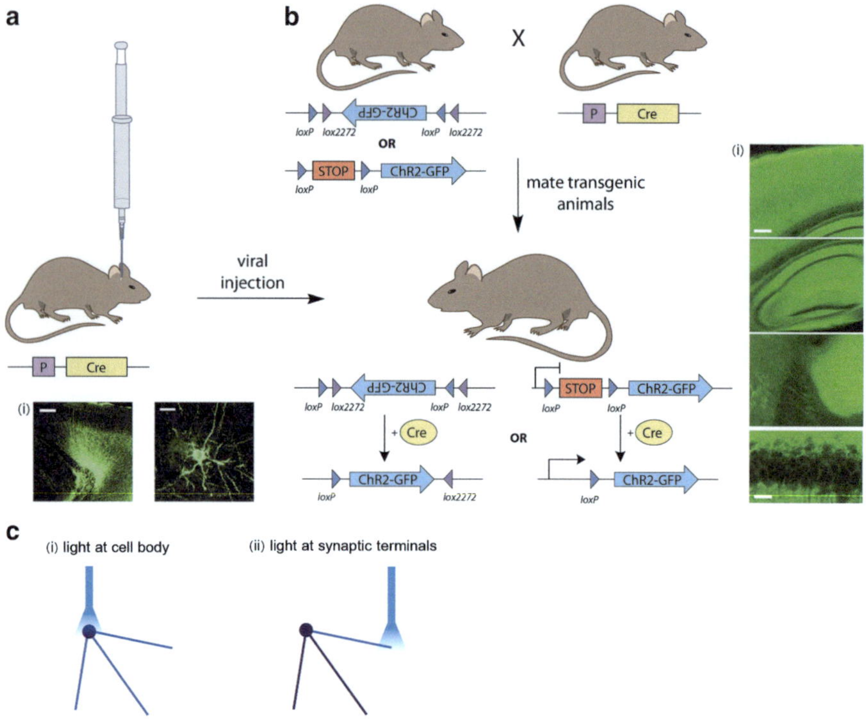

Fig. 9.2 Cell-specific targeting. (**a**, **b**) Cell-specific targeting of opsins with Cre-LoxP recombination [28]. Genes may be delivered virally into Cre-driver line animals (**a**), and Cre-dependent opsin lines and Cre-driver lines can be mated (**b**). **a**-*i* Fluorescence image of Arch-GFP expression in mouse cortex after lentiviral injection. Scale bars, 200 μm (*left*) and 20 μm (*right*). **b**-*i* Native EYFP fluorescence in transgenic ChR2(H134R)-EYFP mice. Scale bars, 200 μm (*top*) and 20 μm (*bottom*). Fluorescence images from [18], © (2010) The Author(s), and from [31], © (2012) The Author(s). (**c**) Anatomical specificity can also be achieved with targeted light delivery to either the cell bodies (*i*) or synaptic terminals (*ii*) of opsin-expressing neurons

9.3.1 Viral Gene Delivery

Lentiviral (LV) and adeno-associated viral (AAV) vectors are commonly used to deliver opsin genes to neurons. Viral methods are frequently used in species where transgenic animals are not readily available, such as in rats [24] and primates [25]. Another benefit of using viruses is high copy numbers and, therefore, high expression of opsins. Strong opsin expression reduces the possibility of tissue damage from higher laser powers.

Viral tropisms are another method by which cell-type-specific expression can be manipulated [26]. Different viruses infect different cell types with varying efficacy due to membrane proteins that are necessary for infection. This can be used to better target opsins to specific cell types. However, viral tropisms also make it difficult to target certain cell types, limiting the full use of optogenetics.

A major limitation of viral methods is the maximum genetic payload that can be packaged into virus. For example, AAV can package up to 4.7 kb of DNA, while lentivirus can package up to ~8 kb. This size cap limits the promoters that can be used with viral vectors; this presents an issue because short promoters may non-specifically target cells [27]. This can be ameliorated using the Cre-LoxP recombinase system of expression in conjunction with viral delivery (discussed below; Fig. 9.2a [28]).

9.3.2 Transgenic Animals

In animals, where a transgenic approach is available, it is a powerful method of targeting specific cell types. Transgenic mouse lines are crossed using the Cre-LoxP recombination to control the expression of transgenes in specific cells (Fig. 9.2b; [29, 30]). For optogenetics, two basic approaches are used: (1) LoxP sites flank a stop codon that is removed in the presence of Cre and (2) the opsin of interest is flanked by the recombination sites and flipped into the coding orientation by Cre. In both cases, the gene is transcribed in the presence of Cre and introduced by viral delivery or in transgenic mice expressing Cre in target cells. ChR2, eNpHR3.0, and Arch Cre-dependent lines have been created that, when crossed with Cre reporter lines, demonstrate long-term expression at suitable copy numbers for optogenetic excitation or silencing [31].

Many transgenic mouse lines have been created [31]. This has made optogenetics accessible; all that is needed is for researchers to breed Cre-dependent opsin lines with Cre-driver lines to create cell-specific opsin expression in neurons of interest.

9.3.3 Anatomical Specificity

In addition to using cell-type-specific expression of opsins, spatial specificity can be achieved through several methods: virus injected at a specific anatomical area, anterograde or retrograde labeling of neurons, and restricted light delivery.

Labeling projections to or from a brain area of interest could be possible with anterograde or retrograde viruses. Lentivirus EIAV pseudotyped with rabies glycoprotein was found to allow retrograde labeling of neurons [32], but has not yet been used for optogenetics. Additionally, a glycoprotein-deficient pseudotyped rabies virus limits the number of synapses the virus can cross [33], allowing for cleaner optogenetic control of the neurons of interest. Herpes simplex virus (HSV) and certain serotypes of AAV may also be used to target projections [34].

Light can be delivered to target either the cell bodies (affecting all postsynaptic targets) or the synaptic terminals of neurons alone (Fig. 9.2c). This requires sufficient opsin expression at the terminals, but the concept has been demonstrated in vivo with ChR2 [35, 36]. Another potential concern is antidromic stimulation of the targeted fibers, which may lead to unintentional activation of other areas that the neurons project to.

9.4 Light Delivery

Lasers and LEDs are the primary sources of light used in optogenetic experiments. These light sources emit light within a narrow band of wavelengths and also allow for precise temporal control of the light. They either shine light directly on the tissue or are coupled with fiberoptics for insertion into the brain to target deeper tissues. As described above, the method of light delivery to the brain is a crucial element of optogenetics. The amount and wavelength of light used determines how many neurons are affected by the light and how responsive those neurons are to the light. Additionally, optimal light conditions are determined by the type of opsin used and the experimental conditions desired (in vivo vs. in vitro, prolonged vs. rapid perturbations, etc.).

9.4.1 Optical Properties of the Brain

The amount of light that reaches a neuron depends on factors such as distance from the light source, absorption and scattering, and wavelength; factors that are taken account into Monte Carlo simulations of light propagation [37]. These models predict that light decreases nonlinearly to a mere 1 % at locations 1 mm away from the light source [18, 38], numbers consistent with experimental results [39]. The major factor contributing to this steep falloff is that most opsins are activated by light in the visible wavelengths (450–600 nm), which falls into the range where hemoglobins absorb the majority of light. This is the impetus behind efforts to develop red- and far-red (>650 nm)-shifted opsins that can be activated with far lower light intensities [8].

9.4.2 Tissue Damage from Device Insertion and Heat

Device insertion as well as heat generated by light can damage tissue. To reduce mechanical damage from device insertion, ideally one would use devices as small as possible. For example, multiple thin fibers or fiber arrays are helpful in reducing mechanical tissue damage [40]. In a fiber array, multiple thin fibers are used to illuminate an entire area, especially in larger rodents such as rats, and in primates, where the light from one fiber might only reach a fraction of the brain area of interest.

Heat-generated damage is difficult to evaluate. Sharp temperature increases, especially at the tip of the fiber, could damage tissues. Even if the heating is not sufficient to cause a lesion, the diffusion of heat could affect cellular metabolism [41]. In addition, heating alone has been shown to influence neurons' responses, as demonstrated by pulsed infrared light stimulation [41] or possibly by pulsed ultrasound stimulation [40, 42]. Optogenetics using light of a few mW of total light power in vivo seems to be safe. It is important to perform controls experiments that address these issues.

9.4.3 Electrode Improvement to Reduce Optical Artifact

A major advantage of optogenetics is that it is orthogonal to many readout technologies, such as electrophysiology, fMRI, behavior, and cellular imaging. However, it was found that laser light produces a strong voltage deflection artifact when directed to the tip of metal electrode [20, 39, 43], but not when directed to glass microelectrodes where the silver/silver chloride electrode in the glass filled with solution was not exposed to light. The light-induced artifact is slow evolving and thus corrupts local field potential (LFP) but not spike waveforms. Thus, cautions have to be made when interpreting LFP results in optogenetic experiments. Optimizing electrode tips, such as coating with conducting polymer indium tin oxide (ITO), may prove useful in reducing this artifact on LFP recordings [44].

9.4.4 Other Side Effects

As discussed briefly above, the extent of nonspecific effects of introducing exogenous proteins at such high concentration in neurons is not known. Activation of Halo changes the $GABA_A$ reversal potential [19], and the acidification of the extracellular space due to Arch activation may have unintended consequences. Simply overexpressing membrane proteins may increase the membrane capacitance, changing neuronal excitability [45].

9.4.5 Hardware for Optogenetics

Specialized hardware for various optogenetic experiments have been well developed in many laboratories, and neurophysiology companies, and are continuously being optimized. For example, thin optical fibers with a size comparable to a recording electrode can be easily handled with conventional methods. Optical fibers can be easily positioned or stereotactically targeted to a brain region of interest or incorporated in a tetrode design in freeing moving rodents. Fiber arrays have been designed to target multiple areas of interest or large areas of interest [40] and may be powered wirelessly [46]. Optical fiber rotary joints are useful devices easily obtained from vendors to prevent tangling of optical fibers, electrophysiological recording cables, or LED power cables.

9.5 Applications

Traditionally, pharmacological manipulations in neuroscience experimentation have been performed with receptor-binding agents. However, the spatial spread and time course of infusion methods have proven problematic in many instances, leading to inconsistent experimental control and ambiguity in the interpretation of results [47]. Similarly, electrical stimulation of cortical sites and subcortical nuclei is difficult to interpret as gross electrical stimulation activates neurons near the electrode indiscriminately [48]. Optogenetics combines consistent and temporally precise activation of spatially aggregated neurons with specific activation of cell types, delivering a high standard of experimental control over neural circuits [11, 49]. Optogenetics research has proven most powerful when performed with dynamic and varied experimental approaches. Utilized in tandem with electrophysiology, microscopy, and behavioral paradigms, targeted light delivery has allowed researchers to manipulate distinct classes of cortical GABAergic interneurons, distinct nuclei and processes of neuromodulator pathways, and neuroanatomically defined cell classes in order to investigate their functional roles in visual processing, memory formation, behavioral learning, stress, mood disorders, and other processes. Here, we discuss a few examples in which optogenetics has been used to reveal the contribution of specific cells in neural computation in transgenic and virally injected animal models.

9.5.1 Parsing the Function of Interneuron Subtypes in Cortical Processing Circuits

Electrophysiological configurations are crucial for understanding the local and large-scale neural behaviors produced by optogenetic manipulations because they are used to characterize and visualize activity at specific brain sites and timescales relevant to experimental hypotheses. For example, single unit recordings, multi-unit recordings, and local field potentials—spiking activity of a single neuron, firing of a small group of neurons, and lower-frequency fluctuations in extracellular field potentials, respectively—are often used to determine the relative activation of neural ensembles in order to quantify the effects of optogenetic manipulations and assess their neurophysiological contribution to behavioral expression.

Several studies published recently highlighted the potential for optogenetics research to effectively glean cortical circuit properties underlying perception and contextual memory. With the use of optogenetic-electrophysiology configurations, researchers have revealed defined processing roles for GABAergic interneuron classes in cortex. Lee et al. [50] used Cre-induced expression of ChR2 to show that activation of parvalbumin (PV)-positive interneurons heightened (PV-Cre) mice's ability to discriminate between oriented spatial gratings concordant with sharpened orientation tuning—measured as a function of firing rate—in neurons of the primary visual cortex (V1). A further work investigated interneuron function in V1 by utilizing multichannel in vivo recording in conjunction with viral induction of halorhodopsin (Halo) and electroporation delivery of ChR2 in somatostatin (SOM)-Cre mice [51]. Since mice were electroporated in utero with ChR2, to be expressed in superficial layers of the maturated cortex, Adesnik et al. [51] were able to show that SOM-positive interneurons suppress surround excitation of pyramidal neurons, suggesting that SOM-positive interneurons receive converging input from nearby cortical neurons and regulate spatial summation of visual input. Finally, while investigating hippocampal circuitry, Royer et al. [52] used Halo to silence PV and SOM interneurons within region CA1 and observed that each cell type plays a specific role in place cell firing with respect to theta-frequency local field potentials, indicating that together, SOM and PV interneurons coordinate temporal aspects of spatial context processing in hippocampus. As discussed in these studies, light delivery methods used in tandem with electrophysiological recording allow researchers to control cortical circuits while observing simultaneous alterations in single cell and neuron population excitability in order to investigate diverse neuronal functions.

9.5.2 Activity-Dependent Expression of Opsins in Behaviorally Relevant Cell Ensembles

A promising aspect of optogenetics research is its capacity to predictably alter animal behavior, allowing the investigation of neural circuits related to behaviors. When used to perturb defined neural populations, optogenetic stimulation and silencing can be used to infer relationships between states of neural activation and underlying cognitive functions. A profoundly useful strategy in identifying and modulating mammalian cell ensembles has been in targeting neural populations by their activity levels during experimental paradigms. This work has largely been pioneered in the laboratory of Susumu Tonegawa at the Massachusetts Institute of Technology [53]. Using the doxycycline system to trigger co-expression of ChR2 with c-fos (an immediate early gene) in mouse dentate gyrus, Liu et al. [53] optogenetically reactivated neurons recruited during contextual fear conditioning, causing mice to reexpress a conditioned response, i.e., freezing. They found that this occurred even when mice were housed in a previously habituated context, indicating that optogenetic reactivation of c-fos-labeled neurons elicits retrieval of contextual memories. In this way, researchers are designing flexible molecular strategies to target and control previously inaccessible neuronal ensembles and opening the door for dynamic, innovative approaches to the examination of *modularity* and cell sparsity within complex neural systems.

9.5.3 Modulating Neuromodulator-Producing Neurons Within Heterogeneous Nuclei

Optogenetics has also proven particularly useful in experimentally accessing neurotransmitter pathways because neuromodulatory neurons are often embedded in heterogeneous nuclei such as the nucleus accumbens (NAc) and the ventral tegmental area (VTA). Researchers often co-express opsins with the metabolic enzymes of neuromodulators, allowing localized activation and silencing within subcortical, midbrain, and brainstem nuclei—inducing or suppressing neuromodulator release. Witten et al. [54] utilized viral expression of Halo in choline acetyltransferase (ChAT)-Cre transgenic mice in order to silence cholinergic interneurons of the NAc. By silencing this sparse population of interneurons bilaterally, they were able to eliminate cocaine conditioning behavior—measured as preference for the location of cocaine delivery. Similarly, Tsai et al. [55] used AAV delivery to express ChR2 in tyrosine-hydroxilase (TH)-Cre transgenic mice, expressing in dopaminergic neurons of the VTA. They observed that light-activated, phasic release of dopamine from VTA, sufficient to induce increased dopamine concentrations in NAc, elicits learned reinforcement responses without delivery of an external reward. In this way optogenetic applications effectively combine spatial specificity—afforded by localized light delivery—and targeted neurotransmitter activation, allowing researchers to unravel the causal roles of anatomically distinct branches of the cholinergic and dopaminergic neurotransmitter cascades in producing animal behaviors.

9.5.4 Interrogation of Long-Range Neuromodulatory Interactions

Finally, an important application of optogenetics has been in the investigation of large-scale circuit interactions. Subcortical and midbrain nuclei engage in complex interactions that determine aspects of volitional behavior and reinforcement conditioning. By taking advantage of neuronal projections between nuclei and cortical regions, researchers can simulate native, large-scale brain activities. In one such study, Chaudhury et al. [56] used retrograde virus tracers to express ChR2 and Halo in NAc and prefrontal cortex (PFC) projections to VTA dopaminergic neurons in order to examine the relationship between reinforcement circuitry and social defeat. They found that the activity of dopamine projections from VTA to NAc and PFC have opposing roles in eliciting susceptibility and resilience to social defeat—indicated by the mouse's capacity for effective reinforcement learning. Utilizing a similar paradigm to deconstruct long-range neuromodulatory interactions during volitional behavior, Warden et al. [57] used viral ChR2 expression in $Ca2^{+}$/calmodulin-dependent protein kinase-α-Cre (CaMKIIα) transgenic mice in order to stimulate projection afferents from PFC within the raphe nucleus of the brainstem. They observed that PFC enhances effortful action under stressful conditions by directly accessing the serotonergic pathway, indicating that PFC interacts with 5-HT to elicit action selection during stress scenarios.

In this way, viral and retroviral expression of opsins can be used to experimentally access the origins and eccentricities of large-scale neurotransmitter systems as they project to and interact between nuclei over extended distances. Indeed, future work with optogenetic co-expression of opsins under CaMKIIα or SERT (a serotonin transporter protein) promoters and dopamine-related promoter regions could be useful in elucidating an integrated neuromodulatory cascade by which PFC interacts with dopamine and 5-HT projecting nuclei, perhaps granting us further insight into a brain-based perspective of reinforcement learning and mood disorder [58].

9.6 Concluding Remarks

Optogenetics, an exciting novel neurotechnology, offers new ways to modify and control neural activity rapidly and reversibly. Three major classes of light-activated opsins, channelrhodopsins (for neural activation), and halorhodopsins, and archaerhodopsins (for neural silencing), when expressed in mammalian neurons, generate sufficient photocurrents to alter neural activities when irradiated with light. By targeting these opsins to specific cells of interest, it is now possible to investigate the casual roles of these neurons in neural computation and animal behaviors. We have described several advances in the application of optogenetics owing to varied experimental approaches: hybridized light delivery and electrophysiological recording configurations, dynamic methods of molecularly induced opsin expression, specific cell targeting within heterogeneous nuclei, and experimental control of neuronal projection systems. Much of the research performed with optogenetics has focused on neocortical, striatal, and hippocampal networks, revealing new windows into the neurophysiology of memory, executive processing, learning, and perception. In this regard, optogenetic applications have displayed extraordinary potential for unraveling neural systems behavior and underlying cognitive functions, as well as furthering our understanding of pathological brain states related to psychiatric disorders and disease. While optogenetics has revolutionized the investigation of specific cells in neural computation with high spatial, temporal, and cell-type specificity, it is important to properly interpret results from optogenetic perturbation studies. Continued progress in generating novel opsins with improved functions or power spectra, along with increasingly available transgenic animal models and associating hardware, will facilitate the continued wide spread use of optogenetics in mapping neural circuits.

References

1. Oesterhelt, D., Stoeckenius, W.: Rhodopsin-like protein from the purple membrane of Halobacterium halobium. Nat New Biol. **233**, 149-152 (1971)
2. Oesterhelt, D., Stoeckenius, W.: Functions of a new photoreceptor membrane. Proc Natl Acad Sci U S A. **70**(10), 2853-2857 (1973)

3. Boyden, E.S., Zhang, F., Bamberg, E., Nagel, G., Deisseroth, K.: Millisecond-timescale, genetically targeted optical control of neural activity. Nat Neurosci. **8**(9), 1263-1268 (2005)
4. Spudich, J. L., Yang, C. S., Jung, K. H., and Spudich, E. N.: Retinylidene proteins: structures and functions from archaea to humans. Annu. Rev. Cell Dev. Biol. **16**, 365–392 (2000)
5. Nagel, G., Szellas, T., Huhn, W., Kateriya, S., Adeishvili, N., Berthold, P., Ollig, D., Hegemann, P., Bamberg, E.: Channelrhodopsin-2, a directly light-gated cation-selective membrane channel. Proc Natl Acad Sci U S A. **100**(24), 13940-13945 (2003)
6. Sineshchekov, O.A., Govorunova, E.G., Spudich, J.L.: Photosensory functions of channelrhodopsins in native algal cells. Photochem Photobiol. **85**(2), 556-563 (2009). doi: 10.1111/j.1751-1097.2008.00524.x
7. Feldbauer, k., Zimmermann, D., Pintschovius, V., Spitz, J., Bamann, C., Bamberg, E.: Channelrhodopsin-2 is a leaky proton pump. P. Natl. Acad. Sci. USA **106**, 12317-12322 (2009)
8. Zhang, F., Prigge, M., Beyriere, F., Tsunoda, S. P., Mattis, J., Yizhar, O., Hegemann, P., and Deisseroth, K.: Red-shifted optogenetic excitation: a tool for fast neural control derived from Volvox carteri. Nat. Neurosci. **11**, 631–633 (2008)
9. Gunaydin, L. A., Yizhar, O., Berndt, A., Sohal, V. S., Deisseroth, K., and Hegemann, P.: Ultrafast optogenetic control. Nat. Neurosci. **13**, 387–392 (2010)
10. Govorunova, E. G., Spudich, E. N., Lane, C. E., Sineshchekov, O. A., and Spudich, J. L.: New channelrhodopsin with a reds shifted spectrum and rapid kinetics from Mesostigma viride. mBio **2**, e00115–00111 (2011)
11. Yizhar, O., Fenno, L. E., Prigge, M., Schneider, F., Davidson, T. J., O'Shea, D. J., Sohal, V. S., Goshen, I., Finkelstein, J., Paz, J. T., Stehfest, K., Fudim, R., Ramakrishnan, C., Huguenard, J. R., Hegemann, P., and Deisseroth, K.: Neocortical excitation/inhibition balance in information processing and social dysfunction. Nature **477**, 171–178 (2011)
12. Matsuno-Yagi, A., Mukohata, Y.: Two possible roles of bacteriorhodopsin; a comparative study of strains of Halobacterium halobium differing in pigmentation. Biochem. Biophys. Res. Commun. **78**, 237–243 (1977)
13. Lanyi, J.K., and Oesterhelt, D.: Identification of the retinal-binding protein in halorhodopsin. J. Biol. Chem. **257**, 2674-2677 (1982)
14. Han, X., and Boyden, E. S.: Multiple-color optical activation, silencing, and desynchronization of neural activity, with single-spike temporal resolution. PloS One **2**, e299 (2007)
15. Zhang, F., Wang, L.-P., Brauner, M., Liewald, J.F., Kay, K., Watzke, N., Wood, P.G., Bamberg, E., Nagel, G., Gottschalk, A., & Deisseroth, K., Multimodal fast optical interrogation of neural circuitry. Nature **446**, 633–639 (2007)
16. Gradinaru, V., Zhang, F., Ramakrishnan, C., Mattis, J., Prakash, R., Diester, I., Goshen, I., Thompson, K.R., Deisseroth, K.: Molecular and cellular approaches for diversifying and extending optogenetics. Cell **141**, 154-165 (2010)
17. Bamberg, E., Tittor, J., and Oesterhelt, D.: Light-driven proton or chloride pumping by halorhodopsin. Proc. Natl. Acad. Sci. U.S.A. **90**, 639–643 (1993)
18. Chow, B. Y., Han, X., Dobry, A. S., Qian, X., Chuong, A. S., Li, M., Henninger, M. A., Belfort, G. M., Lin, Y., Monahan, P. E., and Boyden, E. S.: High-performance genetically targetable optical neural silencing by light-driven proton pumps. Nature **463**, 98–102 (2010)
19. Raimondo, J. V., Kay, L., Ellender, T. J., and Akerman, C. J.: Optogenetic silencing strategies differ in their effects on inhibitory synaptic transmission. Nat. Neurosci. 15, 1102-1104 (2012). doi: 10.1038/nn.3143
20. Han, X., Chow, B. Y., Zhou, H., Klapoetke, N. C., Chuong, A., Rajimehr, R., Yang, A., Baratta, M. V., Winkle, J., Desimone, R., and Boyden, E. S.: A high-light sensitivity optical neural silencer: development and application to optogenetic control of non-human primate cortex. Front. Syst. Neurosci. **5**, 18 (2011)
21. Airan, R.D., Thompson, K.R., Fenno, L.E., Bernstein, H., Deisseroth, K.: Temporally precise in vivo control of intracellular signalling. Nature **458**, 1025–1029 (2009)
22. Oh, E., Maejima, T., Liu, C., Deneris, E., Herlitze, S.: Substitution of 5-HT1A receptor signaling by a light-activated G protein-coupled receptor. J. Biol. Chem. **285**, 30825–30836 (2010)

23. Zhang, F., Gradinaru, V., Adamantidis, A.R., Durand, R., Airan, R.D., et al.: Optogenetic interrogation of neural circuits: technology for probing mammalian brain structures. Nat. Protoc. **5**, 439–456 (2010)
24. Aravanis, A.M., Wang, L.P., Zhang, F., Meltzer, L.A., Mogri, M.Z., et al.: An optical neural interface: in vivo control of rodent motor cortex with integrated fiberoptic and optogenetic technology. J. Neural Eng. **4**, S143–56 (2007)
25. Han, X. Optogenetics in the nonhuman primate. Prog. Brain Res. **196**, 215–233 (2012)
26. Burger, C., Gorbatyuk, O.S., Velardo, M.J., Peden, C.S., Williams, P., Zolotukhin, S., Reier, P.J., Mandel, R.J., Muzyczka, N.: Recombinant AAV viral vectors pseudotyped with viral capsids from serotypes 1, 2, and 5 display differential efficiency and cell tropism after delivery to different regions of the central nervous system. Mol Ther. **10**(2), 302-317 (2004)
27. Nathanson, J. L., Jappelli, R., Scheeff, E. D., Manning, G., Obata, K., Brenner, S., and Callaway, E. M.: Short Promoters in Viral Vectors Drive Selective Expression in Mammalian Inhibitory Neurons, but do not Restrict Activity to Specific Inhibitory Cell-Types. Front. Neural Circuits **3**, 19 (2009)
28. Atasoy, D., Aponte, Y., Su, H. H., and Sternson, S. M.: A FLEX switch targets Channelrhodopsin-2 to multiple cell types for imaging and long-range circuit mapping. J. Neurosci. **28**, 7025–7030 (2008)
29. Sauer, B., and Henderson, N.: Site-specific DNA recombination in mammalian cells by the Cre recombinase of bacteriophage P1. Proc. Natl. Acad. Sci. U.S.A. **85**, 5166–5170 (1988)
30. Tsien, J. Z., Chen, D. F., Gerber, D., Tom, C., Mercer, E. H., Anderson, D. J., Mayford, M., Kandel, E. R., and Tonegawa, S.: Subregion- and cell type-restricted gene knockout in mouse brain. Cell **87**, 1317–1326 (1996)
31. Madisen, L., Mao, T., Koch, H., Zhuo, J. M., Berenyi, A., Fujisawa, S., Hsu, Y. W., Garcia, A. J., 3rd, Gu, X., Zanella, S., Kidney, J., Gu, H., Mao, Y., Hooks, B. M., Boyden, E. S., Buzsaki, G., Ramirez, J. M., Jones, A. R., Svoboda, K., Han, X., Turner, E. E., and Zeng, H.: A toolbox of Cre-dependent optogenetic transgenic mice for light-induced activation and silencing. Nat. Neurosci. **15**, 793–802 (2012)
32. Mazarakis, N. D., Azzouz, M., Rohll, J. B., Ellard, F. M., Wilkes, F. J., Olsen, A. L., Carter, E. E., Barber, R. D., Baban, D. F., Kingsman, S. M., Kingsman, A. J., O'Malley, K., and Mitrophanous, K. A.: Rabies virus glycoprotein pseudotyping of lentiviral vectors enables retrograde axonal transport and access to the nervous system after peripheral delivery. Hum. Mol. Genet. **10**, 2109–2121 (2001)
33. Wickersham, I.R., Lyon, D.C., Barnard, R.J., Mori, T., Finke, S., et al.: Monosynaptic restriction of transsynaptic tracing from single, genetically targeted neurons. Neuron **53**, 639–647 (2007)
34. Lima, S.Q., Hromádka T., Znamenskiy, P., Zador, A.M. PINP: A New Method of Tagging Neuronal Populations for Identification during *In Vivo* Electrophysiological Recording. PLoS ONE **4**, e6099 (2009). doi:10.1371/journal.pone.0006099
35. Gradinaru, V., Mogri, M., Thompson, K.R., Henderson, J.M., Deisseroth, K.: Optical deconstruction of parkinsonian neural circuitry. Science **324**, 354–59 (2009)
36. Tye, K. M., Prakash, R., Kim, S. Y., Fenno, L. E., Grosenick, L., Zarabi, H., Thompson, K. R., Gradinaru, V., Ramakrishnan, C., and Deisseroth, K.: Amygdala circuitry mediating reversible and bidirectional control of anxiety. Nature **471**, 358–362 (2011)
37. Mobley, J., and Vo-Dinh, T.: Optical properties of Tissue. In Biomedical Photonics Handbook, pp 1–72, CRC Press, Boca Raton, FL (2003)
38. Bernstein, J. G., Han, X., Henninger, M. A., Ko, E. Y., Qian, X., Franzesi, G. T., McConnell, J. P., Stern, P., Desimone, R., and Boyden, E. S.: Prosthetic systems for therapeutic optical activation and silencing of genetically-targeted neurons. Proc. Soc. Photo-Opt. Instrum. Eng. **6854**, 68540H (2008)
39. Han, X., Qian, X., Bernstein, J. G., Zhou, H. H., Franzesi, G. T., Stern, P., Bronson, R. T., Graybiel, A. M., Desimone, R., and Boyden, E. S.: Millisecond-timescale optical control of neural dynamics in the nonhuman primate brain. Neuron **62**, 191–198 (2009)

40. Bernstein, J.G., Garrity, P.A., Boyden, E.S.: Optogenetics and thermogenetics: technologies for controlling the activity of targeted cells within intact neural circuits. Curr Opin Neurobiol **22**, 61-71 (2012)
41. Wells, J., Kao, C., Mariappan, K., Albea, J., Jansen, E. D., Konrad, P., and Mahadevan-Jansen, A.: Optical stimulation of neural tissue in vivo. Opt. Lett. **30**, 504–506 (2005)
42. Tufail, Y., Yoshihiro, A., Pati, S., Li, M.M., Tyler, W.J.: Ultrasonic neuromodulation by brain stimulation with transcranial ultrasound. Nature Protocols **6**, 1453-1470 (2011)
43. Ayling, O.G., Harrison, T.C., Boyd, J.D., Goroshkov, A., Murphy, T.H.: Automated light-based mapping of motor cortex by photoactivation of channelrhodopsin-2 transgenic mice. Nat Methods **6**, 219-224 (2009)
44. Zorzos, A. N., Dietrich, A., Talei Franzesi, G., Chow, B., Han, X., Fonstad, C. G., Boyden, E.S.: Light-proof neural recording electrodes. In: Society for Neuroscience, Annual Meeting of. Chicago, 17-21 October (2009)
45. Zimmermann, D., Zhou, A., Kiesel, M., Feldbauer, K., Terpitz, U., Haase, W., Schneider-Hohendorf, T., Bamberg, E., and Sukhorukov, V.L.: Effects on capacitance by overexpression of membrane proteins. Biochem. Biophys. Res. Commun. **369**, 1022–1026 (2008)
46. Wentz, C. T., Bernstein, J. G., Monahan, P., Guerra, A., Rodriguez, A., Boyden, E. S.: A Wirelessly Powered and Controlled Device for Optical Neural Control of Freely-Behaving Animals. Journal of Neural Engineering **8**(4), 046021 (2011)
47. Sendelbeck, L.S., Urquhart, J.: Spatial distribution of dopamine, methotrexate and antipyrine during continuous intracerebral microperfusion **328**, 251-258 (1984)
48. Tehovnik, E.J.: Electrical stimulation of neural tissue to evoke behavioral responses. Journal of Neuroscience Methods **65**, 1-17 (1996)
49. Fenno, L., Yizhar, O., et al.: The Development and Application of Optogenetics." Annual Review of Neuroscience **34**, 389-412 (2011)
50. Lee, S., Kwan, A. C., et al.: Activation of specific interneurons improves V1 feature selectivity and visual perception. Nature **488**, 379-385 (2012)
51. Adesnik, H., Bruns, W., et al.: A neural circuit for spatial summation in visual cortex. Nature **490**, 226-231 (2012)
52. Royer, S., Zemelman, B. V., et al.: Control of timing, rate and bursts of hippocampal place cells by dendritic and somatic inhibition. Nature Neuroscience **15**(5), 769-775 (2012)
53. Liu, X., Ramirez, S., et al.: Optogenetic stimulation of a hippocampal engram activates fear memory recall. Nature **484,** 381-385 (2012)
54. Witten, I. B., Lin, S., et al.; Cholinergic Interneurons Control Local Circuit Activity and Cocaine Conditioning. Science **330**, 1677-1681 (2010)
55. Tsai, H., Zhang, F., et al.: Phasic Firing in Dopaminergic Neurons Is Sufficient for Behavioral Conditioning. Science **324**, 1080-1084 (2009)
56. Chaudhury, D., J. J. Walsh, et al.: Rapid regulation of depression-related behaviors by control of midbrain dopamine neurons. Nature **493**(7433), 532-536 (2013)
57. Warden, M. R., Selimbeyoglu, A., et al.: A prefrontal cortex-brainstem neuronal projection that controls response to behavioral challenge. Nature **493**, 428-432 (2012)
58. Zhuang, X., Masson, J.: et al.: Targeted gene expression in dopamine and serotonin neurons of the mouse brain. Journal of Neuroscience Methods **143**, 27-32 (2005)

Index

M. De Vittorio et al. (eds.), *Nanotechnology and Neuroscience: Nano-electronic, Photonic and Mechanical Neuronal Interfacing*, DOI 10.1007/978-1-4899-8038-0,

MIX
Papier aus verantwortungsvollen Quellen
Paper from responsible sources
FSC® C105338

If you have any concerns about our products,
you can contact us on
ProductSafety@springernature.com

In case Publisher is established outside the EU,
the EU authorized representative is:
Springer Nature Customer Service Center GmbH
Europaplatz 3, 69115 Heidelberg, Germany

Printed by Libri Plureos GmbH
in Hamburg, Germany